Computer-Aided Design of Electrical Machines

Computer-Aided Design of Electrical Machines

K.M. Vishnu Murthy
B.E; M.Tech (Elec)
(Former Sr.Dy.General Manager, BHEL, Hyderabad)
Associate Professor, EEE Dept.,
G. Narayanamma Institute of Technology & Science
Shaikpet, Hyderabad - 500 008.

BSP **BS Publications**
A unit of **BSP Books Pvt., Ltd.**

4-4-309/316, Giriraj Lane, Sultan Bazar,
Hyderabad - 500 095
Phone : 040 - 23445605, 23445688

Published by :

BS Publications
A unit of **BSP Books Pvt., Ltd.**

4-4-309/316, Giriraj Lane, Sultan Bazar,
Hyderabad - 500 095
Phone : 040 - 23445605, 23445688
e-mail : info@bspbooks.net

ISBN : 978-93-52300-66-2 (HB)

Dedicated
to
My Parents

Preface

The aim of this book is to present the sequential steps for developing the computer programs for the design of electrical machines using well-established design formulae and the data of magnetic and non-magnetic materials used in latest designs established based on techno-economical considerations and competitive aspects in the global market.

To enable the student for answering question-wise, the total program is split into a number of parts and each part is run independently. The total program for total design is obtained by adding all the part of programs sequentially.

"MATLAB" version of "C" language is used for computer programming because of its easiness and simplicity over conventional "C" program. Also, this software is well acquainted and available to all students in undergraduate level itself.

The contents of this book cover the syllabus specified by different Universities mostly for M.Tech/M.E , even though undergraduate level students can refer this for their project works.

Since optimization is the final motive of any design, practical methods for optimizing the design being used in industries are made use of for establishing the computer programs for optimal design of various machines.

Although sufficient care has been taken to make error-free text, possibilities of presence of a few errors cannot be ruled out. The author will be very much grateful if such errors are pointed out by the readers for rectification in subsequent prints.

I expresses my happiness for completing this book in my six months stay in USA with my son and daughter-in-law with invaluable support and encouragement given by them as well as my wife.

I thank the authors of books referred in Bibliography.

I sincerely thank the management of GNITS and colleagues of department of EEE for their inspiration and support given to me in bringing out this book. I also express my thanks to those who have directly or indirectly helped me.

My special thanks to God for his kind blessings for successful comletion of this book.

Finally I thank the publisher for publishing this book in record time.

*- **Author***

Contents

Preface .... (v)

CHAPTER - 1

Concept of Computer-Aided Design and Optimization

1.1 Introduction 1

1.2 Computer-Aided Design 2

1.3 Explanation of Details of Flow Chart 2

1.3.1 Input Data to be Fed into the Program 2

1.3.2 Applicable Constraints/Max or Minimum Permissible Limits 4

1.3.3 Output Data to be Printed After Execution of Program 4

1.3.4 Various Objective Parameters for Optimization in an Electrical Machine 5

1.4 Selection of Optimal Design 5

1.5 Explanation of Lowest Cost and Significance of "Kg/KVA" 5

Flowcharts **3**

CHAPTER - 2

Basic Concepts of Design

2.1 Introduction 13

2.2 Specification 13

2.3 Output Coefficient 14

2.4 Importance of Specific Loadings 14

2.5 Electrical Materials 15

2.5.1 Conducting Materials 15

2.5.2 Insulating Materials 16
2.5.3 Magnetic Materials 18
2.6 Magnetic Circuit Calculations 20
2.7 General Procedure for Calculation of Amp-Turns 22
2.8 Heating & Cooling 24
2.8.1 Heating 24
2.8.2 Cooling 25
2.9 Modes of Heat Dissipation 25
2.10 Standard Rating of Electical Machines 25
2.11 Ventilation Schemes 26
2.11.1 In Static Machines (Transformers) 26
2.11.2 In Rotating Machines 27
2.12 Quantity of Cooling Medium 30
2.13 Types of Enclosures 30
2.14 General Design Procedure 31
2.15 Steps to Get Optimal Design 32

CHAPTER - 3

Armature Windings

3.1 Introduction 33
3.2 Important Terms Related to Armature Windings 33
3.3 Classification of Armature Windings 34
3.4 Winding Pitches 36
3.5 Armature Windings for DC Machines 37
3.5.1 Rules for Design Development for Lap-Winding for Armature 37
3.5.2 Examples for 4 Pole, 16 Slot DC Machine with Progressive Double Layer Lap Winding 38
3.5.3 How to Draw the Lap Winding Diagram 39
3.5.4 How to Represent the Position of Brushes 39
3.5.5 Rules for Design Development of Wave-Winding 41
3.5.6 Example for 4 Pole, 17 Slot DC Machine with Double Layer Wave Winding 41

3.5.7 How to Draw the Wave Winding Diagram 42
3.5.8 How to Represent the Position of Brushers 44
3.5.9 Dummy or Idle Coils 44
3.5.10 Criteria for Decision of Type of Winding 44
3.6 Armature Winding for AC Machines 44
3.6.1 Similarities between DC & AC Machines 44
3.6.2 Differences between DC & AC Machines 45
3.6.3 How to Draw a 3-ph Single Layer Winding Diagram 45
3.6.4 How to Draw a 3-ph Double Layer Winding Diagram 47

CHAPTER - 4

DC Machines

4.1 Introduction 57
4.2 Sequential Steps for Design of Each Part & Programming Simultaneously 57
4.3 Calculation of Armature main Dimensions and flux for pole (Part-1) 58
4.3(a) Computer Program in "C" in MATLAB for Part-1 59
4.4 Design of Armature Winding & Core (Part-2) 60
4.4(a) Computer Program in "C" in MATLAB for Part-2 63
4.5 Design of Poles & Calculation of AT (Part-3) 65
4.5(a) Computer Program in "C" in MATLAB for Part-3 68
4.6 Design of Shunt Field & Series Field Windings (Part-4) 69
4.6(a) Computer Program in "C" in MATLAB for Part-4 71
4.7 Design of Commutator & Brushes (Part-5) 71
4.7(a) Computer Program in "C" in MATLAB for Part-5 72
4.8 Design of Inter-Pole/Compensating-Winding and Overall Performance (Part-6) 73
4.8(a) Computer Program in "C" in MATLAB for Part-6 75
4.9 Computer Program in "C" in MATLAB for Complete Design 76
4.10 Computer Output Results for Complete Design 82
4.11 Modifications to be done in the above Program to Get Optimal Design 84
4.12 Computer Program in "C" in MATLAB for Optimal Design 85
4.12(a) Computer Output Results for Optimal Design 91

CHAPTER - 5

Transformers

5.1 Introduction 93

5.2 Core Type Power Transformer 93

5.2.1 Sequential Steps for Design of Each Part & Programming Simultaneously 93

5.2.2 Design of Magnetic Frame (Part-1) 94

5.2.2 (a) Computer Program in "C" in MATLAB for Part-1 96

5.2.3 No-Load Current (Part-2) 96

5.2.3 (a) Computer Program in "C" in MATLAB for Part-2 97

5.2.4 Design of LV Winding (Part-3) 98

5.2.4 (a) Computer Program in "C" in MATLAB for Part-3 99

5.2.5 Design of HV Winding (Part - 4) 100

5.2.5 (a) Computer Program in "C" in MATLAB for Part-4 102

5.2.6 Performance Calculations (Part - 5) 102

5.2.6 (a) Computer Program in "C" in MATLAB for Part-5 103

5.2.7 Tank Design & Weights (Part-6) 104

5.2.7 (a) Computer Program in "C" in MATLAB for Part-6 105

5.2.8 Computer Program in "C" in MATLAB for Complete Design 105

5.2.8 (a) Computer Output Results for Complete Design 110

5.2.9 Modifications to be done in the above Program to get Optimal Design 112

5.2.10 Computer Program in "C" in MATLAB for Optimal Design 113

5.2.10 (a) Computer Output Results for Optimal Design 117

5.3 Shell Type Power Transformer 118

5.3.1 Sequential Steps for Design of Each Part and Programming Simultaneously 118

5.3.2 Design of Magnetic Frame (Part-1) 118

5.3.2 (a) Computer Program in "C" in MATLAB for Part-1 120

5.3.3 No-Load Current (Part-2) 121

5.3.3 (a) Computer Program in "C" in MATLAB for Part-2 122

5.3.4 Design of LV Winding (Part-3) 123

5.3.4 (a) Computer Program in "C" in MATLAB for Part-3 124

5.3.5 Design of HV Winding (Part-4) 124

5.3.5 (a) Computer Program in "C" in MATLAB for Part-4 125

5.3.6 Performance Calculations (Part-5) 126
5.3.6 (a) Computer Program in "C" in MATLAB for Part-5 127
5.3.7 Tank Design & Weights (Part-6) 127
5.3.7 (a) Computer Program in "C" in MATLAB for Part-6 128
5.3.8 Computer Program in "C" in MATLAB for Complete Design 128
5.3.8 (a) Computer Output Results for Complete Design 133
5.3.9 Modifications to be Done in the above Program to get Optimal Design 135
5.3.10 Computer Program in "C" in MATLAB for Optimal Design 135
5.3.10 (a) Computer Output Results for Optimal Design 139

CHAPTER - 6

Synchronous Machines

6.1 Introduction 141
6.2 Salient Pole Type 141
6.2.1 Sequential Steps for Design of Each Part and Programming Simultaneously 141
6.2.2 Calculation of the Stator Main Dimensions & Flux/Pole (Part-1) 142
6.2.2 (a) Computer Program in "C" in MATLAB for Part-1 143
6.2.3 Design of Armature Winding & Core (Part-2) 144
6.2.3 (a) Computer Program in "C" in MATLAB for Part-2 146
6.2.4 Design of Rotor & Calculation of AT (Part-3) 147
6.2.4 (a) Computer Program in "C" in MATLAB for Part-3 149
6.2.5 Carter Coefficients & Ampere Turns for Various Parts of Magnetic Circuit (Part -4) 150
6.2.5 (a) Computer Program in "C" in MATLAB for Part-4 153
6.2.6 Open Circuit Characteristic Curve (OCC) (Part - 5) 154
6.2.6 (a) Computer Program in "C" in MATLAB for Part-5 156
6.2.7 Field Current at Rated Load & PF (Part-6) 158
6.2.7 (a) Computer Program sin "C" in MATLAB for Part-6 159
6.2.8 Field Winding Design + Losses + Efficiency (Part-7) 159
6.2.8 (a) Computer Program in "C" in MATLAB for (Part-7) 161
6.2.9 Calculation of Temp-Rises & Weights (Part-8) 162
6.2.9 (a) Computer Program in "C" in MATLAB for Part-8 163

6.2.10 Computer Program in "C" in MATLAB for Complete Design 164
6.2.10 (a) Computer Output Results for Complete Design 172
6.2.11 Modifications to be done in the above Program to ger Optimal Design 174
6.2.12 Computer Program in "C" in MATLAB for Optimal Design 174
6.2.12 (a) Computer Output Results for Optimal Design 181
6.3 Non-Salient Pole (Cylindrical Solid Rotor) Type 182
6.3.1 Sequential Steps for Design of Each Part and Programming Simultaneously 182
6.3.2 Calculation of Stator Main Dimensions (Part-1) 182
6.3.2 (a) Computer Program in "C" in MATLAB for Part-1 184
6.3.3 Design of Armature Winding & Core (Part-2) 185
6.3.3 (a) Computer Program in "C" in MATLAB for Part-2 189
6.3.4 Design of Rotor-Winding (Part-3) 190
6.3.4 (a) Computer Program in "C" in MATLAB for Part-3 192
6.3.5 Ampere Turns for Various Parts of Magnetic Circuit at 100%V (Part-4) 192
6.3.5 (a) Computer Program in "C" in MATLAB for Part-4 196
6.3.6 Open Circuit Characteristic Curve (OCC) Part-5 197
6.3.6 (a) Computer Program in "C" in MATLAB for Part-5 199
6.3.7 Calculation of Field AT at Rated Load & PF and Other Calculations (Part-6) 201
6.3.7 (a) Computer Program in "C" in MATLAB for Part-6 202
6.3.8 Computer Program in "C" in MATLAB for Complete Design 203
6.3.8 (a) Computer Output Results for Complete Design 212
6.3.9 Modifications to be done in the above Prgram to get Optimal Design 215
6.3.10 Computer Program in "C" in MATLAB for Optimal Design 216
6.3.10 (a) Computer Output Results for Optimized Design 221

Chapter - 7

Three-Phase Induction Motors

7.1 Introduction 223
7.1.1 Squirrel Cage Motor 223
7.1.2 Sequential Steps for Design of Each Part and Programming Simultaneously 223

7.1.3 Calculation of Stator Main Dimensions & Flux (Part-1) 224

7.1.3 (a) Computer Program in "C" in MATLAB for Part-1 225

7.1.4 Design of Stator Winding (Part-2) 226

7.1.4 (a) Computer Program in "C" in MATLAB for Part-2 229

7.1.5 Design of Squirrel Cage Rotor (Part-3) 230

7.1.5 (a) Computer Program in "C" in MATLAB for Part-3 231

7.1.6 Total AT & Magnetizing Current (Part-4) 232

7.1.6 (a) Computer Program in "C" in MATLAB for Part-4 235

7.1.7 Short-Circuit Current Calculation (Part-5) 236

7.1.7 (a) Computer Program in "C" in MATLAB for Part-5 237

7.1.8 Performance Calculation (Part-6) 238

7.1.8 (a) Computer Program in "C" in MATLAB for Part-6 239

7.1.9 Computer Program in "C" in MATLAB for Complete Design 239

7.1.9 (a) Computer Output Results for Complete Design 247

7.1.10 Modifications to be done in the above Program to get Optimal Design 250

7.1.11 Computer Program in "C" in MATLAB for Optimal Design 250

7.1.11 (a) Computer Output Results for Optimal Design 256

7.2 Slip-Ring Type Induction Motor 257

7.2.1 Sequential Steps for Design of Each Part and Programming Simultaneously 257

7.2.2 Design of Slip-Ring Rotor (Part-3) 257

7.2.2 (a) Computer Program in “C” in MATLAB for Part-3 259

7.2.3 Computer Program in “C” in MATLAB for Complete Design 260

7.2.3 (a) Computer Output Results for Complete Design 267

7.2.4 Modifications to be Done in the above Program to get Optimal Design 270

7.2.5 Computer Program in “C” in MATLAB for Optimal Design 271

7.2.5 (a) Computer Output Results for Optimal Design 276

Chapter - 8

Single-Phase Induction Motors

8.1 Introduction 277

8.2 Sequential Steps for Design of Each Part and Programming Simultaneously 277

8.3 Calculation of Main Dimensions (Part-1) 278
8.3 (a) Computer Program in "C" in MATLAB for Part-1 279
8.4 Design of Main Winding (Part-2) 280
8.4 (a) Computer Program in "C" in MATLAB for Part-2 283
8.5 Design of Rotor (Part-3) 284
8.5 (a) Computer Program in "C" in MATLAB for Part-3 285
8.6 Amp-Turns Calculation (Part-4) 285
8.6 (a) Computer Program in "C" in MATLAB for Part-4 288
8.7 Leakage Reactance Calculation (Part-5) 289
8.7 (a) Computer Program in "C" in MATLAB for Part-5 292
8.8 Design of Auxiliary Winding (Part-6) 293
8.8 (a) Computer Program in "C" in MATLAB for Part-6 296
8.9 Weights, Losses and Performance using Eq.Ckt (Part-7) 297
8.9 (a) Computer Program in "C" in MATLAB for Part-7 300
8.10 Computer Program in "C" in MATLAB for Complete Design 301
8.10 (a) Computer Output Results for Complete Design 309
8.11 Modifications to be done in the above Program to get Optimal Design 312
8.12 Computer Program in "C" in MATLAB for Optimal Design 313
8.12 (a) Computer Output Results for Optimal Design 318

Bibliography **319**

Index **321**

CHAPTER 1

Concept of Computer-Aided Design and Optimization

1.1 Introduction

A design problem has to be formulated considering the various constraints, processes, availability of materials, quality aspects, cost aspects etc. Constraints can be from technical, cost or availability aspects. Technical constraints can be from calculation methods, available process systems, skilled labour, manufacturing facilities, machinery, or tools etc. Sometimes transport facilities to site also pose problems. If suitable quality materials are not available indigenously they may have to be imported, which effects cost and delivery time.

In designing any system, accuracy of prediction, economy, quality and delivery period play a vital role. Basically, a design involves calculating the dimensions of various components and parts of the machine, weights, material specifications, output parameters and performance in accordance with specified international standards. The calculated parameters may not tally with the final tested performance. Hence, design has to be frozen keeping in view the design analysis as well as the previous operating experience of such machines.

The practical method in case of bigger machines is to establish a computer program for the total design incorporating the constraint parameters and running the program for various alternatives from which final design is selected.

Though the final design may meet all the required specifications, it need not be an optimal one as regards the weight and cost of the active materials, and certain performance aspects like efficiency, temperature rise etc.

Various Objective Parameters for Optimization in an Electrical Machine

(a) Higher Efficiency

(b) Lower weight for given KVA output (Kg/KVA)

(c) Lower Temperature-Rise

(d) Lower Cost

(e) Any other parameter like higher PF for induction motor, higher reactance etc.

Here we have to understand that if an optimized design is finalized keeping in view of any of the parameters mentioned above, the design may not be optimum for the above parameters. For example, if a design is selected for higher efficiency, it may not be optimum in other parameters, maybe the cost is high. Normally this is a case where compromise is to be made. This is because more costly materials like higher grade silicon steel stampings are used for armature core to reduce iron losses and hence efficiency is increased. Similar controversies will exist for other options also. One more practical example in case of bigger machines is that a constraint is imposed in weight or volume of the machine to transport it through any road bridge or tunnel to reach the site where the machine has to finally operate. Hence, it is desirable to define clearly the objective function for optimization to which the design should fulfill.

1.2 Computer-Aided Design

In any practical design the number of variables is so high that hand calculations are impossible. The number of constraints is also large and for these to be satisfied by final design, a lengthy iterative approach is required. This is only possible with the help of computer programs (refer Flowchart 1.1).

1.3 Explanation of Details of Flowchart

1.3.1 Input Data to be Fed into the Program

(a) Data

1. Rating of the machine (KW/KVA)
2. Rated Voltage
3. Rated Frequency (for AC only)
4. Rated Speed (RPM)
5. Type of Connection of Phases (Star/Delta) for 3 ph AC only
6. Type of Winding (Lap/Wave)
7. Number of Parallel Paths
8. Shunt/Compound in case of DC Machine
9. Squirrel Cage/Slip Ring type for 3-ph Ind.Motor
10. Rated Slip /Rotor speed for Ind.Motor
11. Salient Pole/Round rotor type for 3-ph Alternators
12. Rated power factor for 3-ph Alternators
13. Core/Shell type for Transformers
14. Ratings of HV/LV for Transformers

(b) Applicable curves in array format like,

1. B/H for magnetic materials used for Core, Poles,
2. Loss Cuves for magnetic materials
3. Hysteresis loss vs. frequency
4. Carters coefficients for slots and vent ducts
5. Apparent Flux density
6. Leakage Coefficients of slots
7. HP vs. output coefficient for 1-ph Ind.Motor

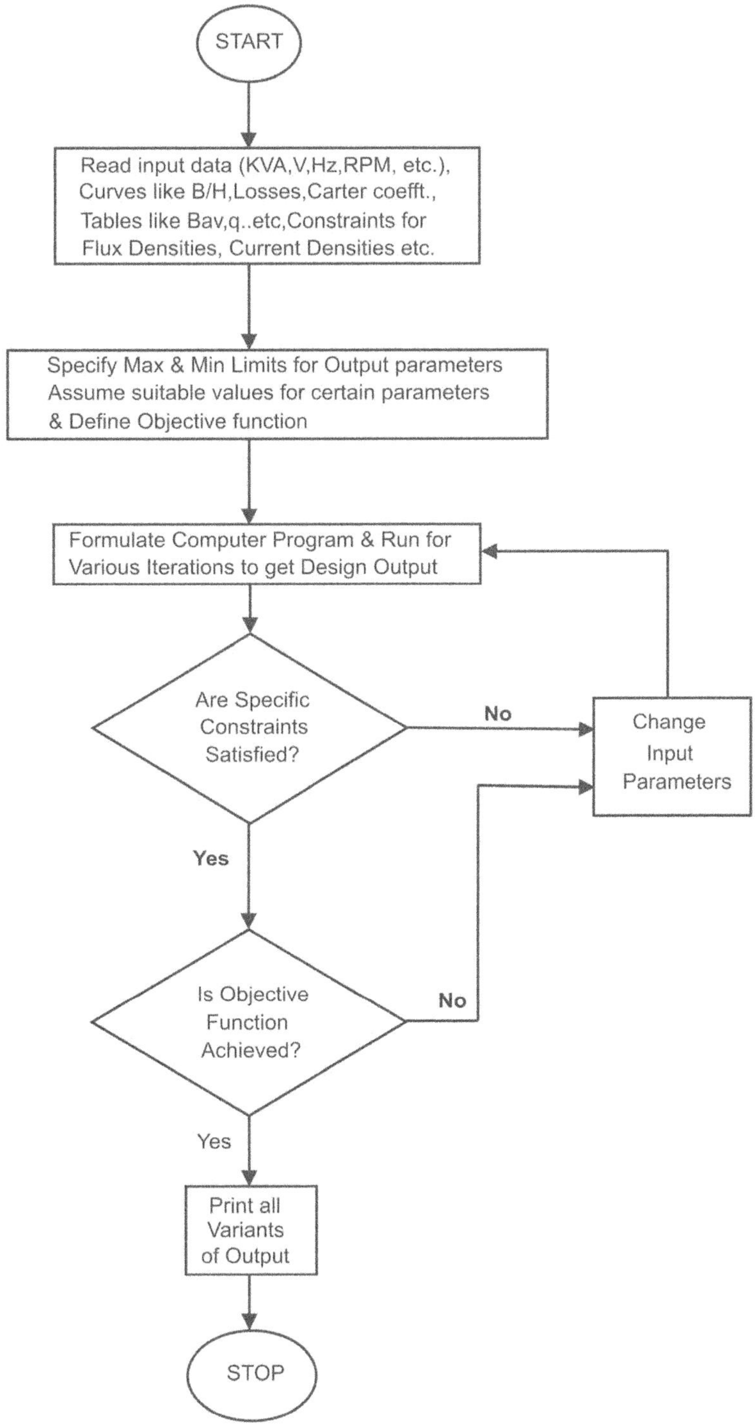

Flowchart 1.1 Computer-aided design.

(b) Applicable tables in array format like,

1. Output vs. Specific Electric Loading(q)
2. Output vs. Specific Magnetic Loading(Bav)
3. Output vs. AT/pole
4. No. of poles vs. Pole pitch
5. Depth of Shunt field winding vs. Arm Dia
6. Standard sizes of brushes
7. Power Factor and Efficiency at different ratings of Ind. Motor
8. Frequency vs. Frequency constant of 1-ph Ind. Motor
9. No. of poles vs. Di/De ratio for 1-ph Ind. Motor
10. Efficiency and PF vs. output for 1-ph Ind. Motor
11. Thickness of Stator winding ins vs. Voltage for Rotating machines
12. Window space Factor vs. KVA for transformers.

1.3.2 Applicable Constraints / Maximum or Minimum Permissible Limits

1. Flux density in core, tooth, yoke
2. Current densities in all windings of the machine
3. Ratio of Pole arc to pole pitch
4. Ratio of Length to pole arc
5. Current Volume per slot of DC armature
6. Peripheral velocity of rotor
7. Frequency of flux reversal in DC armature
8. Current per Brush arm in DC armature
9. Voltage between Commutator segments in DC armature
10. Current per Brush arm in DC armature
11. Polepitch
12. Temperature Rises
13. Power factor in Ind. Motor
14. No load current in Ind. Motor
15. Starting Torque in Ind. Motor
16. Number of Slots in Armature
17. Space factor of the slot in 1-ph Ind. Motor
18. Rotor slots in Ind. Motor
19. Eddy current loss factor in AC machine
20. Short Circuit Ratio of Alternator
21. Leakage reactance on AC Machine
22. Regulation
23. Saturation factor.

1.3.3 Output Data to be Printed after Execution of Program

(a) Applicable Data

1. Main Dimensions and Internal dimensions of the machine
2. No. of slots

3. Turns in all windings
4. Copper sizes in all windings
5. Weights
6. Losses
7. Efficiency
8. Reactances
9. Full load Field current
10. Temperature rise
11. No. of cooling tubes for a transformer
12. Diameter and number of segments in Commutator
13. Full load slip of Ind. Motor.

(b) Applicable Curves

1. Open Circuit, Short Circuit and Load magnetization characteristics of Alternator
2. Slip vs. Torque curves of Ind. Motor.

1.3.4 Various Objective Parameters for Optimization in an Electrical Machine

(a) Higher Efficiency
(b) Lower weight for given KVA output (Kg/KVA)
(c) Lower Temperature-Rise
(d) Lower Cost
(e) Any other parameter like higher PF for Induction motor, higher Reactance etc.

1.4 Selection of Optimal Design

Depending upon the required objective function, the suitable design variant from the printed outputs can be picked up. We can also do it by computation by incorporating a statement to print only the best suited as per the objective.

1.5 Explanation of Lowest Cost and Significance of "*Kg/KVA*"

Normally best objective for optimal design is the lowest cost. We will understand what is the cost?

Total cost = Material Cost + Labour cost + Over head costs

Material cost = Cost of material at manufacturing works + Import duties + Sales taxes

(For imported materials cost depends on foreign exchange rates available at that time)

Labour cost = Payment made to Workers

Over head costs = Supervision charges + Depreciation charges on heavy machinery

We can understand from the above that it is very difficult to get all the information to arrive at total cost in the bookish design. Hence we look at more practically feasible objective from class work point of view which is nothing but **minimum Kg/KVA.** It means that we are able to get output of rated KVA with minimum amount of material.

Flowcharts for optimal designs of all Electrical Machines are as follows. (Flowchart 1.2 to 1.7).

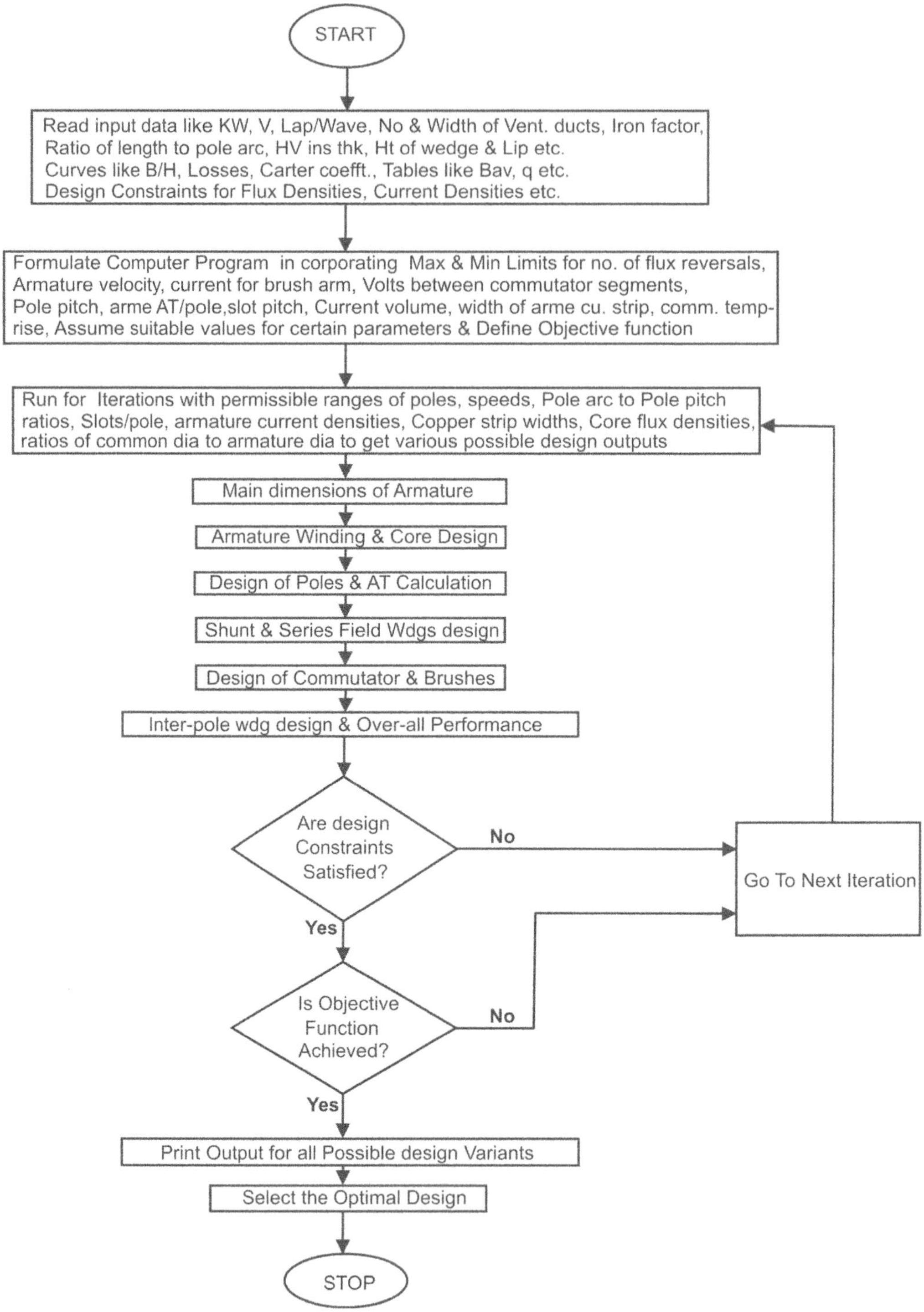

Flowchart 1.2 Flowchart for computer-aided optimal design of DC machine.

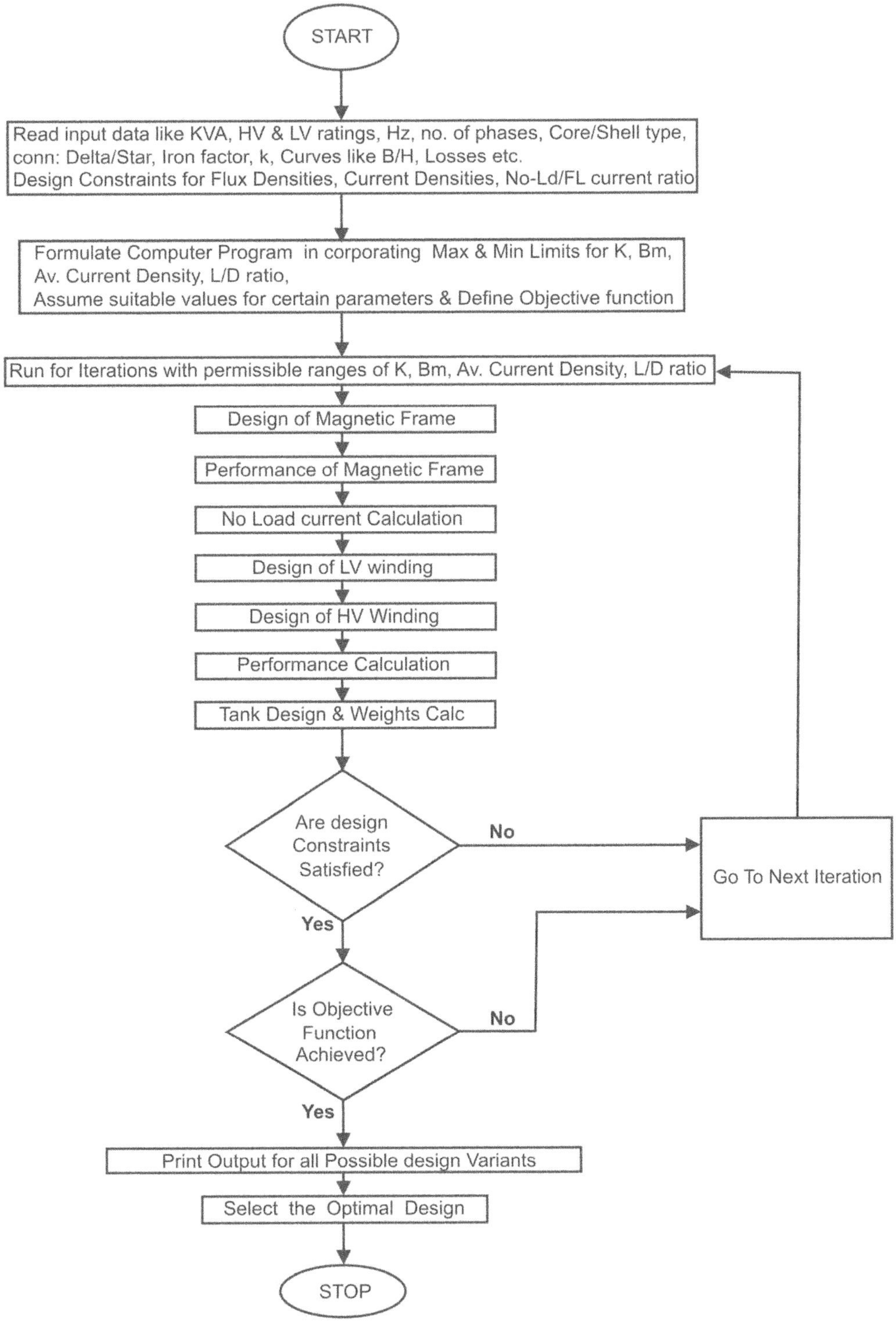

Flowchart 1.3 Flowchart for computer-aided optimal design of transformer.

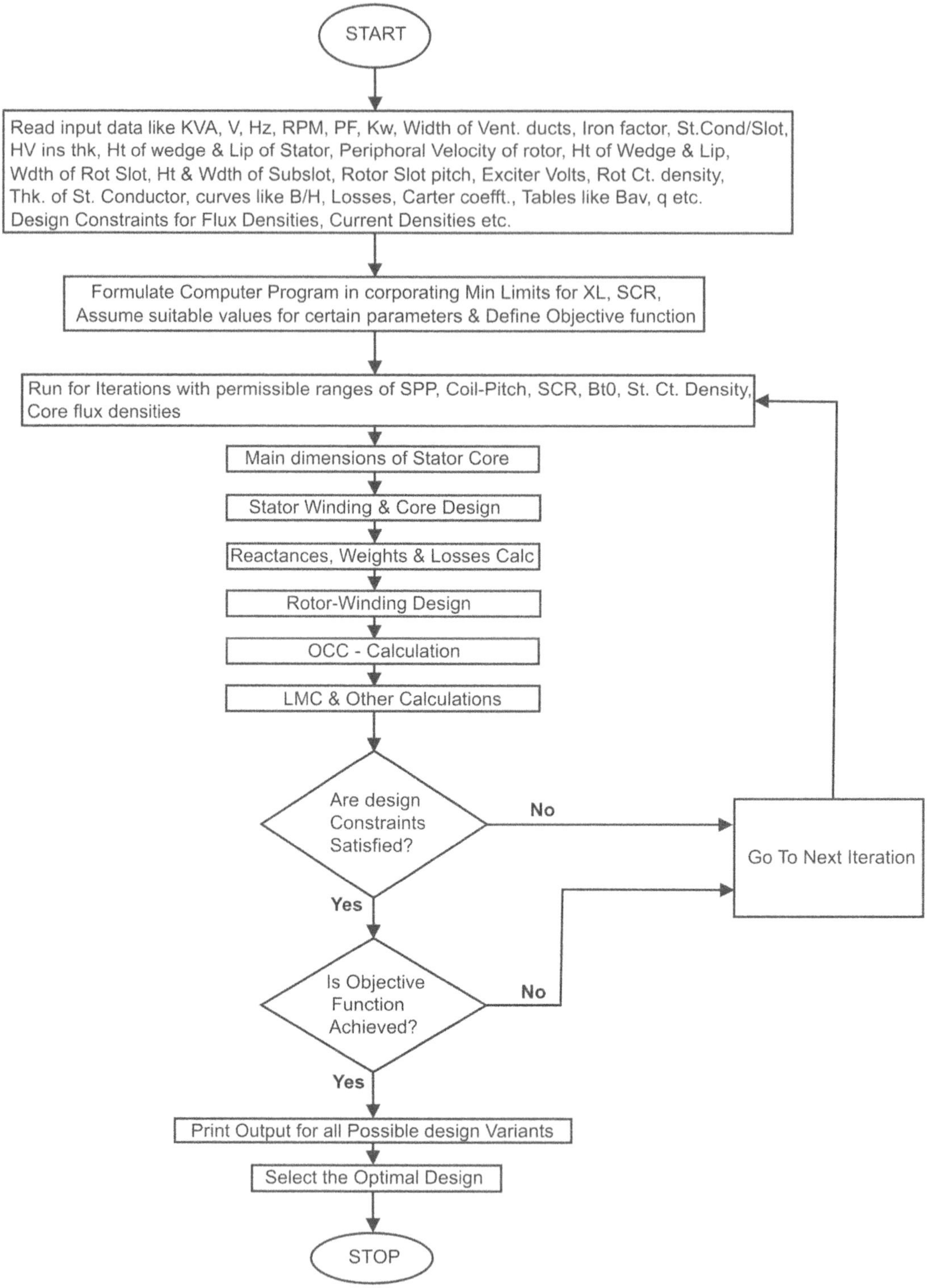

Flowchart 1.4 Flowchart for computer-aided optimal design of non-salient pole generator.

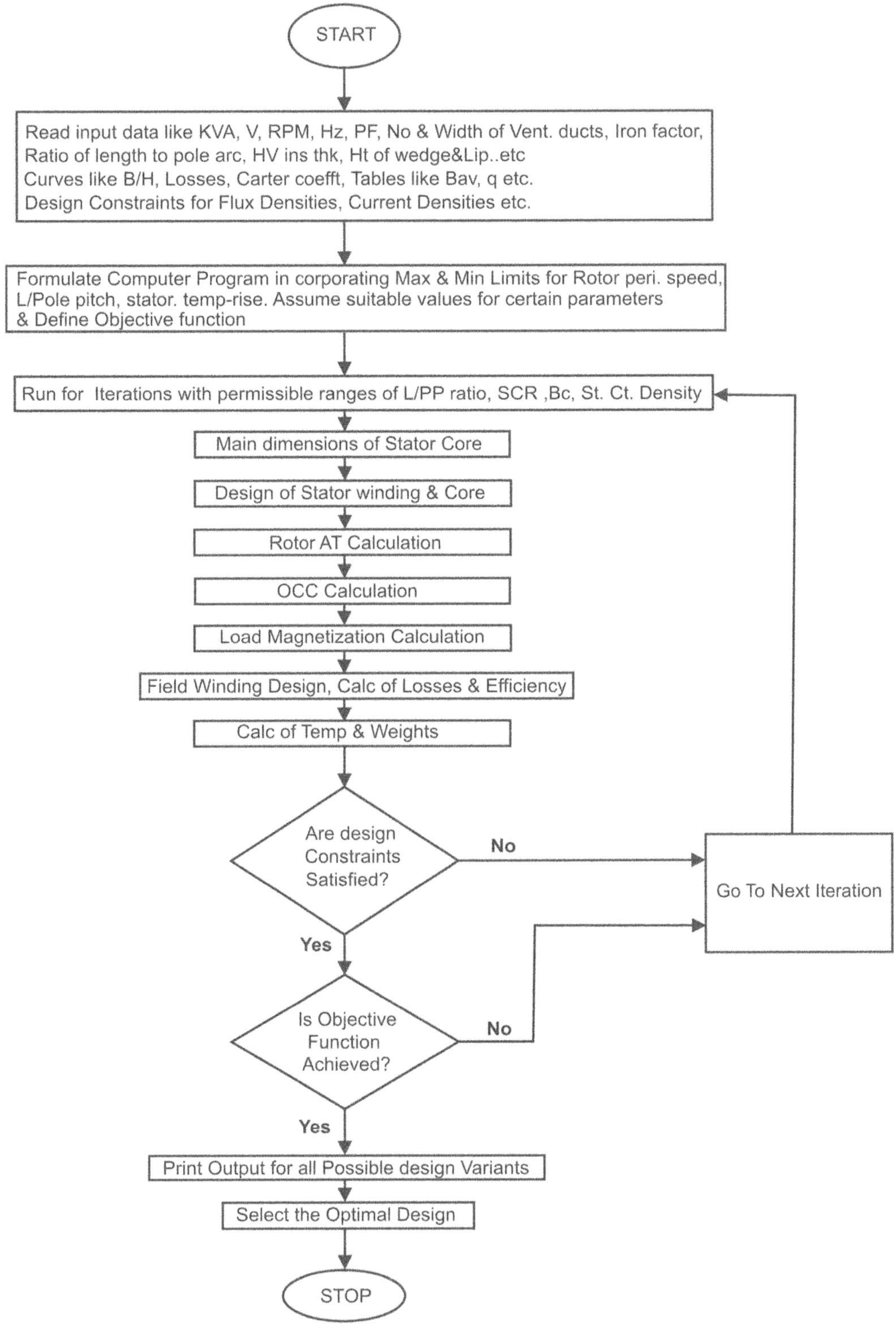

Flowchart 1.5 Flowchart for computer-aided optimal design of salient pole generator.

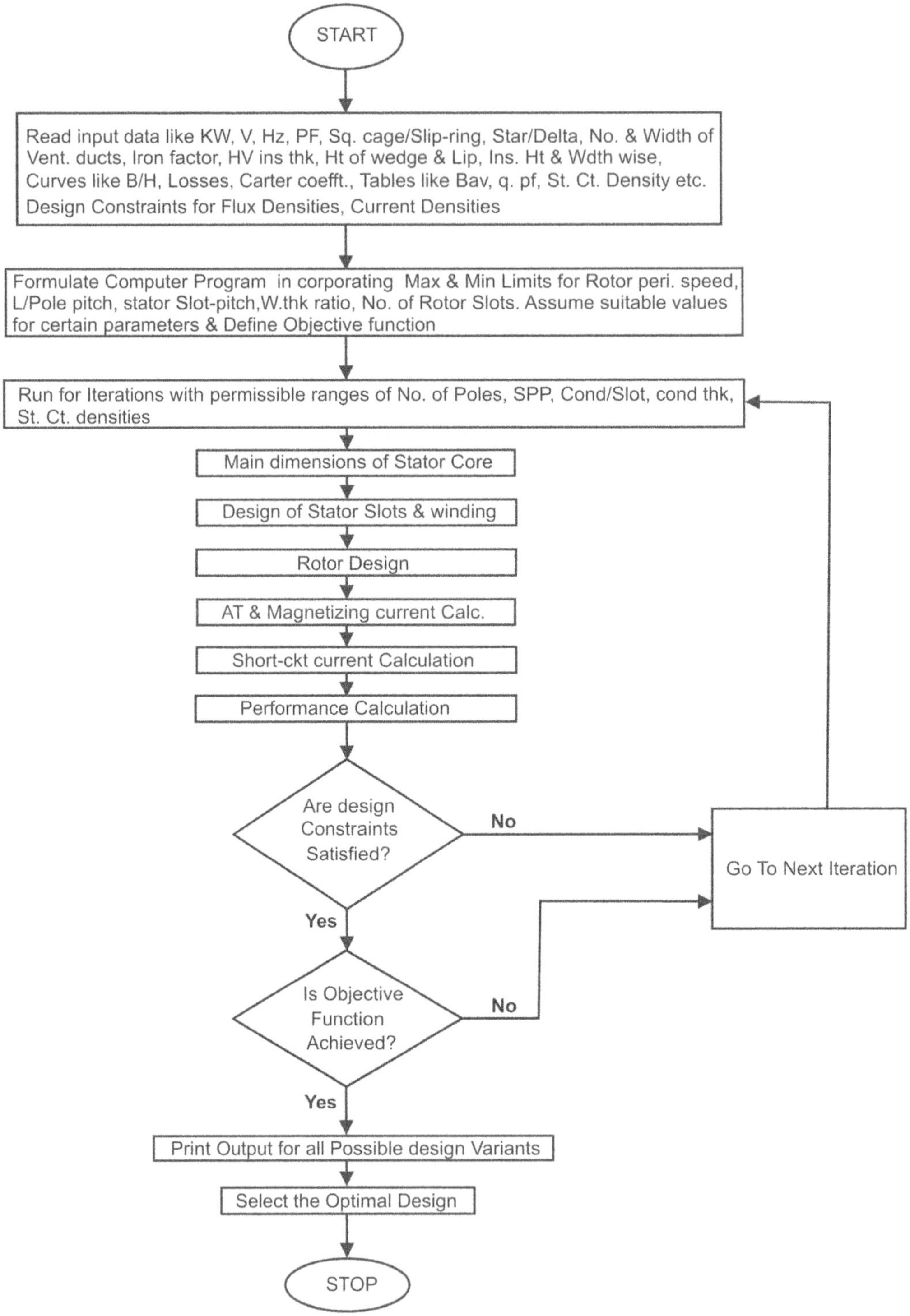

Flowchart 1.6 Flowchart for computer-aided optimal design of 3-ph induction motor.

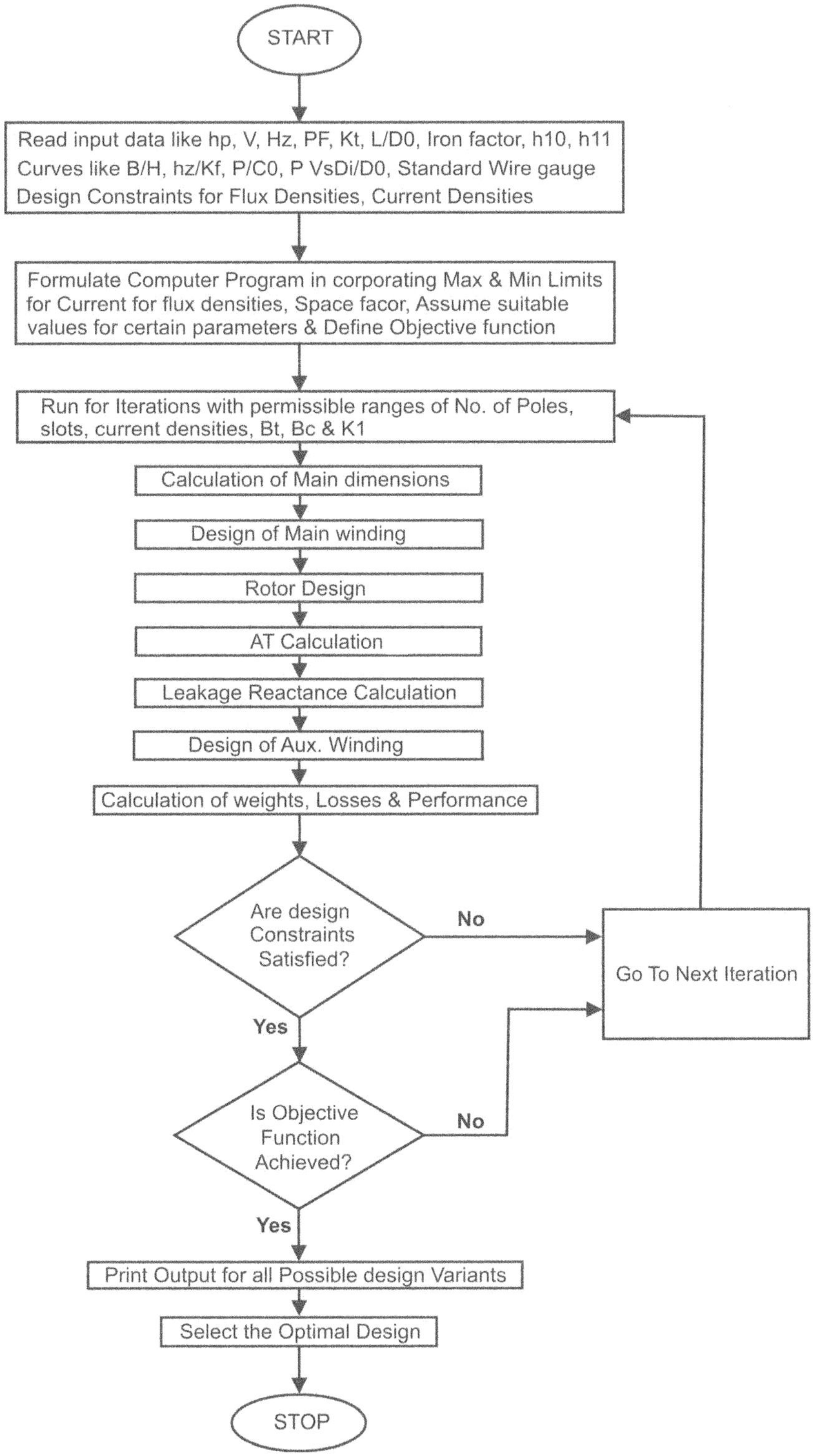

Flowchart 1.7 Flowchart for computer-aided optimal design of 1-ph induction motor.

CHAPTER 2

Basic Concepts of Design

2.1 Introduction

Aim of design is to determine the dimensions of each part of the machine, the material specification, prepare the drawings and furnish to manufacturing units. Design has to be carried out keeping in view the optimizing of the cost, volume and weight and at the same time achieving the desired performance as per specification. Knowledge of latest technological trends to supply a competitive product is a must. Design should conform to stipulations specified by International/National standards.

Design is the most important activity. The designer should be familiar with the following aspects:

(a) Thorough knowledge of international/national standards.

(b) Properties of good electrical materials (like copper), magnetic materials (like silicon steels), insulating materials (like Epoxy mica), mechanical and metallurgical properties of all types of steel.

(c) Governing laws of electrical circuits.

(d) Laws of heat transfer.

(e) Prices of materials used, foreign exchange rates, types of duties levied on products.

(f) Labour rates of both skilled and unskilled labour

(g) Knowledge of competitor's products.

2.2 Specification

Basic inputs for carrying out a design of electrical machine are KVA, KW, PF, Voltage, Speed, Frequency, No. of phases, Type of cooling, class of insulation, permitted temperature rise, type of winding connections, cooling medium temperature, any stipulations imposed by customer etc. In absence of any input data, the same may be taken from relevant standards (Table 2.1).

Table 2.1 Applicable Indian Standard Nos.

	Transformers	Ind. Motors	Sync. Mc/s
Design & Performance	10561	325 & 9628 2223 & 1231	4722/5422
Testing	2026/13956	4029	7132

2.3 Output Coefficient

Starting point for design is the output coefficient (K) where K = $\dfrac{P}{D^2 \times L \times N}$

Where P → Output of machine in KVA

D → Diameter of armature (m)

L → Gross length of armature (m)

N → Speed of the machine (RPM)

For larger machines, output coefficient is high.

By providing a fan and improved cooling, output coefficient can be increased.

If output coefficient (K) is higher, product of "D^2LN" is lower. i.e. either "D^2L" (Volume) is lower or Speed (N) is lower for same KVA output.

That means volume of a better cooled machine is lower for same output and speed.

2.4 Importance of Specific Loadings

Specific magnetic Loading (Bav) = $\dfrac{P \times \Phi}{\pi \times D \times L}$

Where P → No of poles

Φ → Flux per pole

D → Diameter of armature (m)

L → Gross length of armature (m)

Specific Electric Loading (q) = $\dfrac{Ia \times Za}{\pi \times D}$

Where *Ia* → Armature Current (A)

Za → No of Armature Conductors

D → Diameter of armature (m)

Advantages and Disadvantages due to Higher Specific Magnetic Loadings

S. no.	Advantages	Disadvantages
1	Lower Arme-Copper Loss	Higher Saturation and draws higher AmpTurns
2	Reduction in Volume	Higher Iron Losses
3	Reduction in Size	Increased Field Current Losses in Sync and DC Mc/s
4	Reduction in Weight	Higher Magnetizing Current and poor PF in Ind. Mc/s
5	Lower material Cost	Higher Flux densities in Tooth and Core
6	Lower Overall Price	Increased Temp-Rises
7	Higher Stability in Sync. Mc/s	Increased Magnetic hum and noise
8		Reduced Leakage Reactance and hence higher Currents during Short Circuits
9		Lower Efficiency

Advantages and Disadvantages due to Higher Electric Loadings

S. no.	Advantages	Disadvantages
1	Reduction in Volume	Higher Arme-Copper Losses
2	Reduction in Size	Higher Leakage Reactance and hence poor Voltage Regulation
3	Reduction in Weight	Increased Temp-Rises
4	Lower material Cost	Increased Arme-Reaction AT and higher field current in DC and Sync. Mc/s
5	Lower Overall Price	Lower Stability in Sync. Mc/s
6		Lower Efficiency

Considering the above aspects, suitable values of Specific Magnetic and Specific Electrical Loadings are to be selected. In fact by assuming different values of Bav and q falling within the permissible range, many design variants are to be worked out with the help of computer programs and an optimized variant has to be arrived at.

2.5 Electrical Materials

Materials used in Electrical machines are classified into three types:

1. Conducting; 2. Insulating and 3. Magnetic

Design of electrical machines depends mainly on quality of materials used. If low quality materials are used, the machine will be less efficient, more bulky, higher weight and higher cost. Operational running cost will also be higher. A designer should have perfect knowledge of properties and cost of these materials so that the design can be both efficient and cost-effective.

2.5.1 Conducting Materials

Conducting materials are of two categories

1. *Material of low conductivity (high resistivity):* Used for heating devices, thermo couples, resistance etc.

2. *Material of high conductivity (low resistance):* Used for windings of electrical machines and equipments. Material with lowest resistance should be selected so that it contributes lowest Ohmic losses to enhance efficiency and to reduce Temp-rise.

Requirements of high conductive materials:

(a) Highest Possible conductivity (least Resistance)
(b) Least possible temperature coefficient of resistance
(c) Adequate resistance to corrosion
(d) Adequate mechanical strength and high tensile strength
(e) Suitable for jointing by brazing/soldering/welding so that the joints are highly reliable contributing lowest resistance.
(f) Suitable for rollability, drawability, so that conductors of required shape (wire/strip) are easily manufactured.

Best conducing material is silver. Next best is copper and then aluminium. Properties of these are compared in the following table.

S. no.	Property	Unit	Silver	Copper	Aluminium
1	Conductivity	--	1.0	0.975	0.585
2	Resistivity	μΩ-cm	1.46	1.777	2.826
3	Temp-Coefft	% per °C	0.337	0.393	0.4
4	Cost	---	Prohibitively high	Medium	Low

3. *Super conducting materials:* Materials whose resistivity sharply decreases to practically zero value when the temperature is brought down below ***transition*** temperature are called super conductors. Due to practically zero resistance, copper losses will be almost zero. Hence, machines with these conductors can be designed with very high value of current density reducing drastically the size of the machine.

   However, these machines are not in commercial use due to practical limitations.

2.5.2 Insulating Materials

Insulating materials are used to provide an electrical insulation between parts at different potentials. Insulating materials are classified as per the following Table.

Required properties of good insulating materials:

(a) High Insulation Resistance
(b) High Dielectric strength
(c) Low Dielectric Losses and Low Dielectric Loss angle(tan δ)
(d) No moisture absorption
(e) Capable of withstanding without deterioration a repeated heat cycle
(f) Good heat conductivity
(g) Good mechanical strength to withstand vibrations and bending
(h) Solid material should have a high melting point
(i) Liquid materials should not evaporate or volatilize.

Comparison of properties of insulating materials

Material	Dielectric Strength (KV/mm) at 50 Hz	Resistance at 20°C in Ω-cm	Permittivity at 50 Hz	Safe temp in °C	Moisture absorption
Cotton	3 to 4	10^9 to 10^{13}	-	90	Absorbant
Paper	6 to 10	10^{15}	2.5	90	Absorbant
Fibre	5	10^9 to 10^{18}	4 to 6	90	Absorbant if Varnish layer is cracked
Rubber	15 to 25	10^6 to 10^{15}	3 to 4	40	Slightly
Mica	40 to 80	5×10^{13}	5 to 8	500	No
Micanite	30	3×10^{13}	6 to 8	130	No
Asbestos	3 to 4.5	10^9 to 10^{15}		500	Absorbant
Backalite	20 to 25	--	5 to 6	200	No
Glass	5 to 15	10^{15}		Room-temp	No
Porcelain	10 to 20	10^{13}	4 to 7	-do-	No
Polythene	30	10^{16}	2.3	70	No

Insulation for Conductor Covering

Copper conductors used in electrical machines are covered with some type of insulating material (usually in the form of tapes) based on thermal grading, dielectric stresses and economy.

Types of insulating materials used for conductor coverings for different temp levels are as given in the following Table.

Types of insulating materials

Ins. class	Max permissible temp. limit (°C)	MATERIALS
A	105	Cotton, silk and paper impregnated in dielectric liquid such as oil.
E	120	Cotton, silk and paper impregnated in dielectric liquid for operating up to 120°C.
B	130	Mica, glass fibre, asbestos with suitable impregnating materials for operating up to 130°C.
F	155	Mica, glass fibre, asbestos with suitable bonding substances capable up to 155°C.
H	180	Silicone elastomers with mica, glass fibre, asbestos. with silicone resins.
C	225	Silicone elastomers with mica, glass fibre, asbestos with silicone resins.

In electrical machines of small ratings, insulation materials of class AandE can be used to reduce cost. But for larger machines they are not suitable since volume and weight of the machine will be higher and efficiency lower. Techno-economical study proved that Class-B and Class-F insulations are most appropriate for machines of medium and large ratings respectively for commercial use.

The latest trend is to design large machines with Class-F insulation and utilize for Class-B temp-rises. Advantages of Class-F insulation is that it possesses excellent properties as indicated above and gives a reliable performance for a longer life.

Class-H and Class-C insulations are costly and hence used in compact machines required for special applications like submarines, space craft etc., where economical aspect is not prime criteria.

Insulating Resin and Varnish

In electrical machines, resins and varnishes are used for impregnation, coating and adhesion. These resins and varnishes have the following additional insulating properties.

(a) Quick drying properties

(b) Chemical stability even under strong oxidizing influence

(c) Should not attack the base insulating material or the copper conductor

(d) Should set hard and good surface.

2.5.3 Magnetic Materials

Magnetic materials play a vital role in electrical machines, since magnetic circuit is created by these materials.

A good magnetic material should possess the following qualities (refer Figs. 2.1 and 2.2)

(a) High magnetic permeability (μ) so that for required flux density it draws minimum no. of amp turns ($H = B/\mu$)

(b) High electrical resistivity to reduce the eddy current losses

(c) Hysteresis loop should be narrow to reduce hysteresis loss.

Silicon Steel

Magnetic properties of permeability and resistivity of steel are greatly improved by adding a certain percentage of silicon. But, if the percentage of silicon increases 4%, steel becomes brittle. These silicon steels are made into laminations of normal thickness of 0.35mm and 0.5 mm either by cold working or hard working.

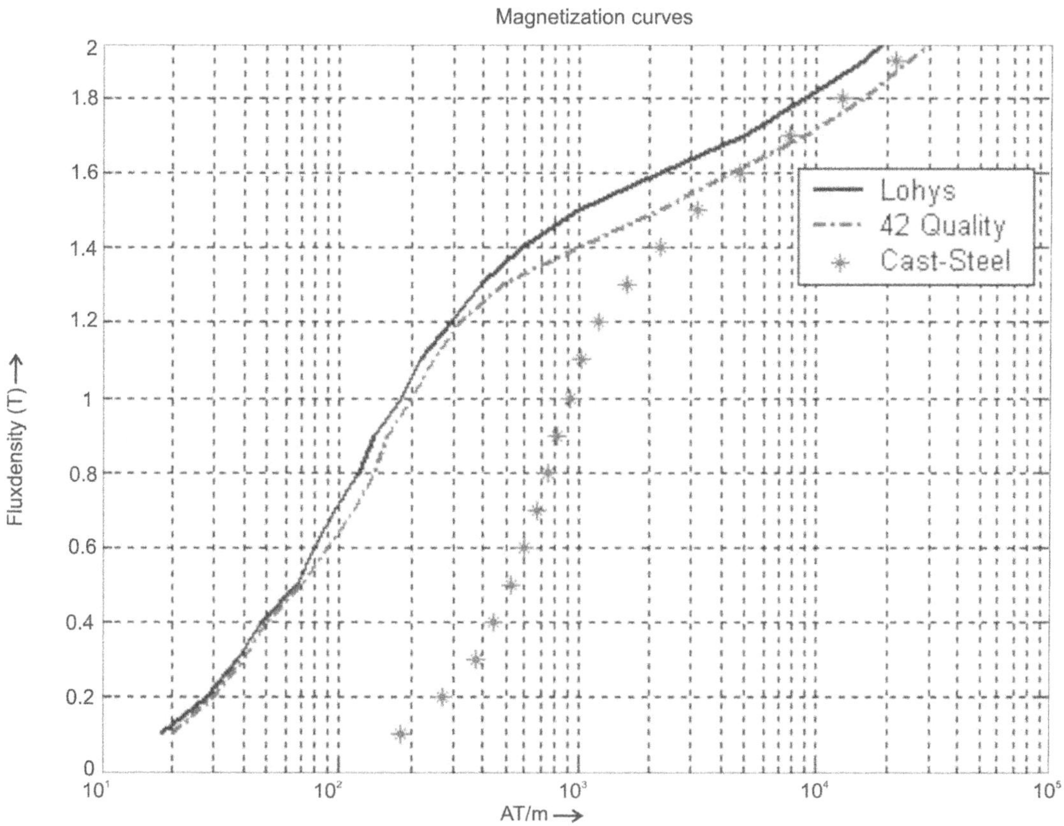

Fig. 2.1 "Lohys" is the best material since permeability is the highest.

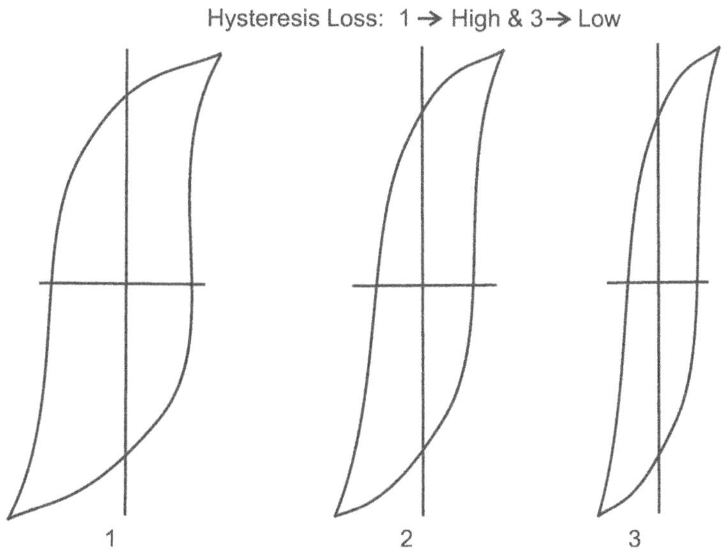

Fig. 2.2 Material : 3 will be with lowest hysteresis loss.

In the cold reduced process, the laminations are annealed at about 1100°C to get grain-oriented material. The cold rolled grain-oriented (CRGO) steels will have superior magnetic properties and can be worked at much higher flux densities. By using these in the transformers, we get the advantage of reduction in size and weight with increased efficiency. Curves given in Fig. 2.3 indicate the loss in Watts/Kg for CRGO and NGRO (Non-grain-oriented) sheet steels.

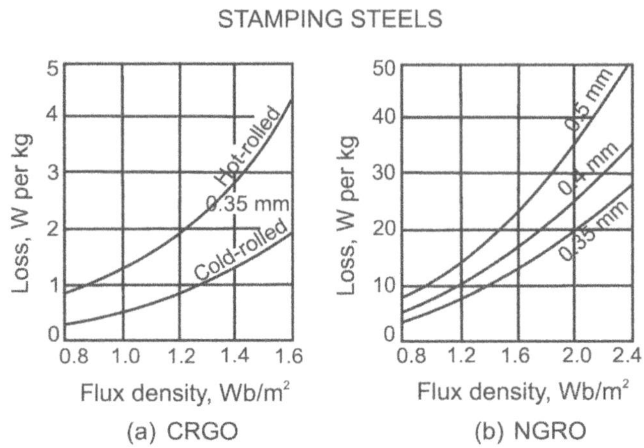

Fig. 2.3 Core loss at 50 c/s.
(a) Transformers (b) Machines

A designer should get the magnetic characteristics, loss curves of different materials from the supplier, study well and decide a suitable material for design.

2.6 Magnetic Circuit Calculations

In any Electro magnetic machine, both the magnetic and electric circuits exist. Magnetic circuit is a closed path like electric circuit. Let us understand a magnetic circuit (Fig. 2.5) by comparing an electrical circuit (Fig. 2.4).

S. no	Electric-Circuit	Magnetic-Circuit
1	Resistance (R) = ρ × L/a	Reluctance (S) = L/(μ × a) = L/(μ0 × μr × a)
2	Voltage (V)	Mmf (Ampere –Turns)
3	Current (I) = V/R	Flux(Φ) = mmf / Reluctance
4	Voltage(V) = I × R	Mmf (AT) = Φ × S
5	If there are 3 resistances in series, V = V1 + V2 + V3 = I × (R1 + R2 + R3)	If there are 3 different cross-sections in series, AT = AT1 + AT2 + AT3 = Φ × (S1 + S2 + S3)

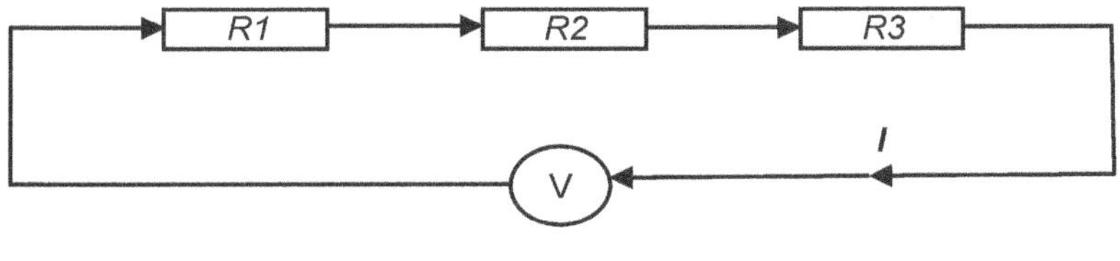

Fig. 2.4 Electrical Circuit with 3-Resistance Elements.

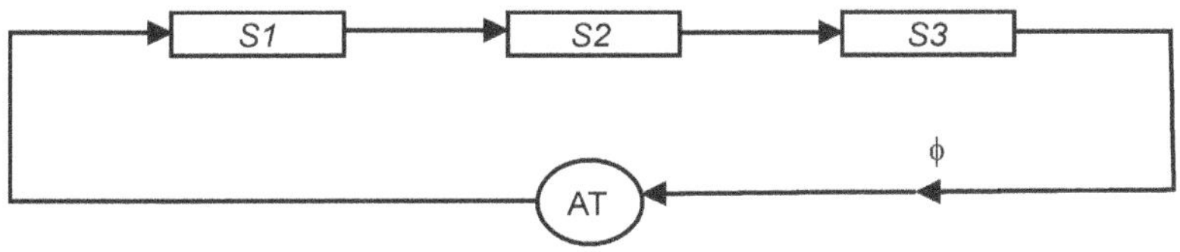

Fig. 2.5 Magnetic Circuit with 3 parts with different Reluctances.

We take the example of a DC generator. EMF induced $= \dfrac{\Phi ZNP}{60A}$ α "Φ" at rated speed of "N" RPM since other parameters are constant. When the field circuit is connected to Voltage Supply (by connecting across the armature or voltage Mains) field current is produced. This field current passing through field turns produces mmf (AT) and the mmf produces flux in inverse relation to total circuit reluctance (S).

Refering to above table, if the values of S1, S2 and S3 are less, then Ampere Turns (AT) drawn by the circuit to produce rated flux "Φ" is lesser. If "AT" are lesser, field current (If) drawn will be lesser since If = AT/T, where T = No. of field coil turns which are constant. Then the CS (cross section) area of copper required is less and weight of copper will be less.

Since total reluctance,

$$(S) = S1 + S2 + S3 = \frac{L1}{\mu 0 \times \mu r1 \times a1} + \frac{L2}{\mu 0 \times \mu r2 \times a2} + \frac{L3}{\mu 0 \times \mu r2 \times a3}$$

It will be lesser if the permeabilities of the materials of the three parts are higher for given lengths and CS areas. Magnetic circuit design is good if the total reluctance is reduced and field current drawn is minimized.

Also ampere turns in any part (AT1) = L1 × H1 = L1 × B1/ (μ0 × μr), where "L1" is the length (m), "H1" is AT/m and "B1" is the flux density in that part.

In a DC machine, Path of Magnetic flux is as shown in Fig. 2.6.

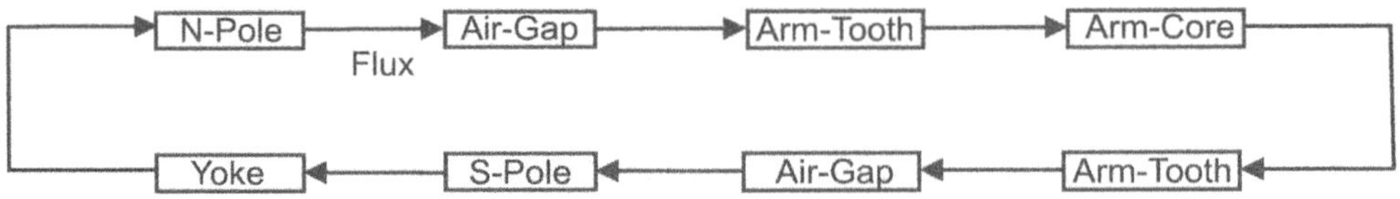

Fig. 2.6 Magnetic-Circuit in a DC machine.

2.7 General Procedure for Calculation of Amp-Turns

(a) Flux/pole (Φ) value is calculated using the formula

(b) Flux density (B) in each part is calculated ($B = \frac{\text{Flux}}{\text{Area of fluxpath}}$)

(c) From the B-H characteristic of the material used for that part (ref Fig. 2.1) value of AT/m (H) is read out

(d) Length of the magnetic path (L) in meters of that part is calculated

(e) AT required for that part = H × L (If the cross-section of the path is not uniform ex:- armature tooth), the path is again split into no of sub-parts, BandH values of each sub-part is estimated and average value of H is arrived at

(f) Same way AT required for each part is calculated

(g) Algebraic sum of AT required for each part gives the total AT required for the Circuit.

S. no	Part	Length (L)	Area of CS (a)	Flux density (B)	AT/m (H)	AT = H × L
1	Air-Gap					
2	Arm -Tooth					
3	Arm -Core					
4	Pole					
5	Yoke					
	Total AT					= ΣAT

Further Details for Calculations

1. For Air-Gap

Amp turns for Air-Gap (ATg) = $0.796 \times Bg \times Kg \times Lg \times 10^6$

Here the additional term "Kg" is the Total Air-Gap Coefficient, where Kg = Kgs × Kgv.

(a) The factor "Kgs" is the gap coefficient for the slot and

$$Kgs = \frac{SlotPitch}{SlotPitch - SlotWidth \times Cg}$$

where "Cg" is the Carter Gap Coefficient and depends on ratio of slot width to Air gap which is to be read from given curve.

(b) The factor "Kgv" is the gap coefficient for ducts and

$$Kgv = \frac{L}{L - nv \times bv \times Cv} \text{ where}$$

L → Gross length of Armature

Nv → No of vent ducts

bv → Width of vent duct

Cv → Carter's coefft (read from given curves)

2. For Armature Tooth

(a) Flux density at one-third section from the narrow end of the tooth is calculated and corresponding value of "H" is read and therefrom "AT" are calculated.

(b) Flux from air gap entering the armature gets divided into 2 parts: (1) majority through tooth (2) slightly through slot. The flux density obtained with the ratio of total flux to the tooth cross section is the apparent Flux density (Bappt). To get the value of "H", one has to use the curves drawn for apparent flux density vs. AT/m for various values of "Ks" where $Ks = \frac{SP \times GL}{TW \times Li}$ and

SP → Slot pitch

GL → Gross length of Armature

TW → Tooth width at ⅓ from narrow end of tooth

Li → Net iron length

Magnetic Circuit in a Synchronous Machine is as shown in Fig. 2.7.

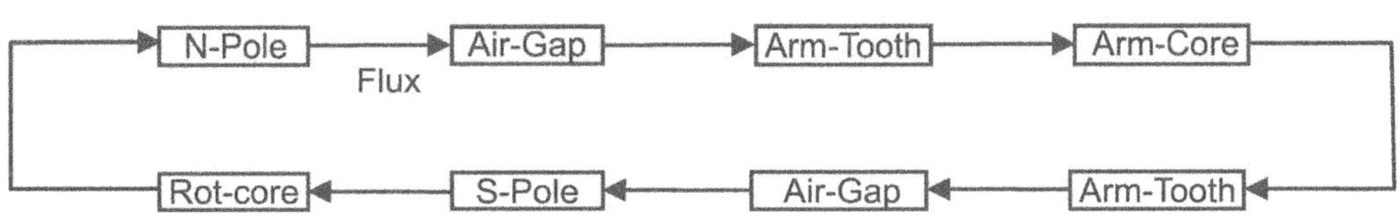

Fig. 2.7 Magnetic-Circuit in a Synchronus machine.

Magnetic Circuit in an Induction Motor is as shown in Fig. 2.8.

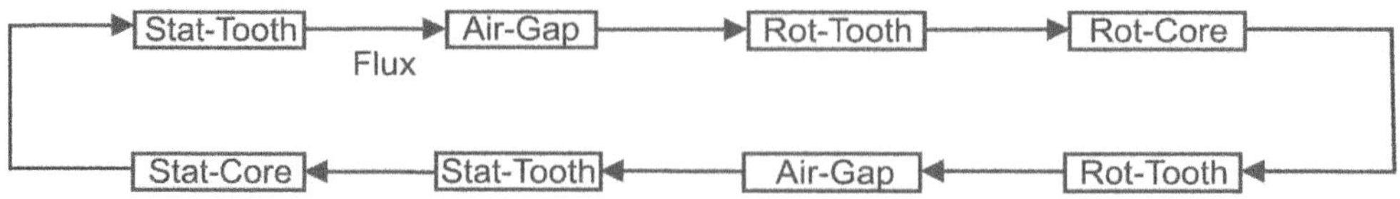

Fig. 2.8 Magnetic-Circuit in an Induction Motor.

2.8 Heating and Cooling

2.8.1 Heating

In electrical machines, heating is the main criterion for design. Electrical machines are designed and manufactured with a selected class of insulation which can withstand a certain temperature. If overheating occurs, insulation will get weakened and results in short circuits leading to the damage of the machine.

In general excess temperatures can change the following insulation properties :

(a) Decrease in Resistance

(b) Decrease in Electrical Strength

(c) Increase in Dielectric Loss angle

(d) Increase in Dielectric losses

(e) Decrease in tensile strength.

As result of these, life of the machine reduces.

In all electrical equipments, various losses produce heat which increases the temperature. If this heat produced is not dissipated, temperature goes on increasing resulting in cracking the insulation and failure of the machine. This dissipation of heat occurs in three modes (i.e.) Conduction, Convection and Radiation. If a perfect cooling medium is designed the heat produced is continuously dissipated so that temperature stabilizes at some value (θmax) and there will not be further increase at that particular load. "θmax " value at rated load is the criterion for deciding class of insulation and further design aspects of the machine.

When a machine is loaded at time t = 0, and when the temp rise is zero (Temp rise = M/c temp – amb temp), the temp-rise gradually increases exponentially with respect to time and and after certain time it attains the steady state value (θmax) governed by the equation $\boldsymbol{\theta = \theta max\ (1 - e^{-t/\tau})}$, where θ → Temperature rise at any time (t) and τ → heating time constant of the machine which is calculated from formula

$$\tau = \frac{GS}{A\lambda}$$

where

G → Weight of the machine (Kg)

S → Average specific heat (Watt-sec/Kg/°C)

A → Area of cooling surface (sq.m)

λ → Specific heat dissipation from the cooling surface (Watts/m^2/°C)

2.8.2 Cooling

When load is removed and the machine is stopped, temperature-rise gradually decreases and cools down exponentially with respect to time, as governed by the formula

$$\theta c = \theta h \times e^{-t/\tau c}$$

where

θh → Temp-rise of the machine just before stopping and load removed

θc → Temperature rise at any time(t)

τc → cooling time constant of the machine

Effect of frequent heating and cooling on a machine

If a machine is subjected for loading and un-loading many times in day, thermal expansion and contraction of insulation occurs and results in early cracking down of insulation and reduction in its life period. Hence, design of such machines needs additional care.

2.9 Modes of Heat Dissipation

Heat is dissipated by three modes: 1. Conduction 2. Convection and 3. Radiation

Since Transformer is a static device, no rotational losses. Heat produced by core and windings are to be dissipated by tank.

In most power transformers watt loss per Kg in the iron and the watt loss per Kg in the copper will be nearly equal. This means that the total losses to be dissipated vary as the weight or volume of the material (L^3), where "L" represents the dimensions of transformer. On the other hand, cooling surface provided by the tank of the transformer varies as (L^2). As the size of the transformer is increases, the ratio of heat generation volume to surface for dissipation (L^3/ L^2) becomes large.

For small transformers a smooth case readily dissipates the heat as can be seen from the above by Convection and Radiation. But in case of rotating machines, rotational losses exist and heat transfer takes place by Conduction and Convection with negligible radiation.

2.10 Standard Ratings of Electrical Machines

Rating of Electrical Machine is mainly decided by the Temp-Rise and Class of Insulation used in the machine. Let us take an example of a electric motor already manufactured with a given class of insulation. If it is loaded continuously at the rated load, let the stabilized temp-rise attained be "T". Now if it is over-loaded by, say, 50% and allowed for continuous operation, the stabilized temperature would be much higher than "T" at which the insulation may fail. Now instead of allowing continuously at this load, if it is loaded for a short time (say, 30 sec) so that the temp-rise will not exceed "T", then the short-time rating of the machine said to be 150% and short time is 30 sec. Similarly if same temp-rise is attained when run at 125% load for 1mt, the one minute rating is 125%.

Based on this understanding, ratings are classified as follows:

(a) Continuous rating

(b) Short time rating

(c) Intermittent-periodic rating (Cyclic loading, ex: 15 mts "ON" and 30 mts "OFF"). For this type of load, there will be initial temp-rise before every start.

(d) RMS horse power rating: When the load on the motor changes in cyclic order in some of the industrial applications such as rolling mills, cranes, hoists etc., in these applications the motor may be required to deliver a particular constant load for a fixed period after which it may deliver another value of constant load for another fixed time, which may be followed by a no-load period. The heating of motor in such type of loadings is proportional to HP delivered to the load. Thus for such load cycles

$$\text{RMS Horse Power rating} = \sqrt{\frac{(HP_1^2 \times t1) + (HP_2^2 \times t2) + (HP_3^2 \times t3) + \ldots}{(t1 + t2 + t3 + \ldots)}}$$

2.11 Ventilation Schemes

2.11.1 In Static Machines (Transformers)

Since transformer is a static device, no heat transfer by conduction and hence cooling is more difficult than a rotating machine and the problem of getting rid of the heat in large transformers is more difficult. This will explain the progressive design with increasing transformer size, fluted tanks, tubular construction (to increase the surface area) and finally in the largest sizes the necessity of artificial cooling.

Natural Radiation (AN): Small transformers using for metering and power are cooled by natural ventilation and convection of heat from their surfaces.

Air Blast (AB): Instead of immersing the transformer in oil the heat is dissipated by a blast of air forced through special ventilating ducts in core and between sections of winding. This method of cooling requires a supply of clean air, fans, and special construction to assure its correct distribution. Its advantage lies in reduced fire and explosion risks.

Oil-Immersed, Self Cooled (ON):- The transformer is immersed in a tank filled with oil, which acts as an insulator also. Heated oil due to its lower density rises up through the circulating ducts of winding giving away its heat to the sides of the tank from which it is radiated to surrounding air. The oil becomes cold and because of its higher density flow down, thus creating a natural circulating path.

Large capacity transformers require corrugations on the surface of the tank or radiating jackets to increase the surface area.

Oil-Immersed, Forced Oil Cooled (OFN): Oil is circulated by a pump

Oil-Immersed, Forced Oil Cooled (OFB): Air is blasted by a fan

Oil-Immersed Water cooled (OFW): Instead of depending entirely upon the transfer of heat from the oil to the outside surface of the tank, some of the heat can be dissipated from the oil (circulated by a pump) in the coiled tubes in the top of the tank. Circulating water is forced through these coils. Occasionally the oil is circulated and cooled outside the transformer.

2.11.2 In Rotating Machines

In rotating machines heat transfer is much better as the rotating component enhances cooling medium pressure and discharge.

Ventilation inside a Rotating Machines is of four types:

(a) ***Radial Ventilation*:** Core is divided into a number of packets separated by radial ventilating ducts of 1cm normal length. When the rotor rotates, air is pressurized by centrifugal action, augmented by fans mounted on rotor, passes radially through the radial ventilating ducts. This circulation of air takes away the heat generated in the windings and core and dissipates the heat to the frame from where it is radiated.

Disadvantage of this method is that net core length is reduced. Advantage is uniform cooling of the machine. This system is used in motors of medium ratings.

(b) ***Axial Ventilation*:** In some of the small induction motors, axial ducts are provided in rotor alone. In some motors they are provided in stator also. Since no radial ducts are provided length of core increases which is advantageous. But the disadvantage is that heat transfer is not uniform throughout the length. Hence, axial ventilation is used in high speed machines with small output.

(c) ***Combined Radial and Axial Ventilation*:** Used for large electrical motors.

(d) ***Forced-Ventilation (Cooling of Turbo-Generators)***

Depending upon the capacity of generators, cooling methods are as follows:

(a) ***Small Ratings*:** Heat generated in core and windings is transferred to stator frame through discharge of air caused by rotation from where it is radiated to atmosphere. Ventilating ducts are provided in the core.

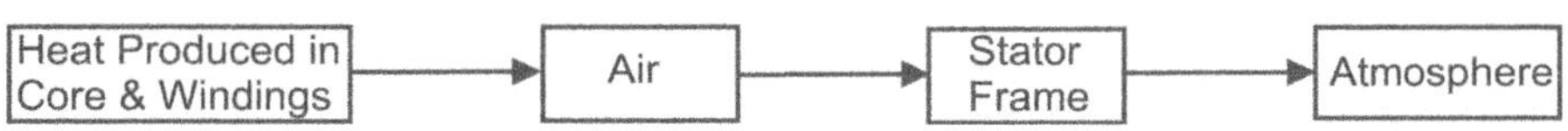

Heat transfer in small ratings.

(b) Medium Ratings

1. **Open Circuit Ventilation** system is employed in medium ratings of machines where cooling medium (Air) pressurized by fan mounted on the rotor, is made to enter the machine from one end, flows through the ventilating ducts and over the over-hang of the winding. Coolant extracts the heat and as a result, its temperature

rises. The hot coolant is now released into surrounding atmosphere through the other end of the machine as shown below. Disadvantage with this system is collection of impurities like dust (present in the atmospheric air) on the parts of the machine even air filters are provided at the inlet.

Heat transfer in medium ratings with open-circuit cooling.

(2a) Closed Circuit Ventilation System is a better method where atmosphere is polluted. In this system, machine is boxed up. Primary coolant (Air) within the machine, which is pressurized by fan mounted on the rotor, is circulated through the ventilating ducts and over the overhang of the winding. Coolant extracts the heat and as a result, its temperature rises. The hot air is now passed through a heat exchanger where it is cooled by a secondary coolant (water) and sucked by the same fan and recirculated. The secondary coolant (water) absorbing heat in the heat exchanger, becomes hotter and then let into atmosphere or downstream of a river. Normally cold water entering the heat exchanger will be taken from a river or a pond pressurized by a centrifugal pump. This water circulation is in open cycle. (refer sketch given below).

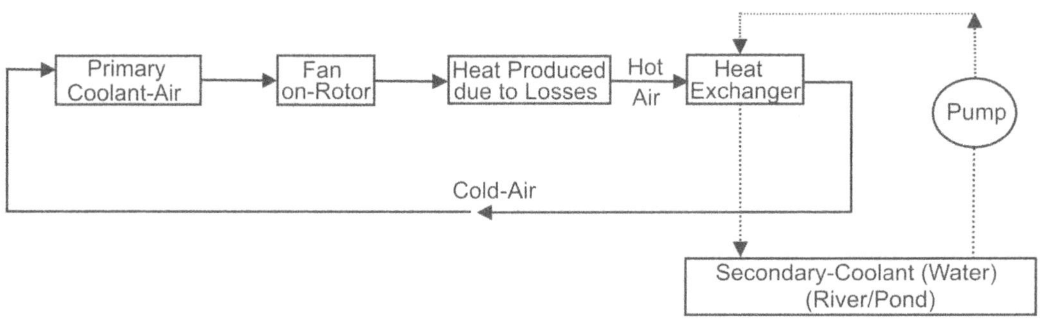

Heat transfer in medium ratings with closed-circuit cooling.

In Larger Ratings of Generators, hydrogen cooling is adopted.

Advantages with Hydrogen Cooling

(a) Hydrogen is 14 times lighter than air and hence ventilation losses are lesser, resulting in higher efficiency.

(b) Heat transfer coefficient is 1.4 times higher than that of air and thermal conductivity is 6.6 times higher than that of air. Hence, for a given volume of machine, KVA output is more.

(c) Since ventilation losses are lower, noise generated will be less.

(d) Life of insulation will be more due to absence of nitrogen. (In air cooled machines, ozone formation due to corona reduces insulating property.)

(e) No formation of dust on machine parts like in air cooled machines.

(f) Since hydrogen is not a supporter of combustion, risk of fire within the generator is eliminated.

Disadvantages

(a) If air content in the hydrogen exceeds 22% the gas will explode. To prevent this, hydrogen cooled generators are to be sealed with seal oil system to prevent leakage of hydrogen. Frame is also to be designed against possible explosion.

(b) Hydrogen can not be directly filled into the generator as air is present. First, air has to be purged out by another inert gas (CO_2) which is later to be replaced by H_2. Reverse process is to be followed during shutting down. These operations increase the time of starting and shutting down.

(c) These three auxiliary systems (H_2, CO_2 and Seal Oil) along with their control panels increase both initial and operational costs in addition to requirement of more space. Hence hydrogen cooling is not economical below certain ratings.

(d) Explosion hazard in the space surrounded by the machine exists with any leakage of hydrogen from the generator.

(2b) Hydrogen Cooled Generator is Always Closed Circuit Ventilation System. In this system, machine is boxed up tightly. Primary coolant (Hydrogen) within the machine is pressurized by fan mounted on the rotor, is circulated through the ventilating ducts and over the overhang of the winding. Coolant extracts the heat and as a result, its temperature rises. The hot hydrogen is now passed through a heat exchanger where it is cooled by a secondary coolant (water) and sucked by the same fan and recirculated. The secondary coolant (water) after absorbing heat in the heat exchanger, becomes hotter and then let into atmosphere or down stream of a river. Normally cold water entering the heat exchanger will be taken from a river or a pond pressurized by a centrifugal pump. This water circulation is in open cycle. (refer sketch given below).

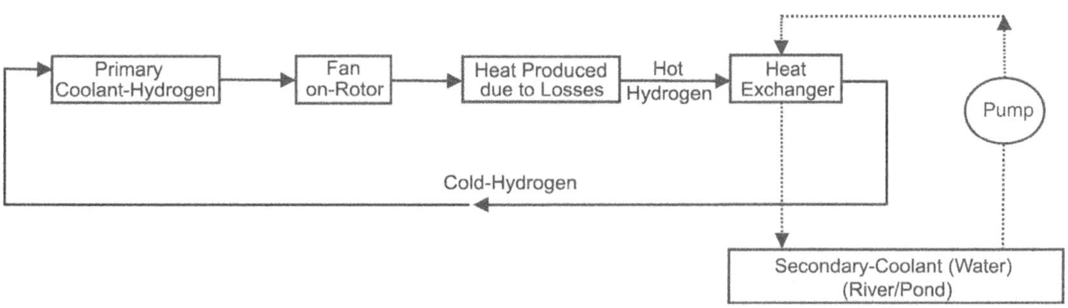

Heat transfer in larger ratings with closed-circuit hydrogen cooling.

3. Water Cooling

In large generators of ratings above 500 MW, water is used as cooling medium as hydrogen cooling is not sufficient. Since specific heat and thermal conductivity of water is much higher than hydrogen, water cooling is techno-economically much advantageous.

2.12 Quantity of Cooling Medium

The Quantity of cooling medium needed to dissipate the heat caused by the losses in the machine is given by the equation $(H) = (Q \times d) \times Cp \times \theta$, where

H = Losses to be dissipated by cooling medium (KW)

Q → Discharge of cooling medium (m^3/sec)

d → Density of cooling medium(Kg/m^3)

Cp → Specific heat (KW-sec)/Kg/°C

θ → Temp-rise (°C)

It is to be noted that losses to be dissipated by cooling medium do not include bearing friction losses which are normally cooled by bearing oil in medium and larger generators.

In medium and larger generators two fans will be mounted on rotor on either end of generator, each fan designed to produce half the discharge (Q/2). The machine is having a ventilation system balanced on both sides with reference to axial centre of the machine.

2.13 Types of Enclosures

Electrical machine is protected by a metallic cover called enclosure against ingress of moisture, dust, atmospheric impurities and any foreign material. The degree of protection varies in different environments. If the machine is provided under a roof, it is safe from certain problems like falling of rain, snow etc. But still protection is required from air born dust etc. If the machine is not having a roof, higher degree of protection is required.

If higher degree of protection is provided, cooling is lower and vice versa.

Depending upon the required degree of protection, enclosures are classified into following types:

(a) *Open Type:* Ends of machine are in contact with atmosphere. Cooling is better. Here it is with lowest degree of protection.

(b) *Protected type:* End covers are provided with holes for ventilation.

(c) *Screen protected type:* A wire mesh to prevent foreign bodies is additionally provided for protected type (b).

(d) *Drip Proof type:* In a damp environment hanging bowls are provided, so that condensed moisture does not enter the machine.

(e) *Totally Enclosed type:* Machines of closed circuit cooling as mentioned above are provided with this type of enclosure.

(f) *Flame type:* Provided for machines working in explosive and fire hazard environments like coal mines etc.

2.14 General Design Procedure

Sequential steps involved in the design and manufacture of any product:

1. Customer specification as per contract, if available, should be read and salient points of design parameters to be highlighted.
2. Latest National/International standards applicable for this design should be referred.
3. Calculation of main dimensions and subsequently dimensions of each part and Performance parameters, using well-proven computer programs, established with equations, scientific formulae, empirical formulae based on previous experience, curves, tables, charts, etc.
4. Ensuring that the volume and weight of the product do not pose any problem for either manufacture at works or transport to site or erection and commissioning at site. Any foreseen problems should be solved before the related activity begins.
5. Preparation of the specifications of each type of materials used in the product.
6. Preparation of drawings of each part and furnishing to manufacturing shops, Purchase department for purchase of raw materials, tools and sub-contracted items.
7. Writing of process (Sequential steps involved): How to manufacture each part, clearly indicating the types of tools, machines, workmen, etc.
8. Writing of process: How to sequentially assemble each part/component.
9. Writing the process how to carry out tests on each component and fully assembled machine to check the quality as specified by standards.
10. Manufacturing the components and carrying out in-process tests.
11. All components are assembled and testing has to be carried out on full machine
12. If it is a new design, additional tests (type tests) to be done over and above the normal routine tests specified.
13. Dispatching the machine to customer site where it is erected and commissioned to keep ready for normal operation.
14. Loading the machine at rated conditions and checking the performance.

2.15 Steps to Get Optimal Design

(a) Input parameters like KW, Voltage, PF, Frequency, and any parameter guaranteed to customer etc., to be kept constant.

(b) Operating range (minimum and maximum values) of various input design parameters to be selected (like Flux densities, Current densities etc.)

(c) Maximum and minimum values of certain output parameters to be incorporated (like number of stator slots, % Regulation etc.)

(d) A well proven computer program is run to print out various possible alternative designs.

(e) Optimization criteria to be identified. (like lower cost, lower weight, lower Kg/KVA, Higher efficiency, lower temp-rise, etc.)

(f) Optimal design is selected to suit the optimization criteria.

CHAPTER 3

Armature Windings

3.1 Introduction

In this chapter of armature windings, details are given how to design the scheme of winding after completion of calculations of number of poles, number of slots, conductors per slot, chording pitch etc.

3.2 Important Terms Related to Armature Windings

(a) *Conductor:* Active length of winding wire or strip in a slot.

(b) *Turn:* A turn consists of two conductors placed inside separate slots on the armature periphery approximately a pole pitch apart as shown in Fig. 3.1. These conductors lie under opposite poles so that emf induced in the turn is additive.

(c) *Coil:* A coil may consist of one or more number of turns connected in series. If it contains two or more number of turns, it is called a multi-turn coil. Refer Fig. 3.1.

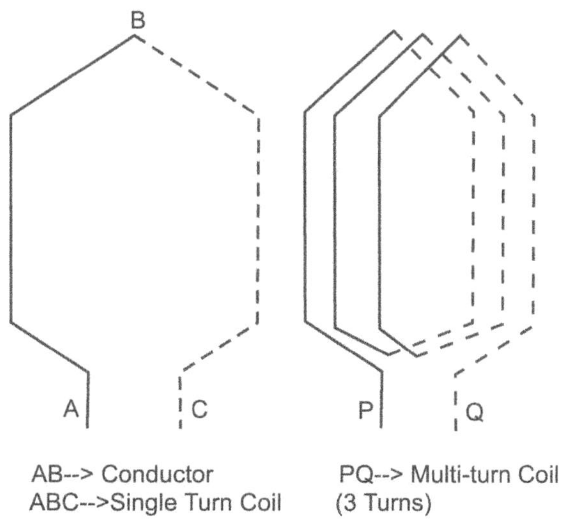

Fig. 3.1

(d) *Coil-side:* A coil consists of two coil sides, upper or lower coil sides placed in two different slots, approximately a pole pitch apart.

(e) *Over-hang:* End portion of coil connecting the two conductors or coil sides is called overhang winding.

(f) *Coil-Span:* Distance measured in a number of slots between two coil sides of a coil placed in slots over the periphery of the armature.

(g) *Full Pitch Coil:* If the coil span is exactly equal to a pole pitch (=slots/pole) then it is full pitched coil.

(h) *Short Pitched coil/Chorded coil:* If the coil span is not equal to a pole pitch (usually less only) then it is called Short pitched coil.

3.3 Classification of Armature Windings

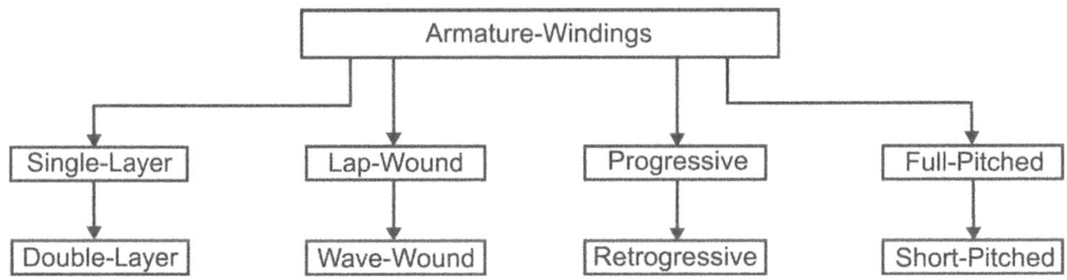

(a) *Single-Layer*: Slot contains only one conductor or one coil side

(b) *Double-Layer*: Slot contains two coil sides. Top/upper coil side of a slot (A) in Fig.3.2 connected to bottom/lower coil side of another slot (B) which is approximately one pole pitch away and bottom/lower side of slot (A) connected to top/upper coil side of another slot (C) which is approximately one pole pitch away.

In Fig. 3.2, A, B, C are three slots located on the armature at approximately at one pole pitch apart, having (1, 2), (3, 4) and (5, 6) as top and bottom conductors respectively. Coil side no.1 (top) of slot A is connected to Coil side.4 (bottom) of slot B and Coil side.2 of slot A(bottom) is connected to coil side 5(top) of slot C.

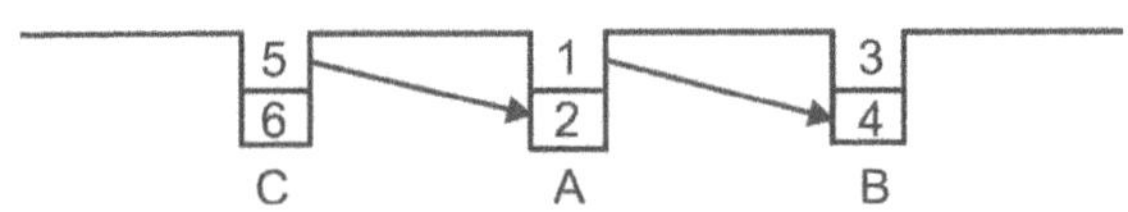

Fig. 3.2

Each turn consists of two coil sides placed in slots separated by approximately one pole pitch apart. Each turn has to be connected with the other to complete the total winding. This connection can be done in two ways, either by Lap winding or Wave winding. Finally each coil end will be connected to one commutator segment.

(c) *Lap-Winding*: If both the coil sides are connected to adjacent commutator segments, say, 1 and 2 as shown in Fig. 3.3, and further connections are made in this manner, it is called simple Lap-winding. Here commutator pitch is unity (±1). It is to be noted that both first and second coils connected in series lie under same set of poles. For machines designed for higher currents more number of parallel paths are required. Then it is called Multiplex Lap winding. For example, if two parallel paths are provided, it is called Duplex Lap winding and commutator pitch is (±2).

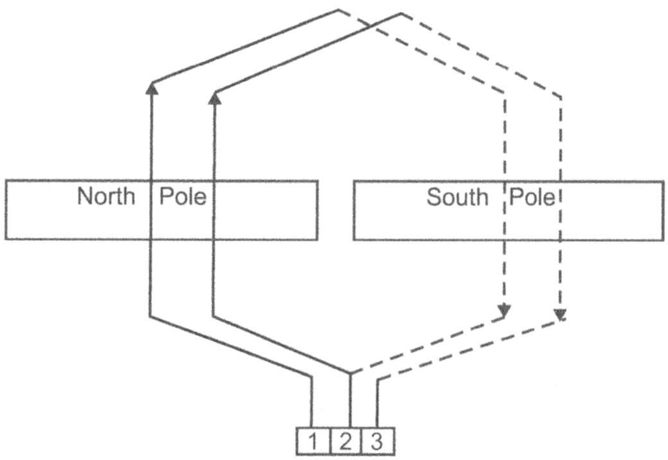

1, 2 & 3 --> Commutator Segments

Fig. 3.3 Lap-winding.

(d) *Wave-Winding*: If both the coil sides are connected to commutator segments which are approximately two poles pitches apart (actually the winding pitch depends on number of poles and number of commutator segments), say, 1 to X as shown in Fig. 3.4 and further connections are made in this manner, it is called simple Wave-winding.

It is to be noted that first and second coils connected in series lie under different set of poles. For machines designed for higher currents more number of parallel paths are required. Then it is called Multiplex Wave winding. For example, if two parallel paths are provided, it is called Duplex Wave winding.

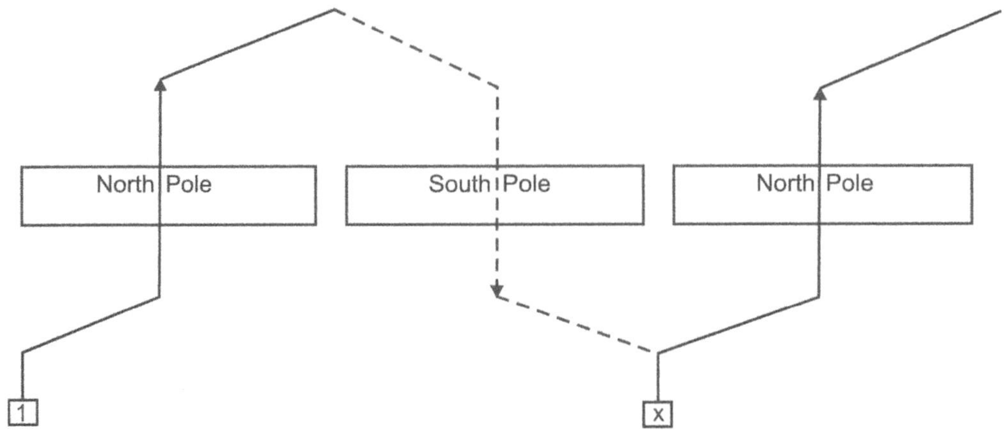

Fig. 3.4 Wave-winding.

(e) *Progressive Winding*: Back pitch (Yb) is higher than the front pitch and winding progresses in clock-wise direction when viewed from commutator end. (Fig. 3.5).

(f) *Retrogressive Winding*: Front pitch (Yf) is higher than back pitch and winding progresses in Anti-clockwise direction when viewed from commutator end.

(g) *Full Pitched Winding*: If the coil span is exactly equal to a pole pitch (=slots/pole) then it is full pitched coil.

(h) *Short Pitched Winding/Chorded Winding*: If the coil span is not equal to a pole pitch (usually less only), then it is called Short pitched coil.

3.4 Winding Pitches (Refer Figures 3.5 and 3.6)

(a) **Back pitch (Yb):** Distance between two coil sides of the coil at the back of the commutator is called Back pitch.

(b) **Front pitch (Yf):** Distance between finish of a coil and the starting of next coil which are connected to same commutator segment, is called Front pitch.

(c) **Resultant pitch:** Distance between starting of two consecutive coils is called resultant pitch (Yr).

For Lap Winding: Yr =Yb – Yf.

For Wave Winding: Yr = Yb + Yf.

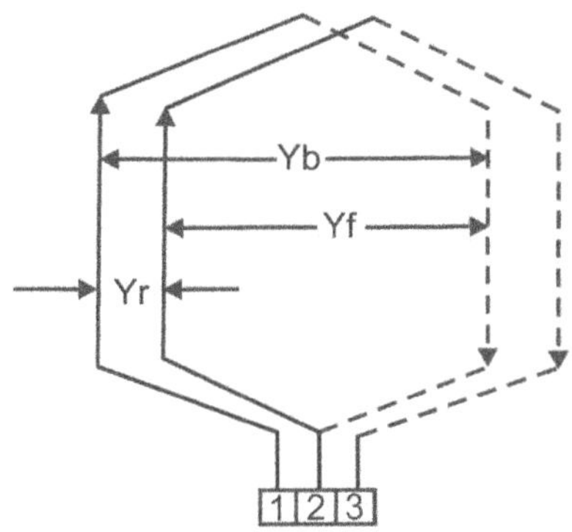

1, 2 & 3 --> Commutator Segments

Fig. 3.5

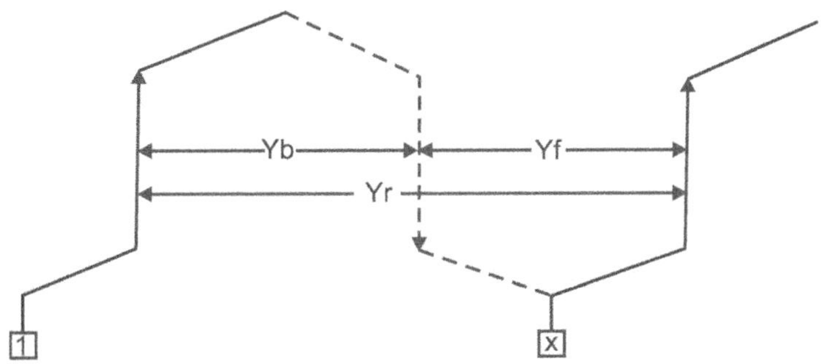

1 & x --> Commutator Segments

Fig. 3.6

3.5 Armature Windings for DC Machines

3.5.1 Rules for Design Development of Lap-Winding for a DC Armature

(a) Back pitch (Yb) and front pitch (Yf) are odd. They are not equal. They differ by 2 or multiples thereof. In general Yb = Yf ± 2m, where "m" is the plex of the winding (1 for simplex, 2 for duplex, 4 for quadraplex etc.)

If Yb = Yf + 2m, it is progressive winding and If Yb = Yf – 2m, it is retrogressive winding.

(b) Both Yf and Yb should be nearly equal to pole pitch.

(c) Average pitch (Ya) = $\frac{Yf + Yb}{2}$ which equals pole pitch $= \frac{Z}{P}$ where "Z" is the total number of conductors and "P" is the number of poles.

(d) Resultant pitch (Yr) = Yb – Yf will always be even since arithmetical difference of two odd numbers is even.

(e) Number of coils is equal to number of commutator segments.

(f) Number of parallel paths in armature is equal to "mP", where "P" is number of poles and "m" is the plex of winding.

(g) Commutator pitch Yc = ± 1m, "+1" for progressive and "–1" for retrogressive windings.

(h) $Yb = \frac{Z}{P} + 1$ and $Yf = \frac{Z}{P} - 1$ for Progressive winding

$Yb = \frac{Z}{P} - 1$ and $Yf = \frac{Z}{P} + 1$ for Retrogressive winding.

3.5.2 Example for 4 Pole, 16 Slot DC Machine with Progressive Simplex Double Layer Lap Winding (Refer Fig. 3.7)

Number of Poles (P) = 4;

Total number of conductors (Z) = slots × conductors/slot = 16 × 2 = 32

For Progressive winding, Back Pitch $(Yb) = \frac{Z}{P} + 1 = \frac{32}{4} + 1 = 9$

and Front Pitch $(Yf) = \frac{Z}{P} - 1 = \frac{32}{4} - 1 = 7$

In Double Layer winding, top conductors in slots are given odd numbers (1, 3,.......31) andbottom conductors are given even numbers (2, 4.........32)

To start with top conductor in slot no.1 is connected to bottom conductor of (1 + Yb) = (1 + 9) = 10 on one end. The other end of conductor "10" is connected to (10 – Yf) = (10 – 7) = 3 and it goes on accordingly as per the following table.

Sr. No	No. of Top Conductor (Zt)	No. of Bottom Conductor (Zb) Zb = Zt + Yb	No of Top Conductor (Zt) Zt = Zb – Yf
1	1	1 + 9 = 10	10 – 7 = 3
2	3	3 + 9 = 12	12 – 7 = 5
3	5	5 + 9 = 14	14 – 7 = 7
4	7	7 + 9 = 16	16 – 7 = 9
5	9	9 + 9 = 18	18 – 7 = 11
6	11	11 + 9 = 20	20 – 7 = 13
7	13	13 + 9 = 22	22 – 7 = 15
8	15	15 + 9 = 24	24 – 7 = 17
9	17	17 + 9 = 26	26 – 7 = 19
10	19	19 + 9 = 28	28 – 7 = 21
11	21	21 + 9 = 30	30 – 7 = 23
12	23	23 + 9 = 32	32 – 7 = 25
13	25	25 + 9 = 34 (= 34 – 32 = 2)	34 – 7 = 27
14	27	27 + 9 = 36 (= 36 – 32 = 4)	36 – 7 = 29
15	29	29 + 9 = 38 (= 38 – 32 = 6)	38 – 7 = 31
16	31	31+ 9 = 40 (= 40 – 32 = 8)	40 – 7 = 33(33 – 32 = 1)

Check: After Sr.no.16 (16th slot) end connection is No.1 conductor, which is same as starting point and hence the winding is OK.

3.5.3 How to Draw the Lap Winding Diagram

Now conductors/pole = Z/P = 32/4 = 8. and no. of brush arms = no. of poles = 4.

Pole No.	1	2	3	4
Polarity	North	South	North	South
Slot Nos. under the Pole	1,2,3,4	5,6,7,8	9,10,11,12	13,14,15,16
Cond. Nos.	1 to 8 (8nos)	9 to 16 (8nos)	17 to 24 (8nos)	25 to 32 (8nos)
Polarity of Induced EMF	+V	-V	+V	-V
No. of coils	4	4	4	4
No. of Comm-Segments	4	4	4	4
No. of Brush arms	1	1	1	1

Sequential Steps:

(a) A graph sheet is taken and 32 conductors are marked and numbered

(b) Represent 4 poles indicating polarities N, S, N, S covering all 32 conductors

(c) Represent direction of induced emf on each conductor by an arrow as per polarity

(d) All 16 commutator segments are represented and numbered

(e) Comm-segment no.1 is connected to one end of top conductor no.1 and its other end is connected to bottom conductor no.10

(f) Other end of bottom conductor no.10 is connected to commutator segment no.2 to which top conductor no.3 is also connected

(g) Other end of top conductor no.3 is connected to bottom conductor no.12

(h) Other end of bottom conductor no.12 is connected to commutator segment no.3 to which top conductor no.5 is also connected

(i) Following this procedure all connections are to be completed as per drawing shown in Fig. 3.7.

3.5.4 How to Represent the Position of Brushes

It is observed that both emfs induced in conductor nos.1 and 8 connected to commutator segment no.1 are entering and adding together making it +ve potential. Hence, a +ve brush arm (A) is kept touching comm-segment no.1. So also commutator segment no.9 is +ve and 2nd +ve brush arm (B) is provided touching it.

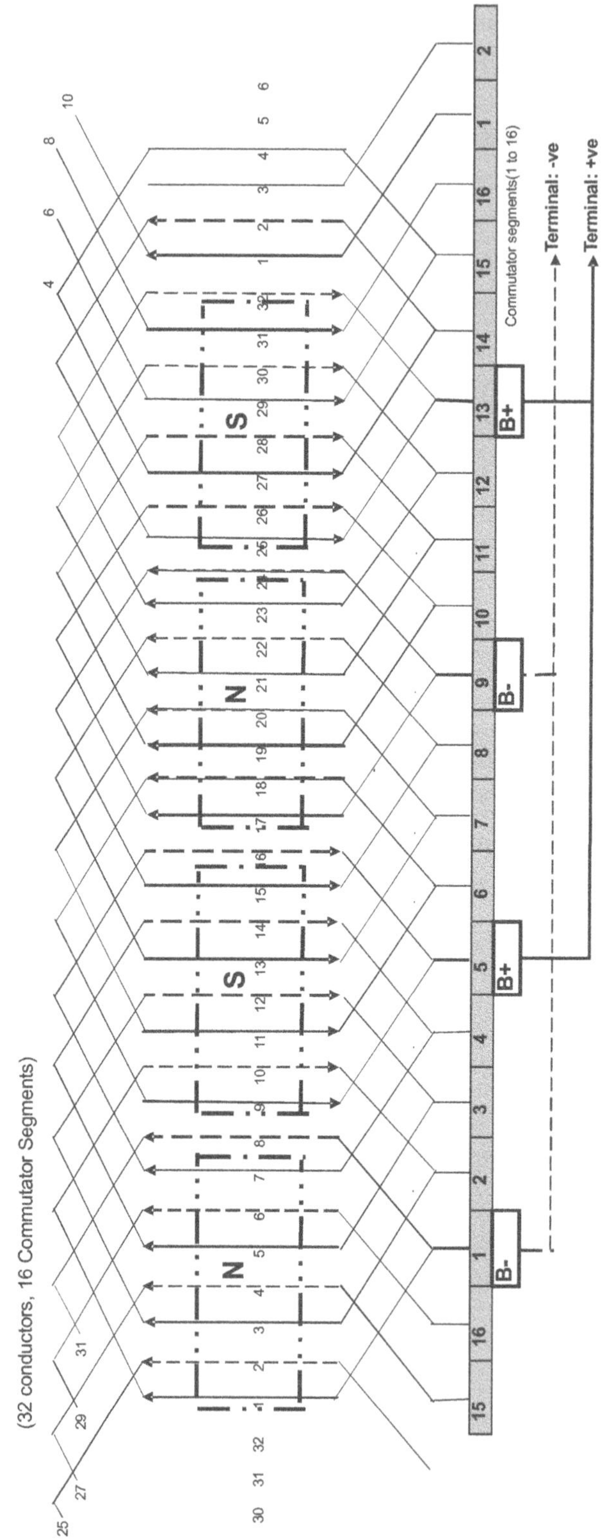

1) Conductors with odd numbers are "TOP" of slot & Conductors with even number are "BOTTOM " of slot

2) Direction of Arrows represent Direction of Induced EMFs

3) B+ --> Positive brush & B-ve --> Negative Brush

Fig. 3.7 Winding diagram of 4 pole, 16 slot, double layer simplex lap winding.

Using the same logic –ve brush arms C and D are positioned touching commutator segments 5 and13 where emfs/currents are leaving.

Both +ve brush arms A and B are connected together making one single positive terminal and both –ve brush arms C and D are connected together making one single negative terminal.

Now number of brushes in each arm = $\dfrac{\text{Current / Parallel Path}}{\text{Current density \& Brush Area}}$

3.5.5 Rules for Design Development of Wave-Winding

(a) Back pitch (Yb) and front pitch (Yf) are odd. They may be equal or differ by 2. Both Yf and Yb should be nearly equal to pole pitch.

(b) Average pitch (Ya) = $\dfrac{Z \pm 2}{P} = \dfrac{Z/2 \pm 1}{P/2} = \dfrac{\text{No. of Commutator Bars} \pm 1}{\text{Pairs of Poles}}$

(c) Resultant pitch (Yr) = Yb + Yf

(d) Commutator pitch Yc = Ya

(e) No. of coils is equal to no. of commutator segments

(f) No. of parallel paths in armature is equal to "2m" where "m" is the plex of winding

3.5.6 Example for 4 Pole, 17 slot DC Machine with Double Layer Wave Winding

No. of Poles (P) = 4;

Total no. of conductors (Z) = slots × conductors/slot = 17 × 2=34

Average Pitch (Ya) = $\dfrac{Z \pm 2}{P} = \dfrac{34 \pm 2}{4}$ = 8 or 9 and Resultant pitch (Yr) = 16 or 18

(Here three options are available: (1)Yf = Yb = 9, (2)Yb = 7 and Yf = 9, (3)Yb = 7 and Yf = 9, all are OK). Now Yf = Yb = 9 option is selected.

Commutator pitch (Yc) = Ya = 9

Since this is a Double Layer winding, top conductors in slots are given odd numbers (1, 3,.......33) and bottom conductors are given even numbers (2,4.........34).

To start with, top conductor in slot no.1 is connected to bottom conductor of (1 + Yb) = (1 + 9) = 10 on one end. The other end of conductor "10" is connected to (10 + Yf) = (10 + 9) = 19 and it goes on accordingly as per the following Table.

Sr. No	No. of Top Conductors (Zt)	No. of Bottom Conductor (Zb) Zb = Zt + Yb	No of Top Conductor (Zt) Zt = Zb + Yf
1	1	1 + 9 = 10	10 + 9 = 19
2	19	19 + 9 = 28	28 + 9 = 37(37 – 34 = 3)
3	3	3 + 9 = 12	12 + 9 = 21
4	21	21 + 9 = 30	30 + 9 = 39 (39 – 34 = 5)
5	5	5 + 9 = 14	14 + 9 = 23
6	23	23 + 9 = 32	32 + 9 = 41 (41 – 34 = 7)
7	7	7 + 9 = 16	16 + 9 = 25
8	25	25 + 9 = 34	34 + 9 = 43 (43 – 34 = 9)
9	9	9 + 9 = 18	18 + 9 = 27
10	27	27 + 9 = 36 (36 – 34 = 2)	2 + 9 = 11
11	11	11 + 9 = 20	20 + 9 = 29
12	29	29 + 9 = 38 (38 – 34 = 4)	4 + 9 = 13
13	13	13 + 9 = 22	22 + 9 = 31
14	31	31 + 9 = 40 (40 – 34 = 6)	6 + 9 = 15
15	15	15 + 9 = 24	24 + 9 = 33
16	33	33 + 9 = 42 (42 – 34 = 8)	8 + 9 = 17
17	17	17 + 9 = 26	26 + 9 = 35 (35 – 34 = 1)

Check : After Sr.no.17 (17th slot) end connection is No.1 conductor which is same as starting point. It is OK.

3.5.7 How to Draw the Wave Winding Diagram

No. of comm-segments = no. of coils = 17 and no. of brush arms = no. of poles = 4

Pole No:	1	2	3	4
Polarity	North	South	North	South
Polarity of Induced EMF	+V	-V	+V	-V
No of Brush arms	1	1	1	1

Sequential Steps

(a) A graph sheet is taken and 34 conductors are marked and numbered.

(b) Represent 4 poles indicating polarities N, S, N, S covering all 34 conductors.

(c) Represent direction of induced emf on each conductor by per polarity.

(d) All 17 commutator segments are represented and numbered.

(e) Comm-segment no.1 is connected to one end of top conductor no.1 and its other end is connected to bottom conductor no.10.

(f) The other end of bottom conductor no.10 is connected to commutator segment no (1 + 9 = 10) to which top conductor no.19 is also connected.

(g) The other end of top conductor no.19 is connected to bottom conductor no.28

(h) The other end of bottom conductor no.28 is connected to commutator segment no: 10 + 9 = 19 (19 – 17 = 2) to which top conductor no.28 is also connected.

(i) Following this procedure all connections are to be completed as per drawing shown in Fig. 3.8.

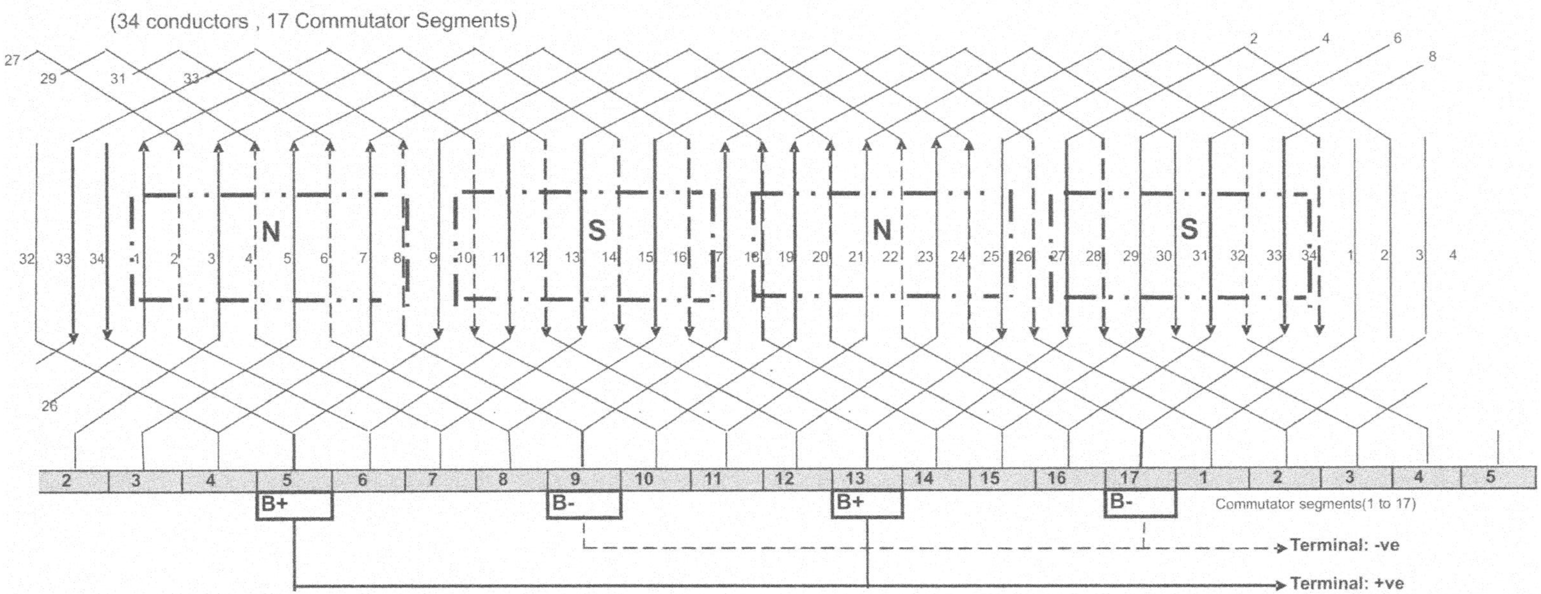

1) Conductors with odd numbers are "TOP" of slot & Conductors with even number are "BOTTOM " of slot

2) Direction of Arrows represents Direction of Induced EMFs

3) B+ --> Positive brush & B-ve --> Negative Brush

Fig. 3.8 Winding diagram of 4 pole, 17 slot, double layer simplex wave winding.

3.5.8 How to Represent the Position of Brushes?

On tracing the trend of emfs through the winding diagram (Fig. 3.8), it is observed that the two emfs meet at one point and separate out at another point. These are the positions of positive and negative brushes respectively. In case these points of meeting and separating the emfs are at the back of the armature, then brushes may be placed at the front (commutator) at positions opposite to these points. It is also seen that there are two parallel circuits in the winding.

3.5.9 Dummy or Idle Coils

These are used with Wave windings and resorted to when the requirements of winding are not met by the standard armature punchings available in armature winding shops. These dummy coils do not influence the electrical characteristics of the winding because they are not connected to commutator. They are only to keep the mechanical balance. They are provided similar to other coils but their ends are cut short and taped.

Example: Suppose a 4 pole, 18 slot, double layer is to be designed with wave winding, $Z = 18 \times 2 = 36$ and $Y_a = \frac{Z \pm 2}{P} = \frac{36 \pm 2}{4} = 9.5$ or 8.5 with which winding is not possible.

Now out of 36 conductors, if 2 conductors (1 coil) is made dummy a winding with 34 conductors is possible as described above.

3.5.10 Criteria for Decision of Type of Winding

For a given number of poles and KW rating a simple wave winding gives double the voltage and half of the current, whereas simple Lap winding gives half the voltage at double the current. (refer the example given below).

Advantage with a wave winding is that equalizer connections are not required.

Winding Type	Voltage(V)	Current(Amps)	Power(Watts)
Wave	2 × V	I/2	P = (2 × V) × (I/2) = VI
Lap	V/2	2*I	P = (V/2) × (2 × I) = VI

Based on a techno-economic study with above considerations, suitable decision is to be made.

3.6 Armature Windings for AC Machines

3.6.1 Similarities between DC and AC Machines

Sr. no.	Parameter	DC - Machine	AC - Machine(Sync)
1	DC magnetic-Flux	Exists	Exists
2	Single and Double Layers	Exists	Exists
3	Lap and Wave Wound	Exists	Exists
4	Progressive and Retrogressive	Exists	Exists
5	Full Pitch and Short pitch	Exists	Exists
6	Wave form of Air-gap flux	Non-Sinusoidal	Non-Sinusoidal
7	Voltage induced in Armature conductors	AC (DC after Commutation)	AC

3.6.2 Differences between DC and AC Machines

Sr. no.	Parameter	DC- Machine	AC - Machine(Sync)
1	Armature	Rotating only	Mostly Stationary (Except in very small ratings)
2	Terminal Voltage	DC	AC
3	Wave form of Voltage	Constant DC	Nearly Sinusoidal
4	Harmonics	Do not Exist	Exists
5	Effect of Chording	Changes the Voltage	Changes magnitude and harmonics
6	Phases	Not applicable	Single/Three Phase
7	Effect of Phase angle	Not applicable	Exists
8	Commutator	Exists	Does not Exist

3.6.2.1 *Additional Terms Applicable for 3 Ph, AC Windings*

(a) Frequency (Hz or cycles/sec)

(b) Phase sequence: R, Y, B displaced at 120° electrical

(c) Phase rotation: R→ Y→ B (or) R→ B → Y

(d) Electrical angle = Pair of Poles × Mechanical Angle

(e) Phase Spread: Electrical Angle in which one half of each phase is spread. Example: For a 3 ph, 2-pole Alternator, $\frac{P/2 \times 360}{3 \times 2} = \frac{2/2 \times 360}{3 \times 2} = 60°E$ (here it is to be understood that 60°E is under N-pole and other 60°E under S-Pole and Total = 120°E for each phase)

(f) Integral Slot Winding: If Slots/Pole/Phase (SPP) is a whole number (Example: for a 72 slots, 4 pole, 3ph armature, SPP = 72/4/3 = 6)

(g) Fractional Slot Winding: If Slots/Pole/Phase (SPP) is not a whole number (Example: for a 66 slots, 4 pole, 3ph armature, SPP = 66/4/3 = 5.5)

(h) Pitch Factor: If the coil span falls short of full pitch by angle by $\alpha°$ (Electrical), then pitch factor = $\cos(n\alpha/2)$, where "n" is the order of Harmonic

(i) Distribution Factor: If $\beta°$ is the slot angle, "m" is slots/phase/pole, then distribution factor = $\frac{\sin(mn\beta/2)}{m\sin(n\beta/2)}$, where "n" is the order of Harmonic.

It is always preferable to use Double Layer, Integral slot, short pitched windings in all 3 phase alternators.

3.6.3 How to draw a 3-ph Single Layer Winding Diagram (Fig. 3.9)

Let us take a simple example of an armature winding for a ***4pole, 3ph, 24 slots, single layer, full pitched Wave wound Alternator.***

It is to be noted here that minimum 36 slots are required in practice. However, the method is the same.

Pole pitch = 24/4 = 6 slots/pole = 180°E.

In one full revolution of rotor a conductor in any slot traverses by 360° mech (= 4/2 × 360 = 720° E). Since there are 24 slots, Slot angle = 720/24 = 30° E.

There are 3 phases R, Y and B separated by a phase angle of 120°. If "R" phase starts at 0°, Y phase at 120° later in the direction of phase rotation and B phase at 240° behind. Since slot angle is 30°, Y phase starts at 120/30 = 4 slots later and B phase starts at 4 more slots later.

In terms of slots if R phase starts at slot.1, Y phase starts at slot no. 1 + 4 = 5 and B phase starts at slot no:5 + 4 = 9.

Since we are assuming full pitched winding, Winding pitch = Pole pitch = 6 slots.

For a single layer winding there will be one conductor per slot and total no. of conductors will be 24*1 = 24.

We designate conductors with numbers: 1, 2, 3, 4, 5,...............24.

Let us Draw the Winding Scheme for "R" Phase First

Conductor (no.1) in first slot is to be connected to one end of conductor (no.1 + 6 = 7) in slot no.7 (1 + winding pitch = 1 + 6 = 7th). The other end of this conductor (no.7) is to be connected to conductor no.7 + 6 = 13 in the 13th slot. The other end of 13th conductor is connected to conductor no.13 + 6 = 19th, thus completing half of the "R" phase. This half connection (R1– R2) is progressive.

The other half connection (R3–R0) is Retrogressive type, starting at conductor no.2 (Terminal.R3), then connected to conductor numbers 2 → 20 (2 + 24 – 6 = 20) → 14 (20 – 6 = 14) → 8(14 – 6 = 8) which the terminal R0.

The purpose of making one half connection progressive and other half to retrogressive is to provide proper mechanical balance of the end winding connections.

Sequential Steps for drawing the Winding Diagram

(a) A graph sheet is taken and 24 conductors are marked and numbered (How ever for convenience represent a few repeated conductors both sides)
(b) Represent 4 poles indicating polarities N, S, N, S covering all 24 conductors
(c) Represent direction of induced emf on each conductor as per polarity (↑ for N pole and ↓ for South pole)
(d) Begin with one end of top conductor (1) and name it R1, starting of 1st half of "R" Phase
(e) The other end of Top conductor no.1 is connected to one end of conductor no.7
(f) The other end of conductor no.7 is connected to one end of conductor no.13
(g) The other end of conductor no.13 is connected to one end of conductor no.19 (which is named as R2)
(h) Begin with one end of conductor (2) and name it R3, starting of 2nd half of "R" Phase
(i) The other end of conductor no.2 is connected to one end of conductor no.20
(j) The other end of conductor no.20 is connected to one end of conductor no.14

(k) The other end of conductor no.14 is connected to one end of conductor no.8 (which is named as R0)

(l) Now R2 and R3 are connected so that all 8 conductors of "R" phase are connected in series to get the rated phase voltage

(m) Now "R1" is phase terminal "R0" will be Neutral terminal of "R" phase

(n) In a similar way "Y" and "B" phase connections are to be made as per the guide lines given below:

(o) Terminals R0, Y0 and B0 are connected together to form the Neutral point of Star Connection.

For "Y" phase

Since "Y" phase starts behind 4 slots, it starts at (1 + 4) = 5th slot. Hence conductor (no.5) in 5th slot is to be connected to conductor (no.5 + 6 = 11) in slot no.11 (5 + winding pitch = 5 + 6 = 11th). The other end of this conductor (no.11) is to be connected to conductor no.11 + 6 = 17 in the 17th slot. The other end of 17th conductor is connected to conductor no.17 + 6 = 23rd, thus completing half of the "Y" phase. This half connection (Y1-Y2) is Progressive type.

The other half connection (Y3-Y0) is Retrogressive type, starting at conductor no. 5 + 1 = 6 (Terminal.Y3), then connected to conductor numbers 6 → 24 (6 + 24 – 6 = 24) → 18 (24 – 6 = 18) → 12 (18 – 6 = 12) which is terminal Y0.

Now Y2 and Y3 are connected so that all 8 conductors of "Y" phase are connected in series to get the rated phase voltage.

Now "Y1" is phase terminal "Y0" will be Neutral terminal of "Y" phase.

For "B" phase

Since "B" phase starts behind 4 more slots, it starts at (5 + 4) = 9th slot. Hence, conductor (no.9) in 9th slot is to be connected to conductor (no.9 + 6 = 15) in slot no.5 (9 + winding pitch = 9 + 6 = 15th). The other end of this conductor (no.15) is to be connected to conductor no.15 + 6 = 21 in the 21th slot. The other end of 21st conductor is connected to conductor no.21 + 6 = 27 (27 – 24 = 3rd) thus completing half of the "B" phase. This half connection (B1-B2) is Progressive type.

The other half connection (B3-B0) is Retrogressive type, starting at conductor no. 9 + 1 = 10 (Terminal.B3), then connected to conductor numbers 10 → 4 (10 – 6 = 4) → 22 (4 + 24 – 6 = 22) → 16 (22 – 6 = 16) which is terminal B0.

Now B2 and B3 are connected so that all 8 conductors of "B" phase are connected in series to get the rated phase voltage.

Now "B1" phase terminal "B0" will be Neutral terminal of "B" phase. Refer Fig. 3.9.

3.6.4 How to Draw a 3-ph Double Layer Winding Diagram

Let us take an example of an armature of a ***2 pole, 3ph, 36 slots, double layer, full pitched, Lap wound Alternator.***

Pole pitch = 36/2 = 18 slots/pole = 180° E = 180° (mechanical) since pairs of poles = 1

In one full revolution of rotor a conductor in any slot traverses by 360°E. Since there ate 36 slots, Slot angle = 360/36 = 10°E

There are 3 phases R, Y and B separated by a phase angle of 120°. If "R" phase starts at 0°, Y phase is at 120° behind and B phase at 240° behind. Since slot angle is 10°, Y phase starts at 120/10 = 12 slots later and B phase starts 24 slots later.

In terms of slots if R phase starts at slot.1, Y phase starts at slot no.1 + 12 = 13 and B phase starts at slot no.13 + 12 = 25.

Since we are assuming full pitched winding Winding pitch = Pole pitch = 18 slots.

For double layer winding there will be 2 conductors per slot and total no of conductors will be 36 × 2 = 72.

We designate top conductors with odd numbers. 1, 3, 5, 7,...............69, 71 and bottom conductors with even numbers. 2, 4, 6, 8..........70, 72. (refer Fig. 3.10.)

Let us Draw the Winding Scheme for "R" Phase First

Top conductor (no.1) in first slot is to be connected to bottom conductor (no.38) in slot no. (1 + winding pitch) = (1 + 18) = 19th slot making the back pitch = (38 - 1) = 37 conductors. The other end of this bottom conductor (no.38) is to be connected to conductor no.3 in the second slot, thus making the front pitch as (38 – 3) = 35 conductors. The average pitch = (37 + 35)/2 = 36 conductors = 18 slots. In the same way winding scheme for total "R" phase is completed as follows:

Phase Group (R)	Sr. No.	No of Top Conductor (Zt)	No. of Bottom Conductor (Zb) Zb = Zt + 37	No. of Top Conductor (Zt) Zt = Zb – 35
I	1	1 (Beginning)	1 + 37 = 38	38 - 35 = 3
	2	3	3 + 37 = 40	40 – 35 = 5
	3	5	5 + 37 = 42	42 – 35 = 7
	4	7	7 + 37 = 44	44 – 35 = 9
	5	9	9 + 37 = 46	46 – 35 = 11
	6	11	11 + 37 = 48	48 → Ending
II	1	37 (Ending)	37 + 37 = 74 (74 – 72 = 2)	74 – 35 = 39
	2	39	39 + 37 = 76 (76 – 72 = 4)	76 – 35 = 41
	3	41	41 + 37 = 78 (78 – 72 = 6)	78 – 35 = 43
	4	43	43 + 37 = 80 (80 – 72 = 8)	80 – 35 = 45
	5	45	45 + 37 = 82 (82 – 72 = 10)	82 – 35 = 47
	6	47	47 + 37 = 84 (84 – 72 = 12)	12 → (Beginning)

Sequential Steps for drawing the Winding Diagram

(a) A graph sheet is taken and 72 conductors are marked and numbered (However for convenience represent repeated conductors on both sides)

(b) Represent 2 poles indicating polarities N, S covering all 72 conductors

(c) Represent direction of induced emf on each conductor as per polarity (↑ for N pole and ↓ for South pole)

(d) Begin with one end of top conductor (1) and name it R1, starting of "R" Phase

(e) The other end of Top conductor no.1 is connected to one end of bottom conductor no.38

(f) The other end of bottom conductor no.38 is connected to one end of top conductor no.3

(g) The other end of top conductor no.3 is connected to one end of bottom conductor no.40

(h) The other end of bottom conductor no.40 is connected to one end of top conductor no.5

(i) Following this procedure all connections are to be completed for the Phase group-I and name the end of last conductor (48) as end of Phase group-I (R2)

(j) Similarly connections for Phase group-II have to be done: Here starting conductor will be in slot no. 1 + Pole Pitch = 1 + 18 = 19th. Top conductor in 19th slot is 37. Hence one end of Phase group-II is 37

(k) Using same logic used for Phase group-I, winding scheme connections for Phase group –II is to be completed as per the above table

(l) The other end of Phase group-II will emerge at conductor no.12

(m) As per the emf direction, emf will enter at 12 and leaves at 37. Hence conductor 12 (R3) is conned to R2 and conductor no.37 will be the exit end of emf for total "R" phase and is named as "R0", which will be Neutral point of Star connection

(n) Refer enclosed drawing (Fig. 3.11).

For "Y" phase

Since "Y" phase starts behind 12 slots, it starts at (1 + 12) = 13th slot. Hence top conductor (no.25) in 13th slot is to be connected to bottom conductor (no.62) in slot no. (13 + winding pitch) = (13 + 18) = 31st. In the same way winding scheme for Phase Group-I of "Y" phase is completed. Here conductor (25) is beginning and named as "Y1". Conductor "72" is the ending and named as "Y2"

Similarly connections for Phase group-II has to be done: Here starting conductor will be in slot no. 13 + Pole Pitch = 13 + 18 = 31st. Top conductor in 31st slot is 61. Hence one end of Phase group-II is 61 and other end emerges as 36th conductor.

Phase Group (Y)	Sr. No	No. of Top Conductor (Zt)	No. of Bottom Conductor (Zb) Zb = Zt + 37	No. of Top Conductor (Zt) Zt = Zb – 35
I	1	25 (Beginning)	25 + 37 = 62	62 – 35 = 27
	2	27	27 + 37 = 64	64 – 35 = 29
	3	29	29 + 37 = 66	66 – 35 = 31
	4	31	31 + 37 = 68	68 – 35 = 33
	5	33	33 + 37 = 70	70 – 35 = 35
	6	35	35 + 37 = 72	72 → Ending
II	1	61 (Ending)	61 + 37 = 98 (98 – 72 = 26)	98 – 35 = 63
	2	63	63 + 37 = 100 (100 -72=28)	100 – 35 = 65
	3	65	65 + 37 = 102 (102 – 72 = 30)	102 – 35 = 67
	4	67	67 + 37 = 104 (104 – 72 = 32)	104 – 35 = 69
	5	69	69 + 37 = 106 (106 – 72 = 34)	106 – 35 = 71
	6	71	71 + 37 = 108 (108 – 72 =3 6)	36 → (Beginning)

As per the emf direction, emf will enter at 36 and leaves at 61 for Phase group-II. Hence conductor 36 (Y3) is conned to 72 (Y2) and conductor no:61 will be the exit end of emf for total "Y" phase and is named as "Y0", which will be Neutral point of Star connection (see Fig. 3.12).

For "B" phase

Since "B" phase starts behind 12 slots to "Y" phase, it starts at (13 + 12) = 25th slot. Hence top conductor (no.49) in 25th slot is to be connected to bottom conductor (no.14) in slot no. (25 + winding pitch) = (25 + 18) = 43rd (43 – 36 = 7th). In the same way winding scheme for phase Group-I of "B" phase is completed. Here conductor (24) is beginning and named as "B2". Conductor "49" is the ending and named as "B1".

Similarly connections for Phase group-II have to be done: Here starting conductor will be in slot no. 25 + Pole Pitch = 25 + 18 = 43rd (43 – 36 = 7th). Top conductor in 7th slot is 13. Hence one end of Phase group-II is 13 and other end emerges as 60th conductor.

Phase Group (B)	Sr. No	No. of Top Conductor (Zt)	No. of Bottom Conductor (Zb) Zb = Zt + 37	No. of Top Conductor (Zt) Zt = Zb – 35
I	1	49(Ending)	49 + 37 = 86 (86 – 72 = 14)	86 - 35 = 51
	2	51	51 + 37 = 88 (88 – 72 = 16)	88 – 35 = 53
	3	53	53 + 37 = 90 (90 – 72 = 18)	90 – 35 = 55
	4	55	55 + 37 = 92 (92 – 72 = 20)	92 – 35 = 57
	5	57	57 + 37 = 94 (94 – 72 = 22)	94 – 35 = 59
	6	59	59 + 37 = 96 (96 – 72 = 24)	24 → Beginning
II	1	13(Ending)	13 + 37 = 50	50 – 35 = 15
	2	15	15 + 37 = 52	52 – 35 = 17
	3	17	17 + 37 = 54	54 – 35 = 19
	4	19	19 + 37 = 56	56 – 35 = 21
	5	21	21 + 37 = 58	58 – 35 = 23
	6	23	23 + 37 = 60	60 → (Beginning)

Here one important point to be noted is that the algebraic sum of emfs entering at Neutral Point is zero. i.e. Er + Ey + Eb = 0 and hence Er + Ey = – Eb. Hence, the direction of induced emf in "B" phase is negative. B0 is + ve and B1 is = –ve.

As per the emf direction, emf will enter at 13 (B0) and leaves at 60 (B3) for Phase group-II. Hence conductor (B3) is connected to (B2). Here 13(B0) is entry point and 49 (B1) is the Exit point of EMF. (see Fig. 3.13).

For all 3 phases

If we superimpose all the above 3 drawings on a single graph sheet, we obtain complete diagram for all 3 phase windings as shown in Fig. 3.14.

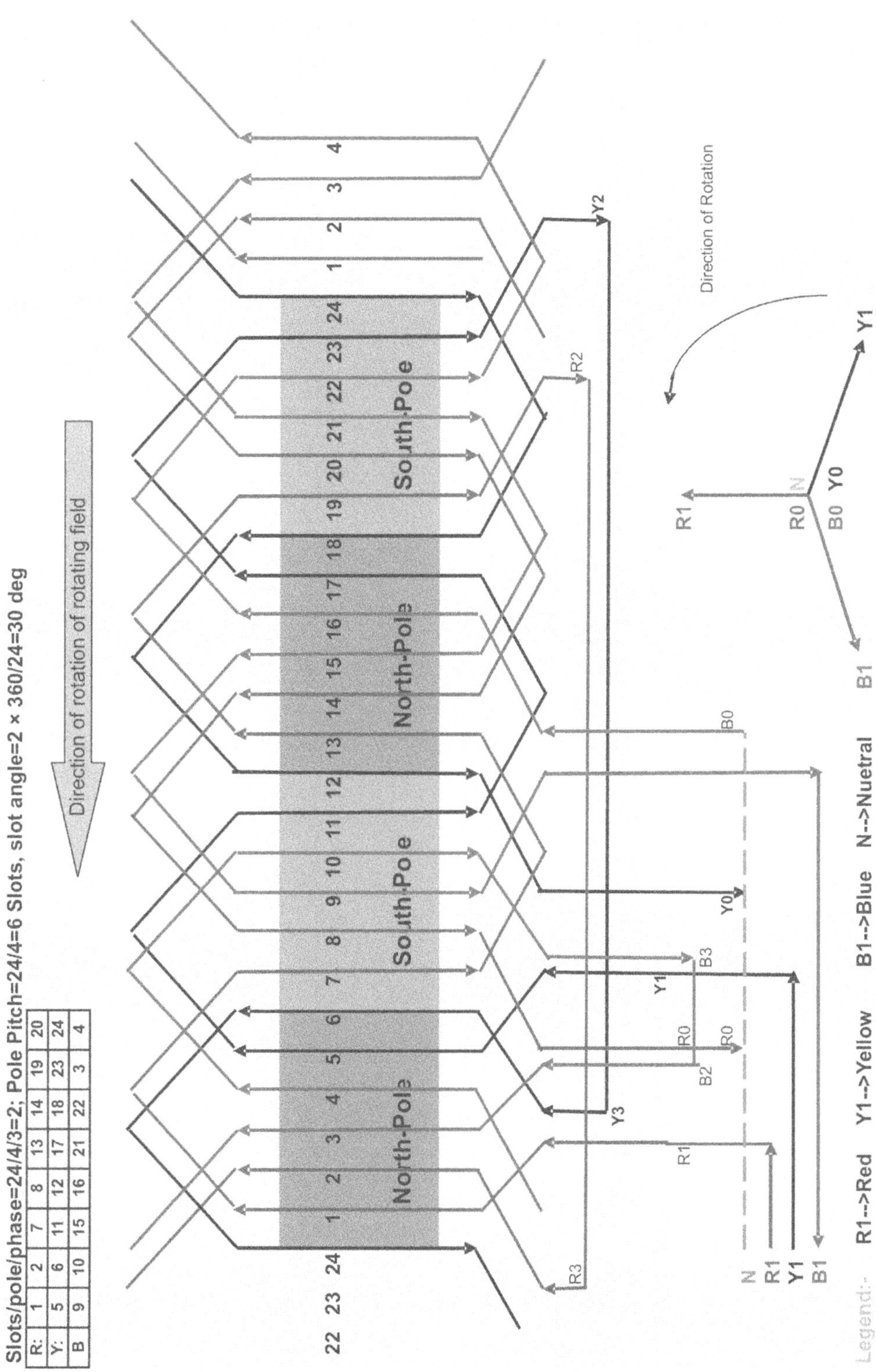

Fig. 3.9 4-Pole, 3 phase, 24 slots, single-layer full pitched Winding of alternator.

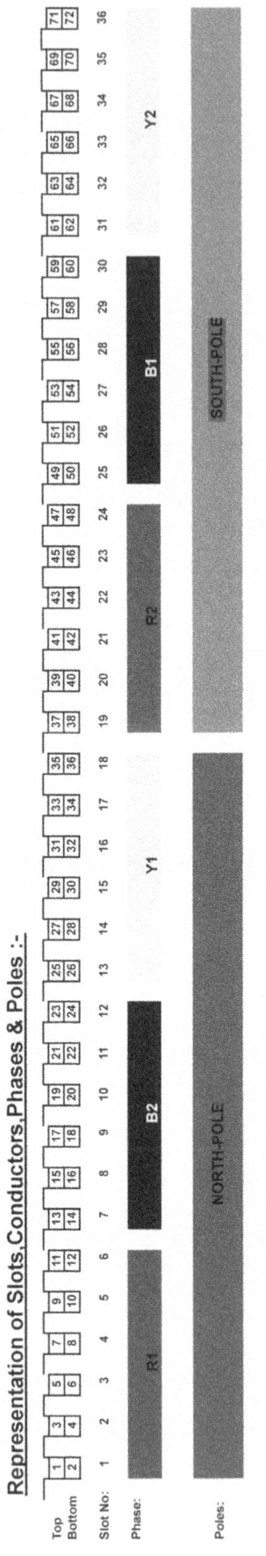

Fig. 3.10 36 Slots, 2 pole, 3 phase double layer winding.

(Slot Pitch=36/2=18Slots=36 Conductors and Phase Spread=36/2/3=6 slots=6×2=12 conductors)

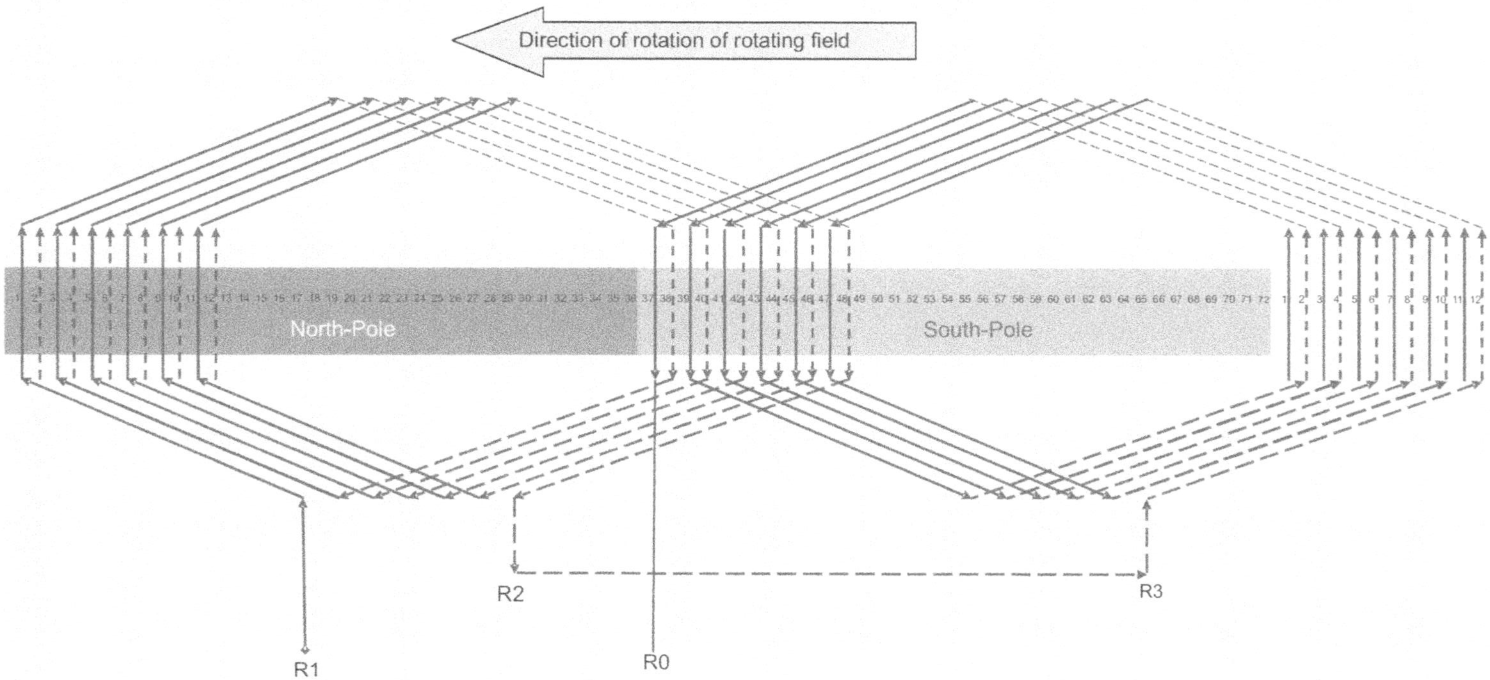

Note(1) :- Odd numbered conductors are on top and Even numbered conductors are on Bottom of the slot
Note(2) :- R1 to R2 --> First Phase group and R3 to R0 --> Second Phase group and R0 to R1 --> Total R Phase

Fig. 3.11 36 slots, 2 pole, 3 ph, double layer full pitched lap widing of an alternator (only R-Phase).

(Slot Pitch=36/2=18Slots=36 Conductors and Phase Spread=36/2/3=6 slots=6 × 2=12 conductors)

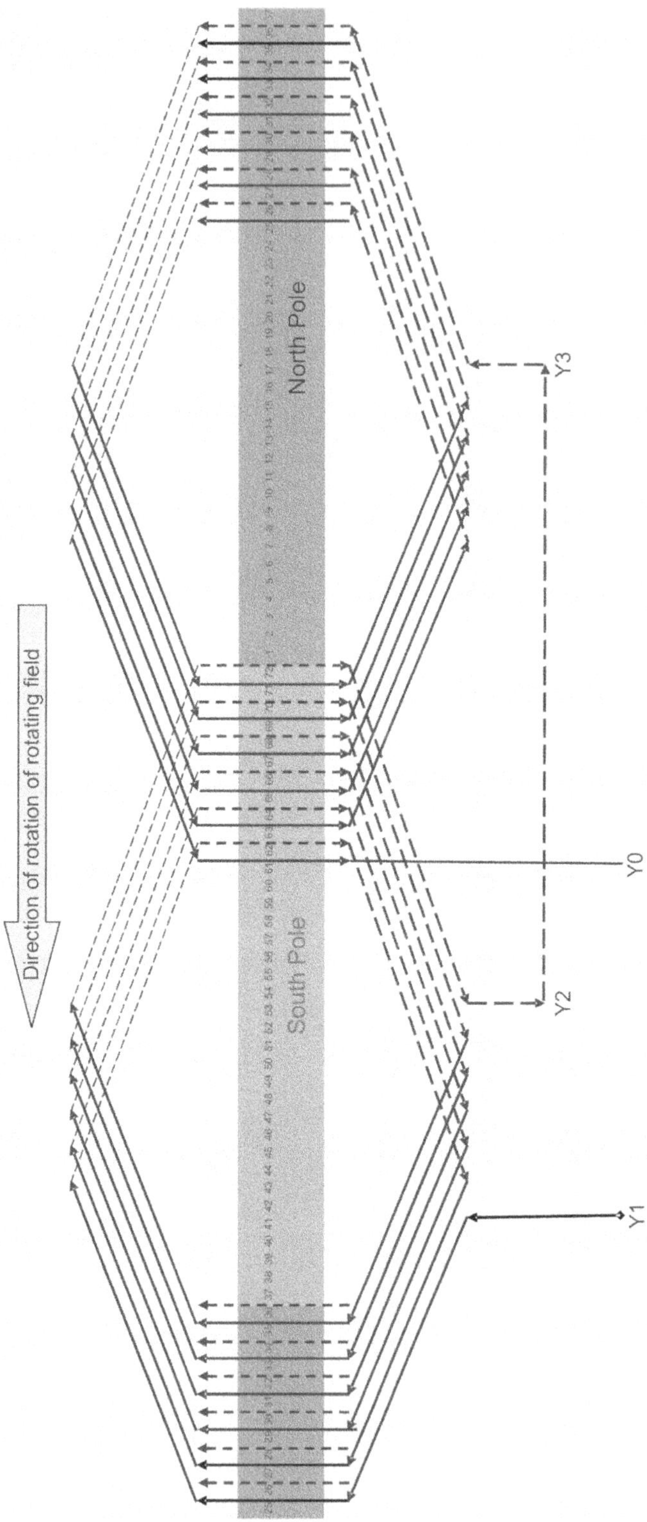

Note(1) :- Odd numbered conductors are on top and Even numbered conductors are on Bottom of the slot

Note(2) :- Y1 to Y2 --> First Phase group and Y3 to Y0 --> Second Phase group and Y0 to Y1 --> Total Y Phase

Fig. 3.12 36 Slots, 2 pole, 3 ph, double layer full pitched lap widing of an alternator (Only Y-Phase).

(Slot Pitch=36/2=18Slots=36 Conductors and Phase Spread=36/2/3=6 slots=6 × 2=12 conductors)

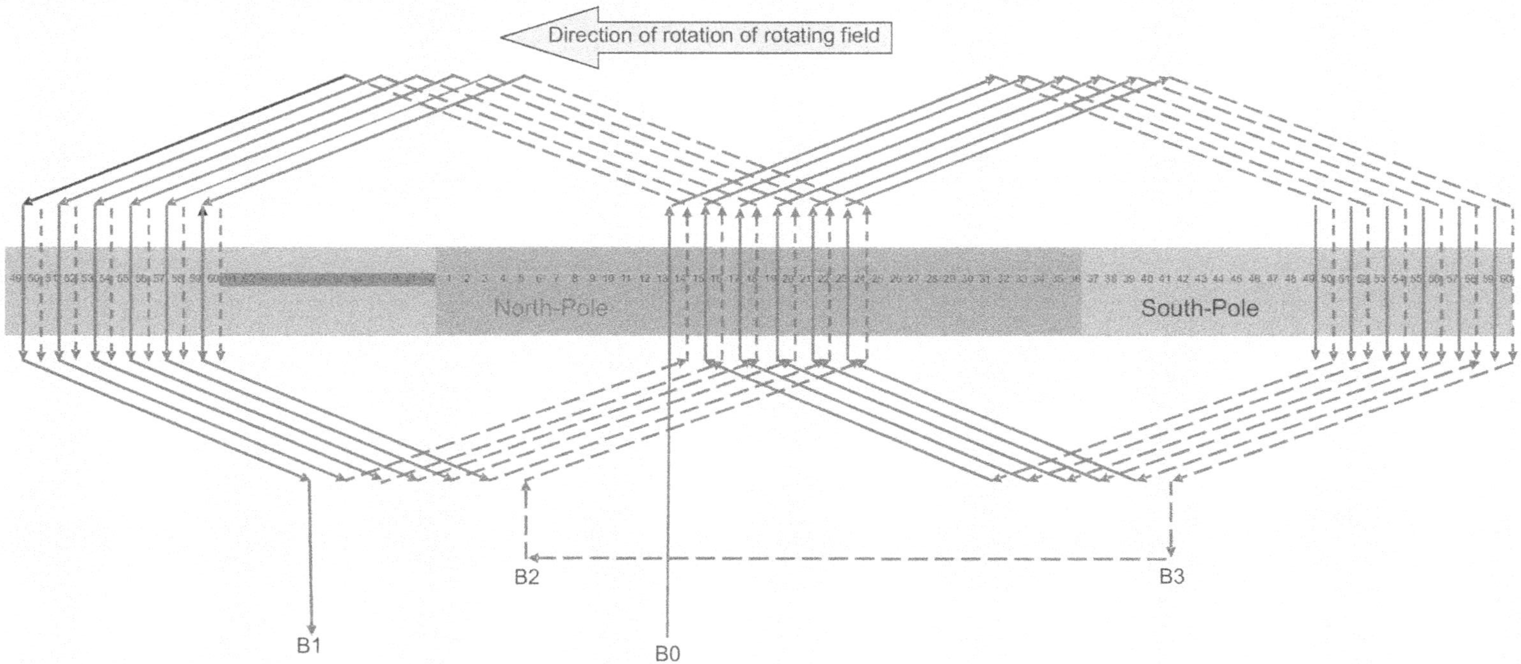

Note(1) :- Odd numbered conductors are on top and Even numbered conductors are on Bottom of the slot
Note(2) :- B1 to B2 --> First Phase group and B3 to B0 --> Second Phase group and B0 to B1 --> Total B Phase

Fig. 3.13 36 Slots, 2 pole, 3 ph, double layer full pitched lap widing of an alternator (Only B-Phase).

(Slot Pitch=36/2=18Slots=36 Conductors & Phase Spread=36/2/3/2=3 slots=3 × 2=6 conductors)

6	7	8	9	10	11	12	&	37	38	39	40	41	42	43	44	45	46	47	48	(R1-->1 & R0-->37)
30	31	32	33	34	35	36	&	61	62	63	64	65	66	67	68	69	70	71	72	(Y1-->25 & Y0-->61)
54	55	56	57	58	59	60	&	13	14	15	16	17	18	19	20	21	22	23	24	(B1-->49 & B0-->13)

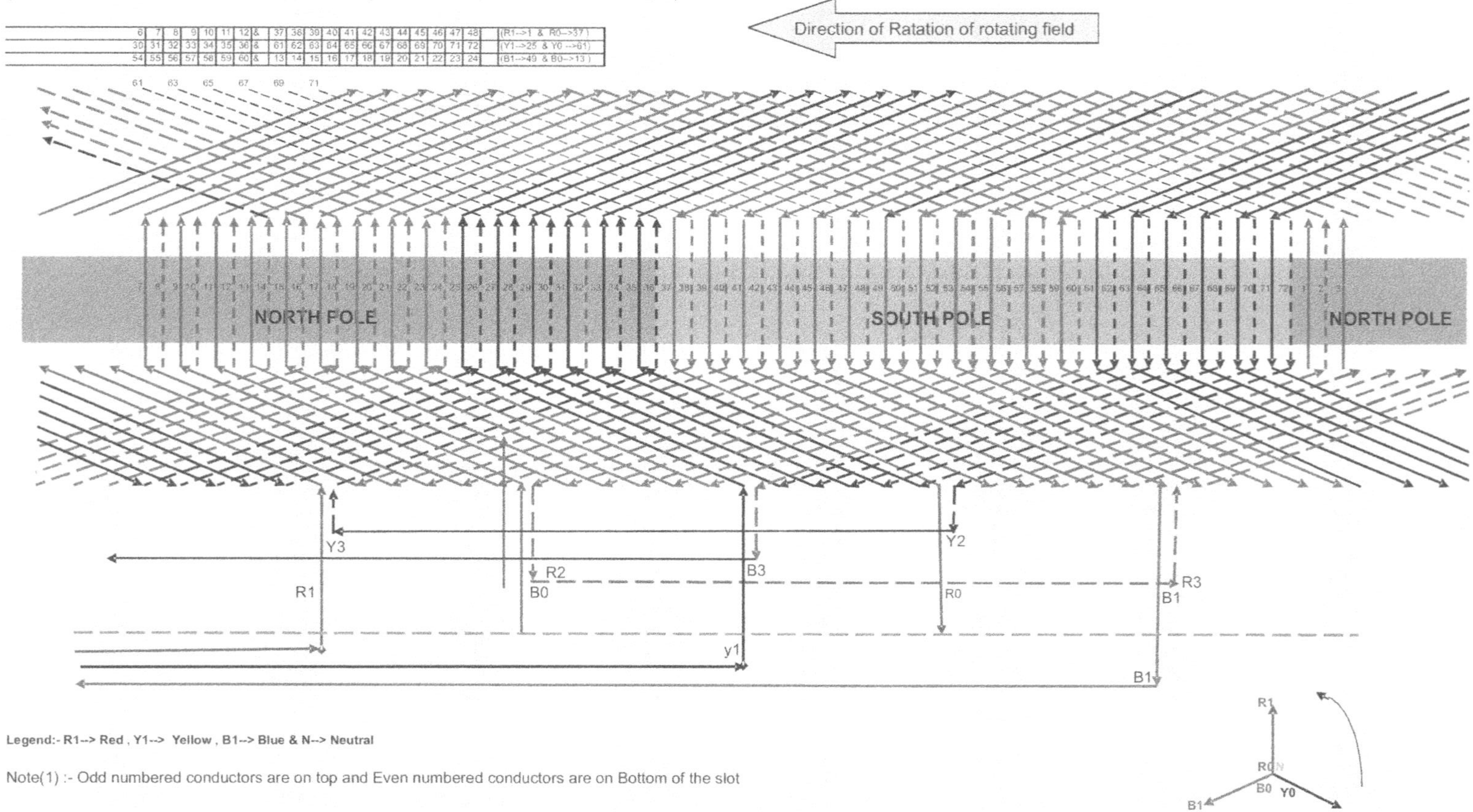

Fig. 3.14 36 slots, 2 pole, 3 ph, double layer full pitched lap winding of an alternator

CHAPTER 4

DC Machines

4.1 Introduction

Based on winding type of armature, DC machines are categorized as lap wound and wave wound. Theory portion of design is not given in this book, but necessary formulae, curves and tables given in standard books are made use of. Since the machine is the same for both operations of Generator and Motor, same design is applicable for both.

Total design is split into six parts in a proper sequence. Design calculations are given for a given Rating of Generator, followed by Computer Program written in "C" language using MATLAB software for each part. Finally all the programs are added together to get the total program by running by which we get the total design. Computer output of total design is given.

This design may not be the optimum one. Now optimization objective and design Constraints are inserted into this total program. When this program is run we will get various alternative feasible designs from which the selected variant based on the Optimization Criteria can be picked up. Computer output showing the important design parameters for various feasible alternatives is given in the end of this chapter along with a logic diagram.

4.2 Sequential Steps for Design of Each Part and Programming Simultaneously

(a) Calculate Output Coefficient, Main dimensions of armature (viz) D,L and Flux/Pole checking the Periphoral velocity and Volts between commutator segments

(b) Calculate no. of slots, size of slot, conductor size, checking current density, current volume, slot balance. Calculate tooth flux density, Height of core, Wt. of iron, Iron losses and Temp rise

(c) Calculate dimensions of poles, Field Coils, Yoke and Amp-Turns required

(d) Calculate Copper size, No. of turns for Shunt and series fields

(e) Determine the diameter of Commutator and no. and size of brushes. Check for Periphoral velocity and gap between brush arms

(f) Calculate the dimensions of Interpoles and Interpole winding. Calculate total losses, efficiency, total weight and Kg/KW.

Note: By adding programs established for each part sequentially, we get the program for complete design.

4.3 Calculation of Armature Main Dimensions and flux/pole (Part-1)

Specific Magnetic Loading & Dimensions & Electric Loading

KW	5	10	50	100	200	300	500	1000	1500	2000	5000	10000
Bav(T)	.43	.45	.56	.6	.62	.64	.66	.68	.7	.72	.74	.76
Amp- Cond/m (*1000)	15	17.5	25	28	32.5	34	36.5	41	42	44	48	51

Following Data is assumed.

No. of ventilating ducts (nv) = 4; Width of ventilating ducts (bv) = 10mm

Pole arc /Pole pitch (PAbPP) = 0.7; Pole Arc = Arme Core Length (L),

Slots/pole (SpP) = 14; Iron factor (k_i) = 0.9; Volage drop on FL = 3%

Values of Specific magnetic loading (B_{av}) = 0.66 T and Specific electric loading (q) = 36500 ac/m → are read from the table given above.

Calculations

$$\text{FL Current (IFL)} = \frac{\text{KW}\times 1000}{\text{V}} = \frac{500\times 1000}{600} = 833.33 \text{ amps}$$

$$\text{Frequency (hz)} = \frac{\text{P}\times\text{N}}{120} = \frac{6\times 700}{120} = 35 \text{ (lies between 25 to 50 and hence OK)}$$

Assuming Field and Armature copper loss = 3% of KW rating,

$$\text{Pa} = 1.03 \times 500 = 515 \text{ KW};$$

$$\text{Output Coefft (C0)} = 0.164 \times \text{Bav} \times \text{q} \times 10^{-3}$$

$$= 0.164 \times 0.66 \times 36500 \times 10^{-3} = 3.9508$$

$$\text{DsqL} = \frac{\text{Pa}}{\text{C0}\times\text{N}} = \frac{515}{3.9508\times 700} = 0.1862 \quad \text{m}^3$$

$$\text{D} = \sqrt[3]{\frac{\text{DsqL}\times\text{P}}{\text{L/PP}\times\pi}} = \sqrt[3]{\frac{0.1862\times 6}{0.7\times\pi}} = 0.798 \text{ m}$$

$$= 798\text{mm} \approx 800 \text{ mm (Rounded off)}$$

Peripheral speed (vA)

$$= \pi\times\text{D}\times\text{N}/60 = \pi\times 800\times 700/60$$

$$= 29.32\text{m/s} \qquad (<= 30 \text{ and hence OK})$$

$$\text{L} = \frac{\text{DsqL}}{\text{D}^2} = \frac{0.1862}{(800/1000)^2} = 0.291\text{m} = 291 \text{ mm} \approx 300 \text{ mm (Rounded off)}$$

Assuming vent ducts (nv = 4) of width (bv = 10mm) each and iron factor (ki = 0.9),

Net Iron length (L_i) = (L-nv × bv) × k_i = (300 – 4 × 10) × 0.91 = 234 mm

Current per Brush Arm (IBA) = IFL × P/2 = 833.33 × 6/2

= 277.8 A (< = 400 and hence OK)

Max Flux density (Bg) = Bav/0.7 = 0.66/0.7 = 0.9429 T

Voltage between commutator segments (Vseg) = 2 × Bg × L/1000 × vA

= 2 × 0.9429 × 300/1000 × 29.32 = 16.59 V (< = 20 and hence OK)

$$\text{Pole Pitch (PP)} = \frac{\pi \times D}{P} = \frac{\pi \times 800}{6} = 418.88\text{mm} \;\; (<= 450 \text{ and hence OK})$$

Pole Arc (Parc) = PAbPP × PP = 0.7 × 418.88 = 293.2 mm

$$\text{Flux/Pole(FI)} = \frac{B_{av} \times \pi \times D \times L \times 10^{-6}}{P} = \frac{0.66 \times \pi \times 800 \times 300 \times 10^{-6}}{6} = 0.0788 \text{ Wb}$$

4.3.1 (a) Program in "C" for Part-1→

```
% Design of 500KW, 600V, Lap wound DC Machine

KW = 500; V = 600; nv = 4; bv = 10; ki = 0.9; dVp=3; Zh = 2;
Loss P = 3;LtoP A = 1; InsHs = 3.2; InsWs=1.8;  Hw=3; HL= 1;

PPmax = 450; ATaPmax = 8500;P = 6; A= P; N=700; PAbPP = 0.7;
SpP = 14; cda1 = 5;  wca = 1.9; Bc = 1.3; DcbD = 0.66;

% <---------1) Main Dimensions of Armature (Part-1)-------->

SKW = [5 10 50 100 200 300 500 1000 1500 2000 5000 10000];

Sbav = [.43 .45 .56 .6 .62 .64 .66 .68 .7 .72  .74  .76];

Sq = [15 17.5 25 28 32.5 34 36.5 41 42 44 48 51];

Bav = interp1 (SKW, Sbav, KW, 'spline');

q = interp1 (SKW, Sq, KW, 'spline')*1e3;

IFL = KW*1000/V; hz = P*N/120;

if hz < 25 || hz > 50 continue; end;

Pa  =  (1  +  LossP/100)*KW;          C0  =  0.164*Bav*q*1e-3;
DsqL = 1/C0*Pa/N; D1= DsqL^(1/3)*(P/(pi*PAbPP*LtoPA))^(1/3);

D = ceil(D1*100)*10;

vA = pi*D/1000*N/60; if vA > = 30 continue; end;

L1 = DsqL/D^2*1e9;  L = ceil(L1/10)*10; Li = (L-nv*bv)*ki;

IBA = IFL/P*2;              if IBA > = 400 continue; end;

Bg = Bav/0.7;             Vseg = 2*Bg*L/1000*vA;
```

```
if Vseg > = 20 continue;end;
PP = pi*D/P;                    if PP > = PPmax continue; end;
Parc = PAbPP*PP;FI1 = Bav*pi*D*L/P*1e-6;
```

4.4 Design of Armature Winding and Core (Part-2)

Arme AT per Pole (ATaP) $\dfrac{\pi \times D \times q \times 10^{-3}}{2 \times P} = \dfrac{\pi \times 800 \times 36500 \times 10^{-3}}{2 \times 6} = 7644.5$

Induced Emf on NO load(E) = (1 + dVp/100) × V = (1 + 3/100 × 600 = 618 V

For Lap wound Machine, No. of parallel paths (A) = P = 6

Total no. of Conductors (Z) = $\dfrac{E \times 60 \times A}{P \times FI \times N} = \dfrac{618 \times 60 \times 6}{6 \times 0.0788 \times 700} = 638.7$

Assuming Slots/Pole (SpP) = 14; No of Slots(S) = SpP × P = 14 × 6 = 84

Slot pitch (sp) = $\dfrac{\pi \times D}{S} = \dfrac{\pi \times 800}{84} = 29.92$ mm

Conductors/Slot (Zs) = Z/S = 638.7/84 = 7.6 ≈ 8 mm (Rounded off even Integer)

Corrected no of conductors (Z) = Zs × S = 8 × 84 = 672

Corrected value of Flux/Pole (FI) = $\dfrac{E \times 60 \times A}{P \times Z \times N} = \dfrac{618 \times 60 \times 6}{6 \times 672 \times 700} = 0.0788$ wb

Current Volume/slot (CV) = $\dfrac{IFL \times Zs}{A} = \dfrac{833.33 \times 8}{6} = 1111$ (<= 1500 and hence OK)

Assuming Shunt Field Current(Ish) = 3 A,

Armature Current (Ia) = IFL + Ish = 833.33 + 3 = 836.33 A

Assuming current density (cda) = 5 A/mm^2;

Cond Area of CS (Aca) = $\dfrac{Ia}{A \times cda} = \dfrac{836.33}{6 \times 5} = 27.878\ mm^2$

Assuming Conductor width (wca) = 1.9mm,

Conductor ht (hca) = $\dfrac{Aca}{Wca} = \dfrac{27.878}{1.9} = 14.67 \cong 15$ mm (Rounded off)

Assuming a factor of 0.98 for rounding at the conductor edges,

Revised Area of conductor (Aca) = 0.98 × hca × wca = 0.98 × 15 × 1.9 = 27.93 mm^2;

Corrected current density (cds) = $\dfrac{Ia}{A \times Aca} = \dfrac{836.33}{6 \times 27.93} = 4.991 A/mm^2$

Assuming no. of conductors ht-wise ia slot (Zh) = 2;

No of cond width-wise in slot (Zw) = Zs/Zh = 8/2 = 4

Assuming cond insulation thk = 0.4 mm and

width-wise Insulation (InsWs) = 1.8 mm,

Slot Width (Ws) = Zw × (wca + 0.4) + InsWs = 4 × (1.9 + 0.4) + 1.8 = 11 mm

Assuming ht -wise Insulation (InsHs) = 3.2 mm, Wedge ht (Hw) = 3 mm and

Lip ht (HL) = 1mm,

Height of Slot (Hs) = Zh × (hca + 0.4) + InsHs + Hw + HL

$$= 2 \times (15 + 0.4) + 3.2 + 3 + 1 = 38 \text{ mm};$$

Cross-Section of Armature Slot (dimensions in mm)

Height-Wise			
Lip			1.0
Wedge			3.0
Ins.Under Wedge			0.5
Cu.Conductor	bare+insulation	15+0.4=15.4	
Top-Layer	for 1 conductor	1*15.4=15.4	15.4
Ins.between Layers			1.0
Bottom- Layer	for 1 conductor	1*15.4=15.4	15.4
Bottom Insulation			0.5
Slack			1.2
Total height			**38.0**

Width-Wise			
Cu.Conductor	bare+insulation	1.9+0.4=2.3	
	for 4 conductors	4* 2.3=9.2	9.2
Insulation on sides	2*0.5=1		1.0
Slack			0.8
Total width			**11.0**

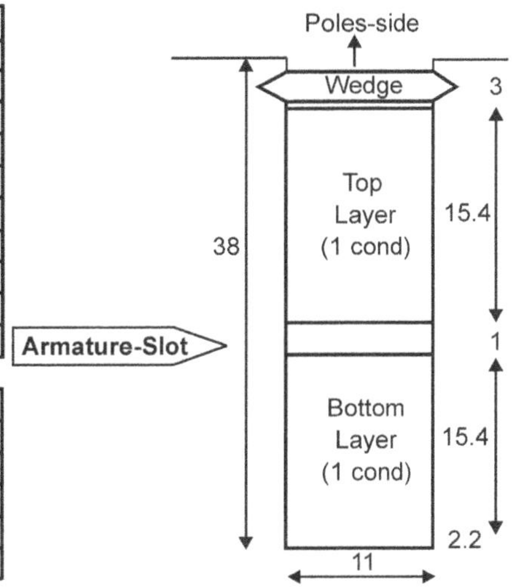

Dia of Arm. teeth at ⅓ height from root (D1b3) = D – 2 × 2 × Hs/3

$$= 800 - 4 \times 38/3 = 749.33 \text{ mm}$$

$$\text{Slot Pitch at above dia (sp1b3)} = \frac{\pi \times D13}{S} = \frac{\pi \times 749.33}{84} = 28.025 \text{ mm}$$

Width of tooth at above dia (Wt1b3) = (sp1b3 – Ws) = (28.025 – 11) = 17.025 mm

Area of CS of tooth at above dia (At1b3)

$$= \frac{Wt1b3 \times Li \times S}{P} = \frac{17.025 \times 234 \times 84}{6} = 55774 \text{ mm}^2$$

Flux density in the tooth (Btav) = $\frac{0.0788 \times 10^6}{\text{At1b3}} = \frac{0.0788 \times 10^6}{55744} = 1.4133\ \text{T}$

Carter Coefft (Ks) = $\frac{\text{sp1b3} \times \text{L}}{\text{Wt1b3} \times \text{Li}} = \frac{28.025 \times 300}{17.025 \times 234} = 2.11$

Max tooth Flux density (Btm) = $\frac{\text{Btav}}{\text{PAbPP}} = \frac{1.4133}{0.7} = 2.019\ \text{T}$

Free Length (Overhang) of conductor (Lfr) = 140 + 1.15 × PP

= 140 + 1.15 × 418.88 = 621.72 mm

Length of Arme-Conductor (Lac) = L + Lfr = 300 + 621.72 = 921.7 mm = 0.9217 m

Arme Wdg Res (Ra) = $\frac{0.021 \times \text{Lac} \times \text{Z}}{\text{Aca} \times \text{A}^2} = \frac{0.021 \times 0.9217 \times 672}{27.93 \times 6^2} = 12.9363\ \text{m}\Omega$

Assuming 20% Eddy Current Losses,

Arm Copper Loss (Pcua) = 1.2 × Ia^2 × Ra = 1.2 × 836.33^2 × 12.9363 × 10^{-3} = 10858 W

Wt of Arme-Copper (Wcua) = Lac × Aca × Z × 8.9/1000

= 0.9217 × 27.93 672 × 8.9/1000 = 153.97 Kg

Flux in core (FIc) = $\frac{Fi}{2} = \frac{0.0788}{2} = 0.0394\ \text{Wb}$

Assuming Flux density in Core (Bc) = 1.3 T, Area of CS in core (Ac)

$= \frac{\text{Fic}}{\text{Bc}} = \frac{0.0394}{1.3} = 30318\ \text{mm}^2$

Height of Core (Hc) = $\frac{\text{A}_c}{\text{L}_i} = \frac{30318}{234} = 129.56\ \text{mm}$

Inner diameter of Core (D_i) = D-2 × (Hs + Hc) = 800 – 2 × (38 + 129.56)

= 464.9 ≈ 470mm (Ronded off)

Corrected height of Core (Hc) = (D – Di)/2 – Hs = (800 – 470)/2 – 38 =127 mm

Mean dia of core (Dmc) = D – 2 × Hs – Hc = 800 – (2 × 38) – 127 = 597 mm

Weight of Core (Wc) = 7.8 × 10^{-6} × π × Dmc × Ac

$= 7.8 \times 10^{-6} \times \pi \times 597 \times 30318 = 443.5\ \text{Kg}$

Hysterisis loss in core/kg (HysPkgc) at 1.3T

= (2.9/30) × hz = 2.9/30 × 35 = 3.3833 → (Emperical formula);

Hysterisis loss in core (Physc) = Wc × HysPkgc = 443.5 × 3.3833 = 1500.6 W

Using 0.5 mm thick Laminations for core, Eddy current Loss in core (Peddyc)

= 0.0045 × $(\text{Bc} \times \text{hz} \times 0.5)^2$ × Wc = 0.0045 × $(1.3 \times 35 \times 0.5)^2$ × 443.5 = 1033 W

Considering 25% more on Loaded Condition, Iron Loss in core (Pic)

$= (\text{Physc} + \text{Peddyc}) \times 1.25 = (1500.6 + 1033) \times 1.25 = 3167$ W;

Weight of Teeth (Wt) $= 7.8 \times 10^{-6} \times \text{Wt1b3} \times \text{Li} \times \text{S} \times \text{Hs}$

$= 7.8 \times 10^{-6} \times 17.025 \times 234 \times 84 \times 38 = 99.19$ Kg

Hysterisis loss in Teeth/kg(HysPkgt) at 2.019T

$= (8/30) \times \text{hz} = 8/30 \times 35 = 9.3333 \rightarrow$ (Emperical formula);

Hysterisis loss in Teeth (Physt) $= \text{Wt} \times \text{HysPkgt} = 99.19 \times 9.3333 = 925.8$ W

For 0.5 mm thick Laminations, Eddy current Loss in teeth (Peddyt)

$= 0.007 \times (\text{Btm} \times \text{hz} \times 0.5)^2 \times \text{Wt} = 0.007 \times (2.019 \times 35 \times 0.5)^2 \times 99.19 = 866.8$ W

Considering 25% more on Loaded Condition, Iron Loss in teeth (Pit)

$= (\text{Physt} + \text{Peddyt}) \times 1.25 = (925.8 + 866.8) \times 1.25 = 2240.7$ W;

Total Iron Loss (Pi) = Pic + Pit =3167 + 2240.7 = 5407.7 W

$$\text{Iron loss percentage (PibKW)} = \frac{\text{Pi}}{\text{KW} \times 1000} \times 100 = \frac{5407.7}{500 \times 1000} \times 100 = 1.0815\%$$

Losses in Armature (Pa) = Pcua + Pi = 10858 + 5407.7 = 16266 W

Over all length of Arme (Larme) $= \text{L} + 70 + 0.3 \times \text{PP}$

$= 300 + 70 + (0.3 \times 418.88) = 495.66$ mm

Cooling Surface of Armature (SA) $= \pi \times \text{D} \times \text{Larme}/100$

$= \pi \times 800 \times 496.66/100 = 12457\text{cm}^2$

Loss per Sq. cm of Cooling surface (Pd) = Pa/SA = 16266/12457 = 1.3057 W/ cm^2;

Temp-Rise of Arme (Tra)

$$= \frac{270 \times \text{Pd}}{1 + 0.09 \times (\text{vA})^{1.3}} = \frac{270 \times 1.3057}{1 + 0.09 \times (29.32)^{1.3}} = 42.62\,^{\circ}\text{C} \quad (<= 50 \text{ and hence OK})$$

4.4.1 (a) Program in "C" for Part-2→

```
%-------2) Armature-Winding and core (part-2)------------------
->

D = 800; q = 36500; P = 6; ATaPmax = 8500; dVp = 3; V = 600;
FI1=0.0829; N=700; SpP=14; IFL = 833.33; cda1 = 5; % Input Data

wca=1.9; Zh = 2; InsWs = 1.8; InsHs = 3.2; Hw = 3; % Input Data

HL = 1; Li = 234; L = 300; PAbPP = 0.7; PP = 418.88; Bc = 1.3;
hz = 35; KW = 500; vA = 29.32; % Input Data
```

```
Ish1 = 3; %Assumptions

ATaP = pi*D/1000*q/(2*P);if ATaP > = ATaPmax continue; end;

E = (1 + dVp/100)*V;  A = P;  Z1 = E*60*A/(P*FI1*N);

S = SpP*P;   sp = pi*D/S; if sp < 25 || sp > 35 continue; end;

Zs1 = Z1/S;  Zsf=floor (Zs1/2)*2; Zsc = ceil (Zs1/2)*2;

Zs = Zsf;       if Zs1 - Zsf > 1 Zs = Zsc;  end; Z = S*Zs;
FI = E*60*A/(P*Z*N);

CV = IFL/A*Zs;      if CV < 1000 || CV > 1500 continue; end;

Ia = IFL + Ish1;   Aca1 = Ia/A/cda1;   hca1 = Aca1/wca;
hca = ceil(hca1);if hca > = 16 continue; end;

Aca = 0.98*hca*wca;       cda = Ia/A/Aca;

if cda > = 5.01 continue; end;

Zw = Zs/Zh;         Ws = Zw*(wca + 0.4) + InsWs;

Hs = Zh*(hca + 0.4) + InsHs + Hw + HL;

D1b3 = D - 2*(2/3*Hs);      sp1b3 = pi*D1b3/S;

Wt1b3 = sp1b3 - Ws; At1b3 = Wt1b3*Li*S/P; Btav = FI*1e6/At1b3;
Ks = sp1b3*L/Wt1b3/Li; if Ks > = 2.2 continue; end;

Btm = Btav/PAbPP; if Btm > = 2.1 continue; end;

Lfr = 140 + 1.15*PP;  Lac = (L + Lfr)/1e3;

Ra = 0.021*Lac/Aca*Z/A^2; Pcua = 1.2*Ia^2*Ra;

Wcua = Lac*Aca*Z*8.9e-3;  FIc = FI/2; Ac = FIc*1e6/Bc;

Hc1 = Ac/Li; Di1 = D-2*(Hs + Hc1); Di = ceil(Di1/10)*10;

Hc=(D - Di)/2 - Hs;  Dmc = D-2*Hs - Hc;  Wc = 7.8e-6*pi*Dmc*Ac;

HysPkgc = (2.9/30)*hz; % Emperial Formula

Physc = Wc*HysPkgc; Peddyc = 0.0045*(Bc*hz*0.5)^2*Wc;

Pic = (Physc + Peddyc)*1.25;

Wt = 7.8e-6*Wt1b3*Li*S*Hs;HysPkgt = (8/30)*hz;

Physt = Wt*HysPkgt;   Peddyt = 0.007*(Btm*hz*0.5)^2*Wt;

Pit = (Physt + Peddyt)*1.25; Pi = (Pic + Pit);

PibKW = Pi/(KW*1e3)*100; Pa = Pcua + Pi;

Larme = L + 70 + 0.3*PP; SA = pi*D*Larme/100;  Pd = Pa/SA;

Tra = 270*Pd/(1 + 0.09*(vA^1.3));
```

4.5 Design of Poles and Calculation of AT (Part-3)

Length of Pole (Lp) = L – 15= 300 – 15 = 285 mm

Assuming Flux density in Pole (Bp) = 1.6 T,

Assuming 20% Leakage flux, Area of Pole (Ap)

$$= \frac{1.2 \times \text{FI}}{\text{Bp} \times 0.95} = \frac{1.2 \times 0.0788}{1.6 \times 0.95} = 0.0622 \text{ m}^2$$

$$\text{Width of Pole (Wp)} = \frac{\text{Ap} \times 10^6}{\text{Lp}} = \frac{0.0622 \times 10^6}{285} = 218.35 \text{ mm} \approx 219\text{mm (Ronded off)}$$

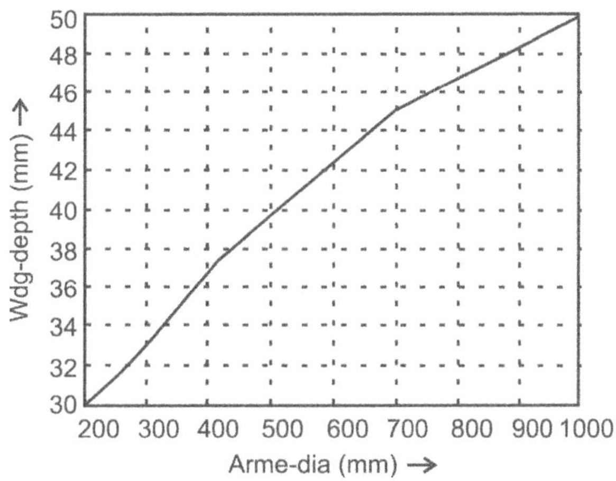

Fig. 4.1 Depth of shunt field winding.

Assuming permissible loss/sq.mt of cooling surface (Pf) = 730 W/m^2,

Copper surface factor for field wdg (Sf) = 0.6;

Depth of shunt field wdg (df) = 47.1

as read from Fig. 4.1 correponding to D = 800 mm,

Amp-Turns for meter length of Field coil (atpm) = $10^4 \times \sqrt{\text{Pf} \times \text{Sf} \times \text{df}/1000}$

$$= 10^4 \times \sqrt{730 \times 0.6 \times 47.1/1000} = 45420$$

$$\text{Armature reaction AT/Pole (ATpp)} = \frac{\text{Ia} \times \text{Z}}{\text{A} \times 2 \times \text{P}} = \frac{836.33 \times 672}{6 \times 2 \times 6} = 7805.8$$

Assuming Field AT (ATf) = ATpp = 7805.8,

$$\text{Height of Field Coil (Hfc)} = \frac{\text{ATf} \times 1000}{\text{atpm}} = \frac{7805.8 \times 1000}{45420} = 171.9 \text{ mm}$$

Assuming height of pole shoe (HPS) = 40 mm,

Height of Pole (Hp) = Hfc + HPS + 0.1 × PP = 171.9 + 40 + 0.1 × 418.88 = 254 mm

Assuming flux density in Yoke (By) = 1.2 T,

$$\text{CS area of yoke (Ay)} = \frac{\text{FI} \times 10^6}{2 \times \text{By}} = \frac{0.0788 \times 10^6}{2 \times 1.2} = 39413 \text{ mm}^2;$$

Assuming axial length of yoke to be 50% more, Lengh of Yoke(Ly)

$= 1.5 \times L = 1.5 \times 300 = 450$ mm

Depth of Yoke (dy) = Ay/Ly = 39413/450 = 87.585 mm (say, 88 mm)

Weight of Main Pole (Wmp) = $7.8 \times 10^{-6} \times Lp \times Wp \times Hp$

$= 7.8 \times 10^{-6} \times 285 \times 219 \times 254 = 123.66$ Kg

Weight of Yoke (Wyoke) = $7.8 \times 10^{-6} \times Ay \times Ly$

$= 7.8 \times 10^{-6} \times 39413 \times 450 = 138.34$ Kg

Calculation of AT for Iron Path

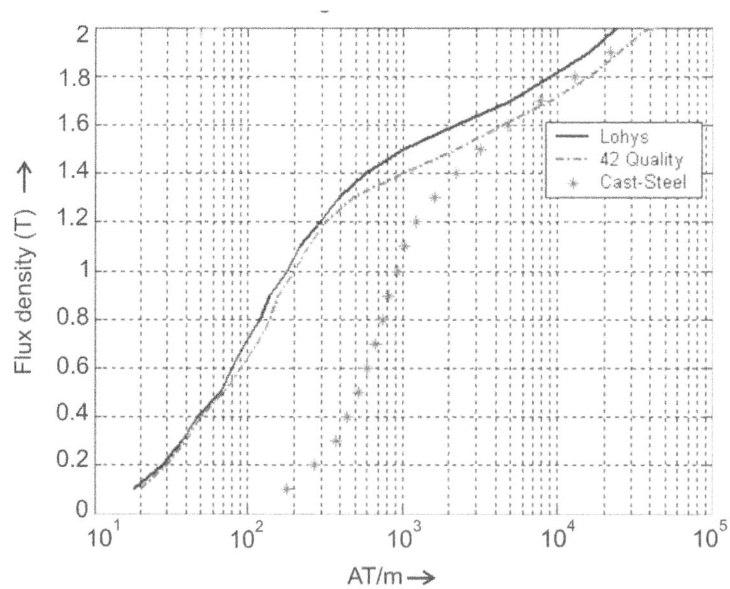

Fig. 4.2 Magnetization curves.

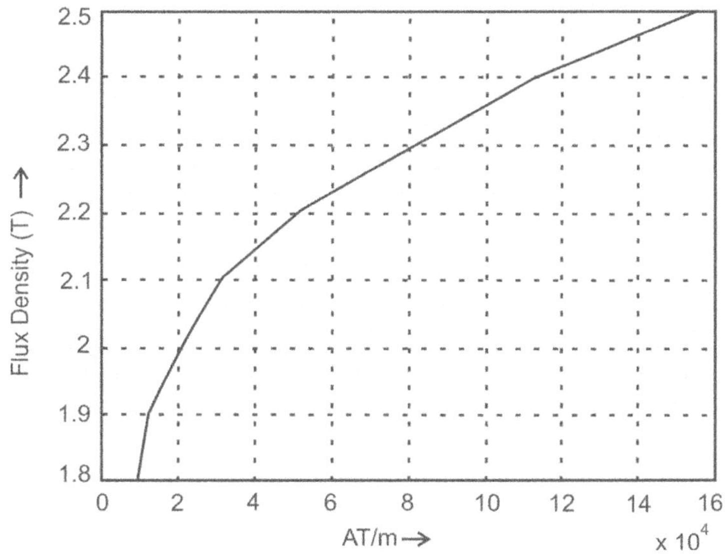

Fig. 4.3 Magnetization curves of teeth at Ks – 2.0

For Pole:

From Fig. 4.2, AT corresponding to Bp = 1.6 T, (atp) = 4400 AT/m (for 42 Quality steel) and AT for Pole (ATp) = $atp \times Hp/1000 = 4400 \times 254/1000 = 1117.6$

For Arm-Teeth:

From Fig. 4.3, AT corresponding to Btm = 2.019 T, (atat) = 21739 AT/m and AT for Teeth (ATt) = $atat \times Hs/1000 = 21739 \times 38/1000 = 826$

For Core:

From Fig. 4.2, AT corresponding to Bc = 1.3 T,

(atac) = 390AT/m (for Lohys steel)

$$\text{Length of Flux Path (Lfpc)} = \frac{\pi \times Dmc}{2 \times P} = \frac{\pi \times 597}{2 \times 6} = 156.29 \text{ mm}$$

AT for Core (ATc) = $atac \times Lfpc/1000 = 390 \times 156.29/1000 = 60.96$

For Yoke:

From Fig. 4.2, AT corresponding to By = 1.2 T, (aty) = 1230 AT/m (for Cast steel)

Mean Dia (Dmy) = $D + 2 \times Hp + dy = 800 + (2 \times 254) + 88 = 1396$;

$$\text{Length of Flux Path (Lfpy)} = \frac{\pi \times Dmy}{2 \times P} = \frac{\pi \times 1396}{2 \times 6} = 365.5 \text{ mm and}$$

AT for yoke (ATy) = $aty \times Lfpy/1000 = 1230 \times 365.5/1000 = 449.5$

Total AT for all parts (ATT) = ATp + ATt + ATc + ATy

=1117.6 + 826 + 60.96 + 449.5 = 2454

Percentage = ATT/ATf × 100 = 2454/7806 × 100 = 31.44 %

AT reqd on no-load (AT0) = 0.85 × ATf = 0.85 × 7806 = 6635

Air-gap AT (ATg) = AT0 – ATT = 6635 – 2454 = 4181

$$\text{AirGapFluxDensity(Bgm)} = \frac{B_{av}}{\text{PAbPP}} = \frac{0.66}{0.7} = 0.9429\text{T}$$

Assuming Air Gap Coefficient (Kg) = 1.15 (Varies from 1.12 to 1.18);

$$\text{AirGapLength(Lg)} = \frac{AT_g}{0.796 \times B_{gm} \times K_g \times 10^6} = \frac{4181}{0.796 \times 0.9429 \times 1.15 * 10^6}$$

= 0.0048 m= 4.85mm

Yoke outer dia (D0) = D + 2 × (Lg + Hp + dy)

= 800 + 2 × (4.85 + 254 + 88) = 1494 mm

4.5.1 (a) Program in "C" for Design of Poles and Calculation of AT-(Part-3)

```
% ←------3) Design of Poles and Calculation of AT-(Part-3) ---
->
L = 300; FI = 0.0788; D = 800; Ia = 836.33; A = 6; Z = 672;
P = 6; PP = 418.88; % Input Data
Btm = 2.019; Hs = 38; Bc = 1.3; Dmc = 597; Bav = 0.66;
PAbPP = 0.7; %Input Data
Bp = 1.6; Pf = 730; Sf = 0.6; By = 1.2; Kg = 1.15; HPS = 40;
%Assumption
DA = [200 300 400 500 700 1000]; dSFW = [30 33 37 40 45 50];
Plot (DA,dSFW); grid; xlabel ('Arme-dia(mm)-->');
ylabel('Wdg-depth(mm)-->');
title ('Depth of Shunt Field Winding');
Lp = L - 15; Ap = 1.2*FI/Bp/0.95; Wp1 = Ap*1e6/Lp;
Wp = ceil(Wp1);
df1 = interp1 (DA,dSFW,D,'spline'); df = ceil (df1*10)/10;
atpm = 1e4*sqrt(Pf*Sf*df/1e3);
Atpp = Ia/A*Z/2/P; Atf = ATpp; Hfc = ATf/atpm*1e3;
Hp1 = Hfc + HPS + 0.1*PP;
Hp = ceil (Hp1); Ay = 1.2*FI*1e6/2/By; Ly = 1.5*L;
```

```
dy1 = Ay/Ly;  dy = ceil(dy1);
Wmp = 7.8e-6*Lp*Wp*Hp;  Wyoke = 7.8*Ay*Ly/1e6;
BB = [.1  .2  .3  .4  .5  .6  .7  .8  .9  1  1.1  1.2  1.3  1.4
1.5  1.6  1.7  1.8  1.9  2.0]; % Tesla
Ha = [18  28  38  48  67  80  95 120 140 180  220  295  390
580 1000 2200 5000  9000 16000 24000]; % for Lohys-steel
H = [20 30 40 50 70 90 115 140 160 200 250 320 480 1000 2200
4400 9000 16000 25000 38000]; % for 42 quality steel
Hcs = [180 270 370 440 520 590 670 750 820 920 1030 1230 1600
2200 3200 4800 7700 13000 22000 40000]; % for Cast-Steel
SBt = [1.8 1.9 2 2.1 2.2 2.3 2.4 2.5];
SATm = [ 9 12 20 31 51 80 113 155]*1e3; % for Ks = 2.0
Semilogx (Ha, BB, '-', H, BB, '-.', Hcs, BB,'*'); grid;
xlabel ('AT/m-->'); ylabel ('Fluxdensity (T)-->');
title ('Magnetization curves');
legend ('Lohys', '42 Quality', 'Cast-Steel');
atp = interp1(BB, H, Bp, 'spline');    ATp = atp*Hp/1e3;
plot (SATm, SBt); grid; xlabel ('AT/m-->');
ylabel ('Fluxdensity (T)-->');
title ('Magnetization curves of teeth at Ks = 2.0');
atat = interp1 (SBt, SATm, Btm, 'spline');   Att = atat*Hs/1e3;
atac  =  interp1(BB,  Ha,  Bc,  'spline');    Lfpc  =  pi*Dmc/2/P;
Atc = atac*Lfpc/1e3;
aty = interp1(BB, Hcs, By, 'spline');    Dmy = D + 2*Hp + dy;
Lfpy = pi*Dmy/(2*P);
ATy = Lfpy*aty/1e3;   TT = ATp + ATt + ATc + ATy;
Perc = ATT/ATf*100;  AT0 = 0.85*ATf; ATg = AT0 - ATT;
Bgm    =    Bav/PAbPP;    Lg    =    ATg/(0.796*Bgm*Kg*1e6)*1e3;
D0 = D + 2*(Lg + Hp + dy);
```

4.6 Design of Shunt Field and Series Field Windings (Part-4)

Assuming 15% voltage drop in Field Regulator,

Voltage across Shunt field Wdg (Vsh) = $0.85 \times V = 0.85 \times 600 = 510$ V

Voltage across each field coil (Vc) = Vsh/P = 510/6 = 85 V

Assuming thickness of insulation on pole body (ti) = 10mm,

Mean Length of turn (Lmt) = $[2 \times (Wp + Lp) + \pi \times (df + 2 \times ti)]/1000$

$= [2 \times (219 + 285) + \pi \times (47.1 + 2 \times 10)]/1000 = 1.2188$ m;

CS area of Conductor (Ash) = $0.021 \times \text{Lmt} \times \text{AT0/Vc}$

$= 0.021 \times 1.2188 \times 6635/85 = 1.9979\ \text{mm}^2$;

Assuming 80% of Field coil ht for Shunt field wdg and

balance 20% for series fld wdg,

Cooling surface of Sh. fld. coil (cssfc) = $2 \times \text{Lmt} \times 0.8 \times \text{Hfc}/1000$

$= 2 \times 1.2188 \times 0.8 \times 171.86/1000 = 0.3351\ \text{m}^2$;

Assuming permissible loss per sq.m of cooling surface = 730 W/ m^2;

Permissible Loss of Sh. Fld Wdg (Psfc) = $730 \times \text{cssfc} \times \text{P}$

$= 730 \times 0.3351 \times 6 = 1467.9$ W;

Sh. Fld Current (Ish) = Psfc/Vsh = 1467.9/510 = 2.8782 A

No. of turns/pole (Tsh) = AT0/Ish $= \dfrac{\text{AT0}}{\text{Ish}} = \dfrac{6635}{2.8782} = 2305$

Selecting a conductor dia (dcu) = 1.6 mm (bare) with ins thk = 0.11mm,

Turns that can be wound in a height of $0.8 \times$ Hfc (T1) $= \dfrac{0.8 \times \text{Hfc}}{\text{dcu} + 0.11} = \dfrac{0.8 \times 171.86}{1.6 + 0.11} = 80.4$ (say, 80)

No. of layers to be provided (NL) $= \dfrac{\text{Tsh}}{\text{T1}} = \dfrac{2305}{80} = 28.8$ (say, 29)

Corrected no. of turns/pole (Tsh) = $\text{T1} \times \text{NL} = 80 \times 29 = 2320$

Depth of winding (dw) = $\text{NL} \times (\text{dcu} + 0.11) = 29 \times (1.6 + 0.11) = 49.59$ mm

CS area of copper cond (Acu)$= \dfrac{\pi \times \text{dcu}^2}{4} = \dfrac{\pi \times 1.6^2}{4} = 2.0106\ \text{mm}^2$

Resistance of Sh.fld winding (Rsh)

$$= \frac{0.021 \times \text{Lmt} \times \text{Tsh} \times \text{P}}{\text{Acu}} = \frac{0.021 \times 1.2188 \times 2320 \times 6}{2.0106} = 177.2\ \Omega$$

Sh. Fld. Copper Loss (Psh) = $\text{Ish}^2 \times \text{Rsh} = 2.8782^2 \times 177.2 = 1468$ W

Wt of Sh. fld. copper (Wcsh) = $\text{Acu} \times \text{Lmt} \times \text{Tsh} \times \text{P} \times 8.9 \times 10^{-3}$

$= 2.0106 \times 1.2188 \times 2320 \times 6 \times 8.9 \times 10^{-3} = 303.6$ Kg

Design of Series Field Winding

AT to be produced by Series Fld Wdg on Full Load (ATse)

$= \text{ATf} - \text{AT0} = 7806 - 6635 = 1171$

Turns /Pole (Tse) = ATse/Ia = 1171/836.33 = 1.4 (say, 2)

Modified AT/Pole (ATse) = $\text{Tse} \times \text{Ia} = 2 \times 836.33 = 1673$

Assuming current density (cdse) = 2.2 A/mm^2,

$$\text{CS area of conductor (Ase)} = \frac{\text{Ia}}{\text{cdse}} = \frac{836.33}{2.2} = 380.15\ \text{mm}^2$$

Copper Strip size selected: 20 X 19 mm

Resistance of Series fld winding (Rse)

$$= \frac{0.021 \times \text{Lmt} \times \text{Tse} \times \text{P}}{\text{Ase}} = \frac{0.021 \times 1.2188 \times 2 \times 6}{380.15} = 0.808\ \text{m}\Omega$$

Series Fld. Copper Loss (Pse) = $\text{Ia}^2 \times \text{Rse} = 836.33^2 \times 0.808/1000 = 565.1$ W

Wt of Series. fld. copper (Wcse) = $\text{Ase} \times \text{Lmt} \times \text{Tse} \times \text{P} \times 8.9 \times 10^{-3}$

$$= 380.15 \times 1.2188 \times 2 \times 6 \times 8.9 \times 10^{-3} = 49.48\ \text{Kg}$$

4.6.1 (a) Program in "C" for Design of Shunt Fld and Series Fld Wdgs (Part-4)

```
% ←--4) Design of Shunt Fieldand series Fld Windings -----→
ti = 10; dcu = 1.6; cdse = 2.2; % Assumptions
V = 600; P = 6; Wp = 219; Lp = 285; df = 47.1; AT0 = 6635;
Hfc = 171. 86; ATf = 7806; Ia = 836.33; % Input Data
Vsh = 0.85*V; Vc = Vsh/P;
Lmt = (2*(Wp + Lp) + pi*(df + 2*ti))/1e3;
Ash = 0.021*Lmt*AT0/Vc;cssfc =2*Lmt*0.8*Hfc/1e3;
Psfc = 730*cssfc*P;  Ish = Psfc/Vsh;  Tsh1 = AT0/Ish;
T1a = 0.8*Hfc/(dcu+0.11);  T1 = floor(T1a);  NL1 = Tsh1/T1;
NL = ceil(NL1);  Tsh = T1*NL;
Dw = NL*(dcu+0.11);  Acu = pi*dcu^2/4;
Rsh = 0.021*Lmt*Tsh*P/Acu; Psh = Ish^2*Rsh;
Wcsh = Acu*Lmt*Tsh*P*8.9*1e-3;  ATse1 = ATf-AT0;
Tse1 = ATse1/Ia; Tse = ceil(Tse1); ATse = Tse*Ia;
Ase = Ia/cdse;   Rse = 0.021*Lmt*Tse*P/Ase;
Pse = Ia^2*Rse;     Wcse = Ase*Lmt*Tse*P*8.9*1e-3;
```

4.7 Design of Commutator and Brushes (Part-5)

Assuming commutator dia is (DcbD = 66%) of Armature dia,

Comm. dia(Dc) = $0.66 \times 800 = 528$ mm

Periphoral speed of Commutator (vC)

$= \pi \times \text{Dc} \times \text{N}/60 = \pi \times 528 \times 700/60 = 19.35$ m/s(<= 20 and hence OK)

No. of Comm Segments (Nc) = Z/2 = 672/2 = 336

Pitch of Comm. Segments (pitC) = $\pi \times Dc / Nc = \pi \times 528 / 336$

$= 4.94$ mm (<=10 and hence OK)

Gap between Brush Spindles (GbSp) = $\pi \times Dc / P = \pi \times 528 / 6$

$= 276.5$mm (lies between 250 to 300 and hence OK)

Selecting Brush of Thickness (Tb) = 10 mm, Width (Wb) = 20 mm and

Height (Hb) = 32 mm and Current density (cdb) = 6 A/cm^2,

No. of Brushes per Spindle (NbPs)

$$= \frac{Ia \times 2}{P \times cdb \times Tb \times Wb / 100} = \frac{836.33 \times 2}{6 \times 6 \times 10 \times 20 / 100} = 23.1, \text{ say } 24$$

Length of Commutator (Lcom) = NbPs × (Wb + 5) + 40 + 30 + 20

$= 24 \times (20 + 5) + 90 = 690$ mm

Assuming Brush Friction Coefft = 0.2 and

Brush pressure on Commutator = 13000 N/m^2,

Brush friction loss (BFL) = $0.2 \times P \times Wb \times Tb \times 10^{-6} \times NbPs \times 13000 \times vC$

$= 0.2 \times 6 \times 20 \times 10 \times 10^{-6} \times 24 \times 13000 \times 19.35 = 1449$ W

Assuming Voltage drop across brush set (Vb) = 2 V,

Brush contact Loss (BCL) = Ia × Vb = 836.33 × 2 = 1673 W

Total Commutator Loss (Pcom) = BFL+BCL = 1449 + 1673 = 3122 W

Wt of Commutator Copper (Wccom) = $Lcom \times \pi \times Dc^2/4 \times 8.9 \times 10^{-6}$

$= 690 \times \pi \times 528^2/4 \times 8.9 \times 10^{-6} = 1345$ Kg

4.7.1 (a) Program in "C" for Design of Commutator and Brushes (Part-5)

```
% <-----------5) Design of Commutator and Brushes ------>
DcbD = 0.66; D = 800; N = 700; Z = 672; P = 6; Ia = 836.33;
% Input data
cdb = 6; Tb = 10; Wb = 20; Hb = 32; Vb = 2; % Assumptions
Dc1 = DcbD*D; Dc = ceil (Dc1); vC = pi*Dc/1000*N/60;
if vC > = 20 continue; end;
Nc = Z/2;  pitC = pi*Dc/Nc;  if pitC > = 10 continue; end;
GbSp = pi*Dc/P;     if GbSp < 250 continue; end;
NbPs1 = (Ia*2/(P*cdb*Tb*Wb/100));  NbPs = ceil(NbPs1);
Lcom = NbPs*(Wb + 5) + 40 + 30 + 20;
BFL = 0.2*P*Wb*Tb*1e-6*NbPs*13000*vC; BCL = Vb*Ia;
Pcom = BFL + BCL;  CSC = pi*Dc*Lcom/100;
Trc = 120*Pcom/CSC/(1 + 0.1*vC);
if Trc > = 50 continue; end;
Wccom = Lcom*pi*Dc^2/4*8.9e-6;
```

4.8 Design of Inter-Pole/ Compensating Winding and Overall Performance

Assuming Interpole length (Lip) = 0.7 × L = 0.7 × 300 = 210 mm

Width of Int. pole shoe (bip) = [{(Zs/2)-1} × pitC + Tb-0.8] × D/Dc

= [{(8/2) – 1} × 4.9368 + 10 – 0.8] × 800/528 = 36.38 mm

Assuming Air gap under Inter-Pole(Lgi) = 1.5 × L = 1.5 × 4.8439 = 7.2659 mm

Ht of Interpole (Hip) = Hp-(Lgi-Lg) = 254 – (7.2659 – 4.8439)

= 251.578 ≈ 252 mm (by rounding off)

Calculation of Flux density under the Inter-Pole (Bip):

Copper ht in Arme-slot (hcai) = 2 × (hca + 0.4) = 2 × (15 + 0.4) = 30.8 mm

Ht above slot in Armaeture (ht) = Hw + HL + InsHs/2 = 3 + 1 + (3.2/2) = 5.6 mm

Periphory of Zs/2 = 4 coils in one layer of slot (perip)

= 2× [{(Zs/2) × (wca+0.4)} + (hca + 0.4)]

= 2 × [{(8/2) × (1.9+0.4)} + (15+0.4)] = 49.2 mm

Leakage coefficients:

Lmdl

$$= 0.4\times\pi\times\left(\frac{\text{hcai}}{3\times\text{Ws}}+\frac{\text{ht}}{\text{Ws}}+\frac{\text{bip}}{6\times\text{Lgi}}\right)=0.4\times\pi\times\left(\frac{30.8}{3\times 11}+\frac{5.6}{11}+\frac{36.38}{6\times 7.266}\right)=2.8612$$

Lmd2

$$=\frac{\text{Lfr}}{\text{L}}\times\left(0.23\times\log\left[\frac{\text{Lfr}}{\text{perip}}\right]+0.07\right)=\frac{621.711}{300}\times\left(0.23\times\log\left[\frac{621.711}{49.2}\right]+0.07\right)=1.3541$$

Total (Lamda) = (Lmd1 + Lmd2) × 10^{-6} = (2.8612 + 1.3541) × 10^{-6}

= 4.2154 × 10^{-6} Wb/m^2 ;

$$\text{Flux density (Bip)} = \frac{\text{L}}{\text{Lip}}\times\frac{2\times\text{Ia}\times\text{Zs}\times\text{Lamda}}{36.3793/1000}=0.3692\text{ T}$$

Design of Inter-Pole Winding:

Amp-turns for Inter-pole (ATip) = ATf + (0.796 × Bip × 1.1 × Lgi × 1000)

= 7806 + (0.796 × 0.3692 × 1.1 × 7.2659 × 1000 = 10154

No. of turns on each Int-Pole (Tip) = ATip/Ia = 10154/836.33

= 12.14 ≈ 13 (by rounding off)

Modified No of AT (ATip) = Tip × Ia = 13 × 836.33 = 10872

Assuming Current density (cdip) = 2.3 A/mm^2,

CS area of conductor (aip) = Ia/cdip = 836.33/2.3 = 363.62 mm^2,

Winding arrangement:

Inter pole ht = 252 mm. Leaving 40 mm for insulation,

Space available for turn (spt) = (Hip – 40)/Tip = (252 – 40)/13 = 16.31 mm

Ht-wise dimension of copper (hcse) = (spt– 0.5) = 16.31– 0.5 = 15.81 (say, 15mm)

Considering 2 layers, thickness of strip (tcse)

$$= \frac{\text{aip}}{2\times\text{hcse}} = \frac{363.62}{2\times 15} = 12.12 \quad \approx 13 \text{ (rounded off)}$$

Modified CS area of cond (aip) = $0.98 \times 2 \times 13 \times 15 = 382.2\ \text{mm}^2$

Modified current density (aip) = Ia/aip = 836.33/382.2 = 2.19 A/ mm^2

Mean length of turn (Lmtip) = [2 × (Lip + bip) + 4 × (tcse + 0.5) + 3]/1000

= [2 × (210 + 36.38) + 4 × (13 + 0.5) + 3]/1000 = 0.5498 m;

Resistance of Int-Pole winding (Rip)

$$= \frac{0.021\times \text{Lmtip}\times \text{Tip}\times \text{P}}{\text{Aip}} = \frac{0.021\times 0.5498\times 13\times 6}{382.2} = 2.3561\,\text{m}\Omega$$

Int-Pole Wdg. Copper Loss (Pip) = $\text{Ia}^2 \times \text{Rip} = 836.33^2 \times 2.3561/1000 = 1576$ W

Wt of Inter-Pole copper (Wcip) = $\text{aip}\times \text{Lmtip} \times \text{Tip} \times \text{P} \times 8.9\times 10^{-3}$

$= 382.2\times 0.5498\times 13\times 6\times 8.9\times 10^{-3} = 145.86$ Kg

Wt of Inter-Poles (Wip) = $7.8\times 10^{-6} \times \text{Lip} \times \text{bip} \times \text{Hip}$

$= 7.8\times 10^{-6} \times 210 \times 36.38 \times 252 = 15.02$ Kg

Assuming Bearing friction and Windage Loss (Pbrg)

$= 0.01\times \text{KW}\times 1000 = 0.01\times 500\times 1000 = 5000$ W

Variable Losses (Pvar) = Pcua + Pse + Pip + BCL

= 10858 + 565.1 + 1648 + 1673 = 14744 W;

Constant Losses (Pconst) = Psh + Pi + BFL + Pbrg

= 1468 + 5408 + 1449 + 5000 = 13325 W

Total Losses (Pt) = (Pvar + Pconst)/1000 = (14744 + 13325)/1000 = 28.068 KW;

$$\text{Efficiency} = \frac{\text{KW}}{\text{KW}+\text{Pt}}\times 100 = \frac{500}{500+28.068}\times 100 = 94.68\%$$

$$\text{Load for Max Efficiency (Ldmxef)} = \text{KW} \times \sqrt{\frac{\text{Pconst}}{\text{PVar}}} = 500\times\sqrt{\frac{13325}{14744}} = 475.33 \text{ KW}$$

Max Efficiency (MaxEff)

$$= \frac{\text{Ldmxef}\times 100}{\text{Ldmxef}+2\times \text{Pconst}/1000} = \frac{475.33\times 100}{475.33+2\times 13325/1000} = 94.69\%$$

Considering Insulation Wt as 1%, Total Weight (WtTot)

= 1.01 × (Wc + Wt + Wcua + Wccom + Wcsh + Wcse + Wcip
+ Wmp + Wyoke + Wip)

= 1.01 × (443.5 + 99.2 + 154 + 1345 + 303.6 + 49.5 + 145.9 + 123.6 + 138.3 + 15)

= 2845.8 Kg

KgPkw = WtTot/KW = 2845.8/500 =5.6908

4.8.1 (a) Program for Design of Inter/Compensating-Pole-Wdg and Overall Performance (Part-6)

```
% <---6) Design of Inter/Compensating-Pole-Wdg and Overall performance ---->
L = 300; Zs = 8; pitC = 4.9368; Tb = 10; D = 800; Dc = 528; Lg = 4.8439; Hp = 254;
hca = 15; % Input data
Hw = 3; HL =1; InsHs = 3.2; wca = 1.9; Ws = 11; Lfr = 621.711; Ia = 836.33;
ATf = 7806; P = 6; % Input data
KW = 500; Pcua = 10858; Pse = 565.1; BCL = 1673; Psh = 1468; Pi = 5408;
BFL = 1449; Wc = 443.5; % Input data
Wt = 99.2; Wcua = 154; Wccom = 1345; Wcsh = 303.6; Wcse = 49.5; Wmp = 123.6;
Wyoke = 138.3;
cdip1 = 2.3; % Assumption
Lip = 0.7*L;  x = Zs/2;  bip = ((x-1)*pitC + Tb-0.8)*D/Dc;  Lgi = 1.5*Lg;
Hip1 = Hp-(Lgi-Lg);       Hip = ceil (Hip1); hcai = 2*(hca + 0.4);
ht = Hw + HL + InsHs/2;  perip = 2*(Zs/2*(wca + 0.4) + (hca + 0.4));
Lmd1 = 0.4*pi*(hcai/3/Ws + ht/Ws + bip/6/Lgi);
Lmd2 = Lfr/L*(0.23*log(Lfr/perip) + 0.07);
Lamda = (Lmd1 + Lmd2)*1e-6; Bip = L/Lip*2*Ia/6*Zs*Lamda/(bip/1e3);
ATip1 = ATf + 0.796*Bip*1.1*Lgi*1e3; Tip1 = ATip1/Ia; Tip = ceil(Tip1);
ATip = Tip*Ia;  aip1 = Ia/cdip1; spt = (Hip-40)/Tip;  hcse1 = (spt-0.5);
hcse = floor(hcse1);  tcse1 = aip1/2/hcse; tcse = ceil(tcse1);   aip = 0.98*2*hcse* tcse;
cdip = Ia/aip;  Lmtip = (2*(Lip + bip) + 4*(tcse + 0.5) + 3)/1e3;
Rip = 0.021*Lmtip*Tip*P/aip;  Pip = Ia^2*Rip;
Wcip = aip*Lmtip*Tip*P*8.9* 1e-3;
Wip = 7.8e-6*Lip*bip*Hip; Pbrg = 0.01*KW*1e3; Pvar = Pcua + Pse + Pip + BCL;
Pconst = Psh + Pi + BFL + Pbrg; Pt = (Pvar + Pconst)/1000;
 Eff = KW/(KW + Pt)*100; Ldmxef = sqrt(Pconst/Pvar)*KW;
 MaxEff = Ldmxef/(Ldmxef + 2*Pconst/1e3)*100;
WtTot = 1.01*(Wc + Wt + Wcua + Wccom + Wcsh + Wcse + Wcip + Wmp +
Wyoke + Wip);
KgPkw = WtTot/KW;
```

4.9 Computer Program in "C" in MATLAB for Complete Design

```
% Design of 500KW, 600V, Lap wound DC Machine
KW = 500; V = 600; nv = 4; bv = 10; ki = 0.9; dVp = 3; Zh = 2;
LossP = 3; Ish1 = 3;
LtoPA = 1; InsHs = 3.2; InsWs = 1.8; Hw = 3; HL = 1;
PPmax = 450; ATaPmax = 8500;
P = 6; A = P; N = 700; PabPP = 0.7; SpP = 14; cda1 = 5;
wca = 1.9; Bc = 1.3; DcbD = 0.66; cdip1 = 2.3; % Assumption
f2 = fopen ('Total_500KW_Output.m','w');
% <------------1) Main Dimensions of Armature-------->
SKW = [5 10 50 100 200 300 500 1000 1500 2000 5000 10000];
Sbav = [.43 .45 .56 .6 .62 .64 .66 .68 .7 .72 .74 .76];
Sq = [15 17.5 25 28 32.5 34 36.5 41 42 44 48 51];
IFL = KW*1000/V; Bav = interp1 (SKW, Sbav, KW, 'spline');
q = interp1 (SKW, Sq, KW, 'spline')*1e3;
hz = P*N/120; if hz < 25 || hz > 50 continue; end;
Pa = (1 + LossP/100)*KW; C0 = 0.164*Bav*q*1e-3;
DsqL = 1/C0*Pa/N;
D1 = DsqL^(1/3)*(P/(pi*PabPP*LtoPA))^(1/3);
D = ceil(D1*100)*10;
L1 = DsqL/D^2*1e9; L = ceil (L1/10)*10; vA = pi*D/1000*N/60;
if vA> = 30 continue; end;
fprintf (f2,'Design of %3d KW,%3d V, LAP wound DC Machine\n',
KW, V);
fprintf (f2,'<-------1) Design of Main Dimensions------>\n');
fprintf (f2, 'Rated KW = % 4.0f\n', KW);
fprintf (f2, 'Rated Volts = % 4.0f\n', V);
fprintf (f2, 'Rated Speed (RPM) = % 4.0f\n', N);
fprintf (f2, 'Rated Current (Amps) = % 4. 1f\n', IFL);
fprintf (f2, 'No of Poles = % 4. 0f\n', P);
fprintf (f2, 'No of ParallelPaths = % 4. 0f\n', A);
fprintf (f2, 'Length/PolearcRatio = % 4. 2f\n', LtoPA);
fprintf (f2,'AvGapFlux-dens(T) = % 4.3f\n', Bav);
fprintf (f2,'q = Sp. ElecLoad (ac/m) = % 6. 0f\n', q);
fprintf (f2, 'Frequency(c/s) = % 4.1f (Range :25to50)\n', hz);
fprintf (f2, 'Polearc/PolePitch = % 4.2f (Range:0.67to0.70)\n',
PabPP);
IBA = IFL/P*2; if IBA > = 400 continue; end;
```

```
Bg = Bav/0.7; Vseg = 2*Bg*L/1000*vA; if Vseg > = 20 continue;
end;
PP = pi*D/P; if PP > = PPmax continue; end;
Parc = PabPP*PP; Li = (L-nv*bv)*ki; FI1 = Bav*pi*D*L/P*1e-6;
ATaP = pi*D/1000*q/(2*P); if ATaP > = ATaPmax continue; end;
E = (1 + dVp/100)*V; A = P; Z1 = E*60*A/(P*FI1*N);
S = SpP*P; sp = pi*D/S; if sp < 25 || sp > 35 continue; end;
Zs1 = Z1/S; Zsf = floor(Zs1/2)*2; Zsc = ceil(Zs1/2)*2;
Zs = Zsf; if Zs1-Zsf  > 1 Zs = Zsc; end; Z = S*Zs;
FI = E*60*A/(P*Z*N);
CV = IFL/A*Zs; if CV < 1000 || CV > 1500 continue; end;
fprintf (f2,'Output Coefft (C0) = % 5.3 f\n',C0);
fprintf (f2,' DsqL Value = % 6.4f\n', DsqL);
fprintf (f2,' Arme dia (mm) = % 4.0 f\n', D);
fprintf (f2,' Arme Length(mm) = % 4.0f\n', L);
fprintf (f2,'Periphoral Vel(m/s) = % 4.2f ( < = 30 )\n', vA);
fprintf (f2, 'PeakGapFlux-dens(T) = % 4.3f\n', Bg);
fprintf (f2, 'Volts/segment = % 4.1f (< = 20)\n', Vseg);
fprintf (f2, 'PolePitch(mm) = % 4.3f\n', PP);
fprintf (f2,' Amps/Brush arm = % 4.0f (< = 400)\n', IBA);
fprintf (f2,' Arm. AT/Pole = % 4.0f\n', ATaP);
fprintf (f2,' Flux/Pole (Wb) = % 6.4f\n', FI);
fprintf  (f2,'  <-------2)Armature-Winding  and  core---------
>\n');
fprintf (f2,' Tot Arm.Conductors = % 4.0 f\n', Z);
fprintf (f2,' No of Slots = % 3.0 f\n', S);
fprintf  (f2,'  SlotPitch  (mm)  =  %  4.2  f  (Range:25  to  35)\n',
sp);
fprintf (f2,' Conductors/Slot = % 4.0 f\n', Zs);
fprintf (f2,' Current (ac/Slot) = % 4.0 f (< = 1500)\n', CV);
Ia = IFL + Ish1; Aca1 = Ia/A/cda1; hca = ceil (Aca1/wca);
if hca > = 16 continue; end;
Aca = 0.98*hca*wca; cda = Ia/A/Aca; if cda > = 5.01 continue;
end;
Zw = Zs/Zh;Ws = Zw*(wca + 0.4) + InsWs;
Hs = Zh*(hca + 0.4) + InsHs + Hw + HL;
D1b3 = D-2*(2/3*Hs); sp1b3 = pi*D1b3/S; Wt1b3 = sp1b3-Ws;
At1b3 = Wt1b3*Li*S/P; Btav = FI*1e6/At1b3;
Ks = sp1b3*L/Wt1b3/Li; if Ks > = 2.2 continue; end;
```

```
Btm = Btav/PabPP; if Btm > = 2.1 continue; end;
Lfr = 140 + 1.15*PP; Lac = (L + Lfr)/1e3;
Ra = 0.021*Lac/Aca*Z/A^2; Wcua = Lac*Aca*Z*8.9e-3;
FIc = FI/2; Ac = FIc*1e6/Bc; Hc1 = Ac/Li; Di1 = D-2*(Hs+Hc1);
Di = ceil(Di1/10)*10;
Hc = (D-Di)/2-Hs; Dmc = D-2*Hs-Hc;
fprintf (f2,' Arm-Cond (HtxW) = % 4.1fmmX % 4.1fmm\n',hca,wca);
fprintf (f2,' Arm-cond (A/mm^2) = % 4.2 f\n', cda);
fprintf (f2,' Arm-Slot (HtxW) = % 4.1 fmm X % 4.1fmm\n',Hs,Ws);
fprintf (f2,' Flux-dens: Av(1/3) = % 4.3 f and Max = % 4.3 f (<
= 2.2T )\n', Btav, Btm);
fprintf (f2,' Arm-Res (m.ohms) = % 6.2 f\n', Ra*1e3);
fprintf (f2,' Wt of ArmCopper (Kg) = % 4.1 f\n',Wcua);
fprintf (f2,' CoreFlux-dens (T) = % 4.3 f\n', Bc);
fprintf (f2,'Ht of Core-Back (m) = % 4.1 f\n', Hc);
fprintf (f2,' Core Inner Dia (m) = % 4.1 f\n', Di);
Pcua = 1.2*Ia^2*Ra; Wc = 7.8 e-6*pi*Dmc*Ac;
HysPkgc = (2.9/30)*hz; % Emperial Formula
Physc = Wc*HysPkgc; Peddyc = 0.0045*(Bc*hz*0.5)^2*Wc;
Pic = (Physc + Peddyc)*1.25;
Wt = 7.8e-6*Wt1b3*Li*S*Hs; HysPkgt = (8/30)*hz;
Physt = Wt*HysPkgt; Peddyt = 0.007*(Btm*hz*0.5)^2*Wt;
Pit = (Physt+Peddyt)*1.25; Pi = (Pic+Pit);
PibKW = Pi/(KW*1e3)*100; Pa = Pcua + Pi;
Larme = L + 70 + 0.3*PP; SA = pi*D*Larme/100;
Pd = Pa/SA; Tra = 270*Pd/(1 + 0.09*(vA^1.3));
Fprintf (f2,' ArmCopper Loss(W) = % 5.0 f\n', Pcua);
Fprintf (f2,' Wt of ArmCore (Kg) = % 4.0 f\n', Wc);
fprintf (f2,' Wt of ArmTeeth (Kg) = % 4.1 f\n', Wt);
fprintf (f2,' Iron Loss (W): Core + Teeth = % 4.0f + % 4.0 f =
% 4.0 f\n', Pic, Pit, Pi);
fprintf (f2,' Temp-Rise of Arme (deg) = % 4.1 f\n', Tra);
fprintf (f2,' <-------3) Design of Poles and AT Calculation----
----->\n');
Bp = 1.6; Pf = 730; Sf = 0.6; By = 1.2; Kg = 1.15; HPS = 40; %
Assumption
DA = [200 300 400 500 700 1000]; dSFW = [30 33 37 40 45 50];
Lp = L-15; Ap = 1.2*FI/Bp/0.95; Wp1 = Ap*1e6/Lp;
```

```
Wp = ceil(Wp1);

df1 = interp1(DA, dSFW, D, 'spline'); df = ceil(df1*10)/10;
atpm = 1e4*sqrt (Pf*Sf*df/1e3);

Atpp = Ia/A*Z/2/P; Atf = ATpp; Hfc = ATf/atpm*1e3;

Hp1 = Hfc + HPS + 0.1*PP; Hp = ceil(Hp1);

Ay = 1.2*FI/2/By; Ly = 1.5*L; dy = Ay*1e6/Ly;

Wmp = 7.8e-6*Lp*Wp*Hp; Wyoke = 7.8*Ay*Ly;

for i = 1:20; BB(i) = i*0.1; end

Ha = [50 65 70 80 90 100 110 120 150 180 220 295 400 580 1000
2400 5000 8900 15000 24000]; % Lohys

H = [20 30 40 50 70 90 120 130 160 200 250 320 480 850 2200
4400 9000 16000 25000 38000]; %, 42 quality

Hcs = [180 270 370 440 520 590 670 750 820 920 1030 1230 1600
2200 3200 4800 7700 13000 22000 40000]; % Cast-Steel

SBt = [1.8 1.9 2 2.1 2.2 2.3 2.4 2.5]; SATm = [9 12 20 31 51 80
113 155]; % for          Ks = 2.0

Atp = interp1(BB, H, Bp, 'spline'); ATp = atp*Hp/1e3;

atat = interp1(SBt, SATm, Btm, 'spline')*1e3;

ATt = atat*Hs/1e3;

atac = interp1(BB, Ha, Bc, 'spline'); Lfpc = pi*Dmc/2/P;

ATc = atac*Lfpc/1e3;

aty = interp1(BB, Hcs, By, 'spline'); Dmy = D + 2*Hp +dy;

Lfpy = pi*Dmy/(2*P*1e3);

ATy = Lfpy*aty; ATTi = ATp + ATt + ATc + ATy;

Perc = ATTi/ATf*100; AT0 = 0.85*ATf; ATg = AT0-ATTi;

By = Bav/0.7;

Lg = ATg/(0.796*By*Kg*1e6)*1e3; D0 = D + 2*(Lg + Hp + dy);

fprintf (f2,' Pole (L*B*H) (mm) = % 4.0 fX % 4.0 fX % 4.0 f\n',
Lp, Wp, Hp);

fprintf (f2,' PoleFlux-dens (T) = % 4.3 f\n', Bp);

fprintf (f2,' Depth of Yoke (mm) = % 4.3 f\n', dy);

fprintf (f2,' YokeFlux-dens (T) = % 4.3 f\n', By);

fprintf (f2,' AT for Iron Parts: Pole + Teeth + Core + Yoke = %
5.0 f + % 4.0 f + % 4.0 f + % 4.0 f = % 4.0 f (% 4.1fperc)\n',
ATp, ATt, ATc, ATy, ATTi, Perc);

fprintf (f2,' Air-gap length(mm) = % 4.2 f\n', Lg);
```

```
fprintf (f2,' <------4) Design of Shunt Fieldand series Fld Windings ------>\n');

ti = 10; dcu = 1.6; cdse = 2.2; % Assumptions

Vsh = 0.85*V; Vc = Vsh/P; Lmt = (2*(Wp + Lp) + pi*(df + 2*ti))/1e3; Ash = 0.021*Lmt*AT0/Vc;

cssfc = 2*Lmt*0.8*Hfc/1e3; Psfc = 730*cssfc*P; Ish = Psfc/Vsh; Tsh1 = AT0/Ish;

T1a = 0.8*Hfc/(dcu + 0.11); T1 = floor(T1a); NL1 = Tsh1/T1;

NL = ceil(NL1); Tsh = T1*NL;

dw = NL*(dcu + 0.11); Acu = pi*dcu^2/4;

Rsh = 0.021*Lmt*Tsh*P/Acu; Psh = Ish^2*Rsh;

Wcsh = Acu*Lmt*Tsh*P*8.9*1e-3;

ATse1 = ATf-AT0; Tse1 = ATse1/Ia; Tse = ceil (Tse1);

ATse = Tse*Ia; Ase = Ia/cdse; Rse = 0.021*Lmt*Tse*P/Ase;

se = Ia^2*Rse; Wcse = Ase*Lmt*Tse*P*8.9*1e-3;

fprintf (f2,' Sh.Field Wdg: Ish = % 4.2fA; Tuns/Coil = % 4.0 f; Rsh = % 5.1 fohms and Psh = % 5.1f W\n', Ish, Tsh, Rsh, Psh);

fprintf (f2,' Series-Field Wdg: Tuns/Pole = % 2.0 f; Rse = % 5.3fm. ohms and Pse = % 5.1 f W\n', Tse, Rse*1e3, Pse);

fprintf (f2, '<------5) Design of Commutator and Brushes------>\n');

cdb = 6; Tb = 10; Wb = 20; Hb = 32; Vb = 2; % Assumptions

Dc1 = DcbD*D; Dc = ceil(Dc1); vC = pi*Dc/1000*N/60;

if vC > = 20 continue; end;

Nc = Z/2; pitC = pi*Dc/Nc; if pitC > = 10 continue; end;

GbSp = pi*Dc/P; % if GbSp < 250 continue; end;

NbPs1= (Ia*2/(P*cdb*Tb*Wb/100)); NbPs = ceil(NbPs1);

Lcom = NbPs*(Wb + 5) + 40 + 30 + 20;

BFL = 0.2*P*Wb*Tb*1e-6*NbPs*13000*vC; BCL = Vb*Ia;

Pcom = BFL + BCL;      CSC = pi*Dc*Lcom/100;

Trc = 120*Pcom/CSC/(1 + 0.1*vC); if Trc > = 50 continue; end;

Wccom = Lcom*pi*Dc^2/4*8.9e-6;

fprintf (f2,' Commutator: Dia = % 4.0 fmm; Per. Vel: Calc = % 4.2 f (Permissible = 20m/s)\n', Dc, vC);

fprintf (f2,' Seg-Pitch: Calc = % 4.2 fmm (Permissible >4mm)\n', pitC);
```

```
fprintf (f2,' Gap bet Spindles (mm): % 4. 0f (Permissible = 250 to 300mm)\n', GbSp);

fprintf (f2,' Brush-Selected: (T*W*H)(mm) = % 3. 0f X %3. 0f X % 3. 0f and Brushes/Spindle = % 3. 0f\n', Tb, Wb, Hb, NbPs);

fprintf (f2,' Commutator: Length = % 4.1 fmm; Losses = % 5. 0f W and Temp-Rise (deg) = % 4.1 f (Max. Perm = 50)\n', Lcom, Pcom, Trc);

fprintf (f2,' <---6) Design of Inter/Compensating-Pole-Wdg and Overall prformance->\n');

Lip = 0.7*L; x = Zs/2; bip = ((x-1)*pitC + Tb-0.8)*D/Dc;

Lgi = 1.5*Lg; Hip1 = Hp-(Lgi-Lg);

Hip = ceil(Hip1); hcai = 2*(hca + 0.4); ht = Hw + HL + InsHs/2; perip = 2*(Zs/2*(wca + 0.4) + (hca + 0.4));

Lmd1 = 0.4*pi*(hcai/3/Ws + ht/Ws + bip/6/Lgi);

Lmd2 = Lfr/L*(0.23*log(Lfr/perip) + 0.07);

Lamda = (Lmd1 + Lmd2)*1e-6;

Bip = L/Lip*2*Ia/6*Zs*Lamda/(bip/1e3);

ATip1 = ATf + 0.796*Bip*1.1*Lgi*1e3;

Tip1 = ATip1/Ia; Tip = ceil(Tip1); ATip = Tip*Ia;

aip1 = Ia/cdip1;    spt = (Hip-40)/Tip; hcse1 = (spt-0.5);

hcse = floor (hcse1); tcse1 = aip1/2/hcse; tcse = ceil(tcse1);

aip = 0.98*2*hcse*tcse; cdip = Ia/aip;

Lmtip = (2*(Lip + bip) + 4*(tcse + 0.5) + 3)/1e3;

Rip = 0.021*Lmtip*Tip*P/aip; Pip = Ia^2*Rip;

Wcip = aip*Lmtip*Tip*P*8.9*1e-3; Wip = 7.8e-6*Lip*bip*Hip;

fprintf (f2,' Inter-Pole (L*B*H)(mm) = % 4.1 f X % 4. 1f X % 4.1 f and Flux. Dens = % 5.4f T\n', Lip, bip, Hip, Bip);

fprintf (f2,' Inter-Pole Wdg: Tuns/Pole = % 2.0 f; Rip = % 5.3 fm.ohms and Pip = % 5.1 f W\n', Tip, Rip*1e3, Pip);

Pbrg = 0.01*KW*1e3; Pvar = Pcua + Pse + Pip + BCL;

Pconst = Psh + Pi + BFL + Pbrg; Pt = Pvar + Pconst;

Eff = KW/(KW + Pt/1e3)*100;  Ldmxef = sqrt(Pconst/Pvar)*KW;

MaxEff = Ldmxef/(Ldmxef + 2*Pconst/1e3)*100;

WtTot = 1.01*(Wc + Wt + Wcua + Wccom + Wcsh + Wcse + Wcip + Wmp + Wyoke + Wip); KgPkw = WtTot/KW;
```

```
fprintf (f2,' Variable Losses (Pcua + Pse +Pip + BCL) = % 4.0 f + % 4.0 f + % 4.0 f + % 4.0 f = % 5.0 f  W \n', Pcua, Pse, Pip, BCL, Pvar);

fprintf (f2,' Constant Losses (Psh + Pi + BFL + Pbg) = % 4.0 f + % 4.0 f + % 4.0 f + % 4.0 f = % 5.0 f W \n', Psh, Pi, BFL, Pbrg, Pconst);

fprintf (f2,' Toal Losses = % 5.0 f W; Efficiency = % 5.2 f perc\n', Pt, Eff);

fprintf (f2,' Load for Max. Eff = % 5.0 f KW and Max. Efficiency = % 5.2 f perc\n', Ldmxef, MaxEff);

fprintf (f2,' WtTot (Kg) = Wc + Wt + Wcua + Wccom + Wcsh + Wcse + Wcip + Wmp + Wyoke + Wip = ');

fprintf (f2,' \n = % 3.0 f + % 3.0 f + % 3.0 f + % 4.0 f + % 3.0 f + % 3.0 f +', Wc, Wt, Wcua, Wccom, Wcsh, Wcse);

fprintf (f2,' % 3. 0f + % 3.0 f + % 3.0 f + % 3.0 f = % 4.0 f\n', Wcip, Wmp, Wyoke, Wip, WtTot);

fprintf (f2,' Wt of Total Machine(Kg) = % 5.0 f and Kg/KW = % 5.2 f \n', WtTot, KgPkw);

fprintf (f2,' <----------------End--------------------->\n\n');

fclose (f2);
```

4.10 Computer Output Results for Complete Design

Design of 500 KW, 600 V, LAP wound DC Machine

<-------1) Design of Main Dimensions------>

Rated KW = 500

Rated Volts = 600

Rated Speed (RPM) = 700

Rated Current (Amps) = 833.3

No .of Poles = 6

No. of ParallelPaths = 6

Length/PolearcRatio = 1.00

AvGapFlux-dens (T) = 0.660

q = Sp.ElecLoad (ac/m) = 36500

Frequency (c/s) = 35.0 (Range :25 to 50)

Polearc/PolePitch = 0.70 (Range: 0.67 to 0.70)

Output Coefft (C0) = 3.951

DsqL Value = 0.1862

Arme dia (mm) = 800

Arme Length (mm) = 300

Periphoral Vel (m/s)=29.32 (<= 30)

PeakGapFlux-dens (T) = 0.943

Volts/segment = 16.6 (< = 20)

PolePitch (mm) = 418.879

Amps/Brush arm = 278 (< = 400)

Arm.AT/Pole = 7645

Flux/Pole (Wb) = 0.0788

<-------2)Armature-Winding and core--------->

Tot Arm.Conductors = 672

No of Slots = 84

SlotPitch (mm) = 29.92 (Range: 25 to 35)

Conductors/Slot = 8

Current (ac/Slot) = 1111 (< = 1500)

Arm-Cond (HtxW) = 15.0mm X 1.9mm

Arm-cond (A/mm^2) = 4.99

Arm-Slot (HtxW) = 38.0mm X 11.0mm

Flux-dens:Av(1/3) = 1.413 and Max = 2.019 (< = 2.2T)

Arm-Res (m.ohms) = 12.94

Wt of ArmCopper (Kg) = 154.0

CoreFlux-dens (T) = 1.300

Ht of Core-Back (m) = 127.0

Core Inner Dia (m) = 470.0

ArmCopper Loss (W) = 10858

Wt of ArmCore (Kg) = 444

Wt of ArmTeeth (Kg) = 99.2

Iron Loss(W):Core + Teeth = 3167 + 2241 = 5408

Temp-Rise of Arme (deg) = 42.6

<-------3) Design of Poles and AT Calculation--------->

Pole (L*B*H) (mm) = 285 × 219 × 254

PoleFlux-dens (T) = 1.600

Depth of Yoke (mm) = 87.585

YokeFlux-dens (T) =0.943

AT for Iron Parts: Pole + Teeth + Core + Yoke = 1118 + 826 + 63 + 449 = 2456 (31.5perc)

Air-gap length (mm) = 4.84

<------4) Design of Shunt Fieldand series Fld Windings ------>

Sh. Field Wdg: Ish = 2.88A; Tuns/Coil = 2320; Rsh = 177.2 ohms and Psh = 1468.0 W

Series-Field Wdg: Tuns/Pole = 2; Rse = 0.808m.ohms and Pse = 565.1 W

<------5) Design of CommutatorandBrushes------>

Commutator: Dia = 528mm; Per.Vel: Calc = 19.35(Permissible = 20m/s)

Seg-Pitch: Calc = 4.94mm (Permissible >4mm)

Gap bet Spindles (mm): 276(Permissible = 250 to 300mm)

Brush-Selected: (T*W*H) (mm) = 10 X 20 X 32 and Brushes/Spindle = 24

Commutator: Length = 690.0mm; Losses = 3122 W and Temp-Rise(deg) = 11.2(Max. Perm = 50)

<---6) Design of Inter/Compensating-Pole-Wdg and Overall prformance->

Inter-Pole(L*B*H)(mm) = 210.0 X 36.4 X 252.0 and Flux.dens = 0.3692 T

Inter-Pole Wdg: Tuns/Pole = 13; Rip = 2.356m. ohms and Pip = 1648.0 W

Variable Losses (Pcua + Pse + Pip + BCL) = 10858 + 565 + 1648 + 1673 = 14744 W

Constant Losses (Psh + Pi + BFL + Pbg) = 1468 + 5408 + 1449 + 5000 = 13325 W

Toal Losses = 28068 W; Efficiency = 94.68 perc

Load for Max. Eff = 475 KW and Max. Efficiency = 94.69 perc

WtTot (Kg) = 1.01*(Wc + Wt + Wcua + Wccom + Wcsh + Wcse + Wcip + Wmp + Wyoke + Wip) = 1.01*(444 + 99 + 154 + 1345 + 304 + 49 + 146 + 124 + 138 + 15) = 2846 Wt of Total Machine(Kg) = 2846 and Kg/KW = 5.69

<----------------------End---------------------------->

4.11 Modifications to be done in the above Program to Get Optimal Design

1. Insert "for" Loops for the following parameters to iterate the total program between min and max permissible limits for selecting the feasible design variants:-

   (a) No. of Poles

   (b) No. of Speeds

(c) Pole arc by Pole pitch ratios
(d) Slots per pole per phase
(e) Arme Wdg Current densities
(f) Conductor widths
(g) Core flux density
(h) Ratio of Commutator dia to Arm dia

2. Insert also minimum or maximum range of required objective functional values as constraint values, for example (a) Efficiency (b) Kg/KW (3) Temp-rise of Armature etc.
3. Run the program to get various possible design variants

Note: From the feasible design variants printed in the output, select that particular design fulfilling the objective of optimal parameter for the design

4.12 Computer Program in "C" in MATLAB for Optimal Design

```
% Design of 500KW, 600V, Lap wound DC Machin

% DC M/C Design 500KW, 600V, for Book

KW = 500; V = 600; nv = 4; bv = 10; ki = 0.9; dVp = 3; Zh = 2;
Loss P = 3;

LtoPA = 1; InsHs = 3.2; InsWs = 1.8; Hw = 3; HL = 1;

PPmax = 450; ATaPmax = 8500;

f2 = fopen ('Optimal_500KW_Output.m','w');

fprintf (f2,' Design of % 3d KW, % 3d V, LAP wound DC
Machine\n', KW, V);

fprintf (f2,' Sn  P   N  hz   Z    S  sp  Zs  CV  hcaXwca cda
Btm  Bc   D   L PA/PP  Tra Dc/Da Dc  Trc  Eff KgPkw\n');

fprintf(f2,'=================================================
=================================================\n');

% <-----------1) Main Dimensions of Armature-------->

SKW = [5 10 50 100 200 300 500 1000 1500 2000 5000 10000];

Sbav = [.43 .45 .56 .6 .62 .64 .66 .68 .7 .72 .74 .76];

Sq = [15 17.5 25 28 32.5 34 36.5 41 42 44 48 51];

IFL = KW*1000/V; Bav = interp1 (SKW, Sbav, KW,' spline');

q = interp1 (SKW, Sq, KW,' spline')*1e3;
```

```
Sn = 0; M1 = 0; M2 =0; M3 = 0; EFFmax = 94; minTra= 50;
minKgPkw = 7; % Assumptions
for P = 4:2:10; for N = 300:50:1000; hz = P*N/120;
if hz < 25 || hz > 50 continue; end;
Pa = (1 + LossP/100)*KW; C0 = 0.164*Bav*q*1e-3;
DsqL = 1/C0*Pa/N;
for PAbPP = 0.68:0.01:0.7;
D1 = DsqL^(1/3)*(P/(pi*PAbPP*LtoPA))^(1/3);
D = ceil(D1*100)*10;      L1 = DsqL/D^2*1e9;
L = ceil (L1/10)*10; vA = pi*D/1000*N/60;
if vA > = 30 continue; end;
IBA = IFL/P*2; if IBA > = 400 continue; end;
Bg = Bav/0.7; Vseg = 2*Bg*L/1000*vA; if Vseg > = 20 continue;
end;
PP = pi*D/P; if PP > = PPmax continue; end;
Parc = PAbPP*PP; Li = (L-nv*bv)*ki; FI1 = Bav*pi*D*L/P*1e-6;
ATaP = pi*D/1000*q/(2*P); if ATaP > = ATaPmax continue; end;
%------------2)Armature-Winding and core------->
E = (1 + dVp/100)*V; A = P; Z1 = E*60*A/(P*FI1*N);
for SpP = 9:15; S = SpP*P; sp = pi*D/S;
if sp < 25 || sp > 35 continue; end;
Zs1 = Z1/S; Zsf = floor(Zs1/2)*2; Zsc = ceil (Zs1/2)*2;
Zs = Zsf; if Zs1-Zsf  > 1 Zs = Zsc; end; Z = S*Zs;
FI = E*60*A/(P*Z*N);
CV = IFL/A*Zs; if CV < 1000 || CV > 1500 continue; end;
Ish1 = 3;  % Assumptions
for cda1 = 4.5:0.1:5;
Ia = IFL + Ish1; Aca1 = Ia/A/cda1;
for wca = 1.6:0.1:2.2;
hca = ceil(Aca1/wca); if hca > = 16 continue; end;
```

```
Aca = 0.98*hca*wca; cda = Ia/A/Aca; if cda > = 5.01 continue;
end;
Zw = Zs/Zh; Ws = Zw*(wca + 0.4) + InsWs;
Hs = Zh*(hca + 0.4) + InsHs + Hw + HL;
D1b3 = D-2*(2/3*Hs); sp1b3 = pi*D1b3/S; Wt1b3 = sp1b3-Ws;
At1b3 = Wt1b3*Li*S/P; Btav = FI*1e6/At1b3;
Ks = sp1b3*L/Wt1b3/Li; if Ks > = 2.12 continue; end;
Btm = Btav/PAbPP; if Btm > = 2.1 continue; end;
Lfr = 140 + 1.15*PP; Lac = (L + Lfr)/1e3;
Ra = 0.021*Lac/Aca*Z/A^2; Wcua = Lac*Aca*Z*8.9e-3;
for Bc = 1.2:0.1:1.3;
FIc = FI/2; Ac = FIc*1e6/Bc; Hc1=Ac/Li; Di1=D-2*(Hs + Hc1);
Di = ceil (Di1/10)*10;
Hc = (D-Di)/2-Hs; Dmc = D-2*Hs-Hc;
Pcua = 1.2*Ia^2*Ra; Wc = 7.8e-6*pi*Dmc*Ac;
HysPkgc = (2.9/30)*hz; % Emperial Formula
Physc = Wc*HysPkgc; Peddyc = 0.0045*(Bc*hz*0.5)^2*Wc;
Pic = Physc + Peddyc;
Wt = 7.8e-6*Wt1b3*Li*S*Hs; HysPkgt = (8/30)*hz;
Physt = Wt*HysPkgt; Peddyt = 0.007*(Btm*hz*0.5)^2*Wt;
Pit = Physt + Peddyt; Pi = (Pic + Pit)*1.25;
PibKW = Pi/(KW*1e3)*100; Pa = Pcua + Pi;
Larme = L + 70 + 0.3*PP; SA = pi*D*Larme/100;
Pd = Pa/SA; Tra = 270*Pd/(1 + 0.09*(vA^1.3));
% <-----------3)Design of Poles and AT Calculation------>\n');
Bp = 1.6; Pf = 730; Sf = 0.6; By = 1.2; Kg = 1.15; % Assumption
DA = [200 300 400 500 700 1000]; dSFW = [30 33 37 40 45 50];
Lp = L-15; Ap = 1.2*FI/Bp/0.95; Wp1 = Ap*1e6/Lp;
Wp = ceil(Wp1);   df1 = interp1(DA, dSFW, D, 'spline');
df = ceil(df1*10)/10; atpm = 1e4*sqrt (Pf*Sf*df/1e3);
Atpp = Ia/A*Z/2/P; Atf = ATpp; Hfc = ATf/atpm*1e3;
```

```
Hp1 = Hfc + 40 + 0.1*PP; Hp = ceil(Hp1);

Ay = 1.2*FI/2/By; Ly = 1.5*L; dy = Ay*1e6/Ly;

Wmp = 7.8e-6*Lp*Wp*Hp; Wyoke = 7.8*Ay*Ly;

for i =1:20; BB(i) = i*0.1; end

Ha = [50 65 70 80 90 100 110 120 150 180 220 295 400 580 1000
2400 5000 8900 15000 24000]; % Lohys

H = [20 30 40 50 70 90 120 130 160 200 250 320 480 850 2200
4400 9000 16000 25000 38000]; % 42 quality

Hcs = [180 270 370 440 520 590 670 750 820 920 1030 1230 1600
2200 3200 4800 7700 13000 22000 40000]; % Cast-Steel

SBt = [1.8 1.9 2 2.1 2.2 2.3 2.4 2.5];

SATm = [ 9 12 20 31 51 80 113 155]; % for Ks = 2.0

Atp = interp1(BB, H, Bp, 'spline'); ATp = atp*Hp/1e3;

Atat = interp1(SBt, SATm, Btm, 'spline')*1e3;

ATt = atat*Hs/1e3;       atac = interp1(BB, Ha, Bc, 'spline');
Lfpc = pi*Dmc/2/P; ATc = atac*Lfpc/1e3;

aty = interp1(BB, Hcs, By, 'spline'); Dmy = D + 2*Hp + dy;

Lfpy = pi*Dmy/ (2*P*1e3);    ATy = Lfpy*aty;

ATTi = ATp + ATt + ATc + ATy; Perc = ATTi/ATf*100;

AT0 = 0.85*ATf; ATg = AT0-ATTi; Bgm = Bav/PAbPP;

Lg = ATg/(0.796*Bgm*Kg*1e6)*1e3;D0 = D + 2*(Lg + Hp + dy);

% <----4)Design of Shunt Fieldand series Fld Windings ----->;

ti = 10; dcu = 1.6;cdse = 2.2; % Assumptions

Vsh=0.85*V; Vc=Vsh/P; Lmt = (2*(Wp + Lp) + pi*(df + 2*ti))/1e3;

Ash = 0.021*Lmt*AT0/Vc;     Cssfc = 2*Lmt*0.8*Hfc/1e3;

Psfc = 730*cssfc*P; Ish = Psfc/Vsh; Tsh1 = AT0/Ish;

T1a = 0.8*Hfc/(dcu+0.11); T1 = floor(T1a); NL1 = Tsh1/T1;

NL= ceil(NL1); Tsh= T1*NL;    dw= NL*(dcu + 0.11);

Acu = pi*dcu^2/4; Rsh = 0.021*Lmt*Tsh*P/Acu; Psh = Ish^2*Rsh;
Wcsh= Acu*Lmt*Tsh*P*8.9*1e-3; ATse1 = ATf-AT0; Tse1 = ATse1/Ia;
Tse=ceil(Tse1);ATse=Tse*Ia;Ase=Ia/cdse;Rse=
0.021*Lmt*Tse*P/Ase;Pse= Ia^2*Rse; Wcse=Ase*Lmt*Tse*P*8.9*1e-3;
```

```
% <-----------5) Design of Commutator and Brushes ------>

cdb = 6; Tb = 10; Wb = 20; Hb = 32; Vb =2; % Assumptions

for DcbD = 0.65:0.01:0.7;

Dc1 = DcbD*D; Dc = ceil (Dc1); vC = pi*Dc/1000*N/60;

if vC > = 20 continue; end;

Nc = Z/2; pitC = pi*Dc/Nc; if pitC > = 10 continue; end;

GbSp = pi*Dc/P; if GbSp < 250 continue; end;

NbPs1 = (Ia*2/(P*cdb*Tb*Wb/100)); NbPs = ceil(NbPs1);

Lcom = NbPs*(Wb + 5) + 40 + 30 + 20;

BFL = 0.2*P*Wb*Tb*1e-6*NbPs*13000*vC; BCL = Vb*Ia;

Pcom = BFL + BCL;   CSC = pi*Dc*Lcom/100;

Trc = 120*Pcom/CSC/ (1 + 0.1*vC); if Trc > = 50 continue; end;

Wccom = Lcom*pi*Dc^2/4*8.9e-6;

%<6)  Design  of  Inter/Compensating-Pole-Wdg  and  Overall
performance--->

cdip1 = 2.3; % Assumption

Lip = 0.7*L; x = Zs/2; bip = ((x-1)*pitC + Tb-0.8)*D/Dc;

Lgi = 1.5*Lg; Hip1 = Hp-(Lgi-Lg);

Hip = ceil(Hip1); hcai = 2*(hca + 0.4); ht = Hw + HL + InsHs/2;
perip = 2*(Zs/2*(wca + 0.4) + (hca + 0.4));

Lmd1 = 0.4*pi*(hcai/3/Ws + ht/Ws + bip/6/Lgi);

Lmd2 = Lfr/L*(0.23*log (Lfr/perip) + 0.07);

Lamda=(Lmd1+ Lmd2)*1e-6; Bip = L/Lip*2*Ia/6*Zs*Lamda/(bip/1e3);

ATip1   =   ATf   +   0.796*Bip*1.1*Lgi*1e3;Tip1   =   ATip1/Ia;
Tip = ceil(Tip1); ATip = Tip*Ia;

aip1 = Ia/cdip1;        spt = (Hip-40)/Tip; hcse1 = (spt-0.5);
hcse = floor(hcse1); tcse1 = aip1/2/hcse; tcse = ceil (tcse1);

aip = 0.98*2*hcse*tcse;cdip = Ia/aip;

Lmtip = (2*(Lip + bip) + 4*(tcse + 0.5) + 3)/1e3;

Rip = 0.021*Lmtip*Tip*P/aip; Pip = Ia^2*Rip;

Wcip = aip*Lmtip*Tip*P*8.9*1e-3; Wip = 7.8e-6*Lip*bip*Hip;
```

```
Pbrg = 0.01*KW*1e3; Pvar = Pcua + Pse + Pip + BCL;
Pconst = Psh + Pi + BFL + Pbrg; Pt = (Pvar + Pconst)/1000;
Eff = KW/(KW+Pt)*100; if Eff < = 94.6 continue; end;
Ldmxef = sqrt(Pconst/Pvar)*KW;
MaxEff = Ldmxef/(Ldmxef + 2*Pconst/1e3)*100;
WtTot = 1.01*(Wc + Wt + Wcua + Wccom + Wcsh + Wcse + Wcip + Wmp
+ Wyoke + Wip); KgPkw = WtTot/KW;
if KgPkw > = 5.7 continue; end;
Sn = Sn + 1; Sn,
if Eff > = EFFmax  EFFmax = Eff; end; if abs (Eff-EFFmax) < =
2e-3 M1 = Sn; end;
if KgPkw < = minKgPkw minKgPkw = KgPkw; end; if abs (KgPkw-
minKgPkw) < = 0.001 M2 = Sn; end;
if Tra < = minTra  minTra= Tra; end; if abs (Tra-minTra) < =
0.0001 M3 = Sn; end;
% fprintf (f2,' %2 d% 3d% 5d% 3.0f% 5.0f% 4d% 5.1f% 3d% 5.0f%
3.0fX% 3.1f% 5.1f% 4.1f% 4.1f% 5.0f% 4.0f% 5.1f% 4.0f% 5.1f%
6.2f% 5.2f\n', Sn, P, N, hz, Z, S, sp, Zs, CV, hca, wca, cda,
Btm, Bc, D, L, Tra, Dc, Trc, Eff, KgPkw);
Fprintf (f2,'% 2d% 3d% 5d% 3.0f% 5.0f% 4d% 5.1f% 3d% 5.0f%
3.0fX% 3.1f% 5.1f% 4.1f% 4.1f% 5.0f% 4.0f% 5.2f% 5.1f% 5.2f%
4.0f% 5.1f% 6.2f% 5.2f\n',...
Sn, P, N, hz,  Z, S, sp, Zs, CV, hca,wca, cda, Btm, Bc, D, L,
PAbPP,Tra,DcbD, Dc, Trc, Eff, KgPkw);
end; end; end; end; end; end; end; end;
fprintf(f2,'====================================================
==========================================\n');
fprintf (f2,' Selection of Design Variant based on Optimization
Criteria:');
fprintf (f2,' \nIf Maximum Efficiency is Required, Select
Variant(Sn) = %3d (%5.2f perc)', M1, EFFmax);
fprintf (f2,'\nIf Minimum Kg/kw is Required, Select Variant
(Sn) = % 3d (%5.2f)', M2, minKgPkw);
fprintf (f2,'\nIf Minimum Arme-Temp-Rise is Reqd, Select
Variant (Sn) = % 3d        (% 4.2f)', M3, minTra);
fclose(f2);
```

4.12 (a) Computer Output Results for Optimal Design

Design of 500 KW, 600 V, LAP wound DC Machine

Sn	P	N	hz	Z	S	sp	Zs	CV	hca×wca	cda	Btm	Bc	D	L	PA/PP	Tra	Dc/Da	Dc	Trc	Eff	KgPkw
1	6	700	35	672	84	30.3	8	1111	15×1.9	5.0	2.1	1.2	810	290	0.69	42.3	0.65	527	11.2	94.71	5.69
2	6	700	35	672	84	30.3	8	1111	15×1.9	5.0	2.1	1.3	810	290	0.69	42.3	0.65	527	11.2	94.71	5.63
3	6	700	35	672	84	30.3	8	1111	15×1.9	5.0	2.1	1.2	810	290	0.69	42.3	0.65	527	11.2	94.71	5.69
4	6	700	35	672	84	30.3	8	1111	15×1.9	5.0	2.1	1.3	810	290	0.69	42.3	0.65	527	11.2	94.71	5.63
5	6	700	35	624	78	32.2	8	1111	15×1.9	5.0	2.1	1.3	800	300	0.70	41.5	0.65	520	11.4	94.80	5.66
6	6	700	35	624	78	32.2	8	1111	15×1.9	5.0	2.1	1.3	800	300	0.70	41.5	0.65	520	11.4	94.80	5.66
7	6	700	35	672	84	29.9	8	1111	15×1.9	5.0	2.0	1.2	800	300	0.70	42.6	0.65	520	11.4	94.69	5.67
8	6	700	35	672	84	29.9	8	1111	15×1.9	5.0	2.0	1.3	800	300	0.70	42.6	0.65	520	11.4	94.69	5.61
9	6	700	35	672	84	29.9	8	1111	15×1.9	5.0	2.0	1.3	800	300	0.70	42.6	0.66	528	11.2	94.68	5.69
10	6	700	35	672	84	29.9	8	1111	15×1.9	5.0	2.0	1.2	800	300	0.70	42.6	0.65	520	11.4	94.69	5.67
11	6	700	35	672	84	29.9	8	1111	15×1.9	5.0	2.0	1.3	800	300	0.70	42.6	0.65	520	11.4	94.69	5.61
12	6	700	35	672	84	29.9	8	1111	15×1.9	5.0	2.0	1.3	800	300	0.70	42.6	0.66	528	11.2	94.68	5.69

Selection of Design Variant based on Optimization Criteria:

If Maximum Efficiency is Required, Select Variant (Sn) = 6 (94.80 perc)

If Minimum kg/kw is Required, Select Variant (Sn) = 11 (5.61)

If Minimum Arme-Temp-Rise if Required, Select Variant (Sn) = 6(41.53).

CHAPTER 5

Transformers

5.1 Introduction

3-ph Transformers are of mainly two types (viz) Core type and Shell type both again of two categories, Power and Distribution. Theory portion of Design is not given in this book, but the necessary formulae, curves and tables given in standard books are made use of.

5.2 Core Type Power Transformer

Total design is split into six parts in a proper sequence. Design Calculations are given for a given Rating of a Transformer, followed by Computer Program written in "C" language using MATLAB software for each part. Finally all the Programs are added together to get the total Program by running which we get the total design. Computer output of total design is given.

This design may not be the optimum one. Now optimization objective and design constraints are inserted into this total Program. When this program is run we will get various alternative feasible designs from which the selected variant based on the Optimization Criteria can be picked up. Computer output showing the important design parameters for various feasible alternatives is given at the end of this chapter along with a logic diagram.

Problem: Design a 800 KVA, 6600/ 440 V, 50 Hz,3 Phase, Delta/Star, Core Type, ON cooled Power Transformer: (limit temp-rise to 50 deg-C).

5.2.1 Sequential Steps for Design of Each Part and Programming Simultaneously

(a) Calculate the dimensions of Magnetic frame consisting of Core, window and Yoke. Calculate Flux densities in those parts and Iron losses.

(b) Calculate Amp-Turns and No load current.

(c) Calculate No. of turns, size of Copper and final dimensions of LV winding.

(d) Calculate No. of turns, size of Copper and final dimensions of HV winding.

(e) Check weather clearances between Magnetic frame and windings are OK.

(f) Calculate Windings copper Losses, total losses, Efficiency, Reactances and % Regulation.

(g) Calculate Dimensions of cooling tank, number and size of cooling tubes, temperature rise, total weight and Kg/KVA.

Note: By adding programs established for each part sequentially we get the Program for complete design.

5.2.2 Design of Magnetic Frame (Part-1)

Assuming 3 stepped core with factor (k) = 0.6 and Value factor (K) = 0.6,

Volts/turn (Et) = $k \times \sqrt{KVA/ph} = 0.6 \times \sqrt{800/3} = 9.798$

Assuming Max. Flux density in Core (Bm) = 1.5 T,

$$\text{CS area of Core (Ai)} = \frac{Et}{4.44 \times f \times Bm} = \frac{9.798}{4.44 \times 50 \times 1.5} = 0.0294\ m^2;$$

Core dia(d) = $\sqrt{Ai/k} = \sqrt{0.0294/0.6} = 0.2214 \approx 0.23m$ (Rounded off)

Corrected (Ai) = $k \times d^2 = 0.6 \times 0.23^2 = 0.0317\ m^2$;

Corrected (Et) = $4.44 \times f \times Bm \times Ai$

$$= 4.44 \times 50 \times 1.5 \times 0.0317 = 10.57 \text{ V/turn}$$

Assuming Average Current density (cdav) = 2.6 A/mm^2;

$$\text{Window Space factor (kw)} \approx \frac{10}{30 + KV} \times 1.15$$

$$\approx \frac{10}{30 + 6600/1000} \times 1.15 = 0.314$$

$$\text{Window Area (Aw)} = \frac{KVA \times 1000}{3.33 \times f \times Bm \times kw \times cdav \times 10^6 \times Ai}$$

$$= \frac{800 \times 1000}{3.33 \times 50 \times 1.5 \times 0.314 \times 2.6 \times 10^6 \times 0.0317} = 0.1235\ m^2;$$

$$\text{Assuming Length of Core (L)} = \frac{Aw}{D - d} = \frac{0.1235}{0.44 - 0.23}$$

$$= 0.5883 \approx 0.59 \text{ m}$$

$$\text{Distance between Core centres (D)} = \frac{Aw}{L} + d$$

$$= \frac{0.1235}{0.59} + 0.23 = 0.4394 \approx 0.44 \text{ m}$$

Length of Yoke (W) = $2 \times D + 0.9 \times d = 2 \times 0.44 + 0.9 \times 0.23 = 1.1$ m

Assuming Iron factor (ki) = 0.92;

CS area of core (Ac) = Ai / ki = 0.0317/ 0.92 = 0.0345 m^2;

Assuming CS area of yoke is 15% more (Ay) = 1.15 × Ac

$$= 1.15 \times 0.0345 = 0.0397 \text{ m}^2;$$

Width of Yoke (by) = 0.9 × d = 0.9 × 0.23 = 0.207 m

Height of Yoke (hy) = Ay/by = 0.0397/0.207 = 0.1917 m;

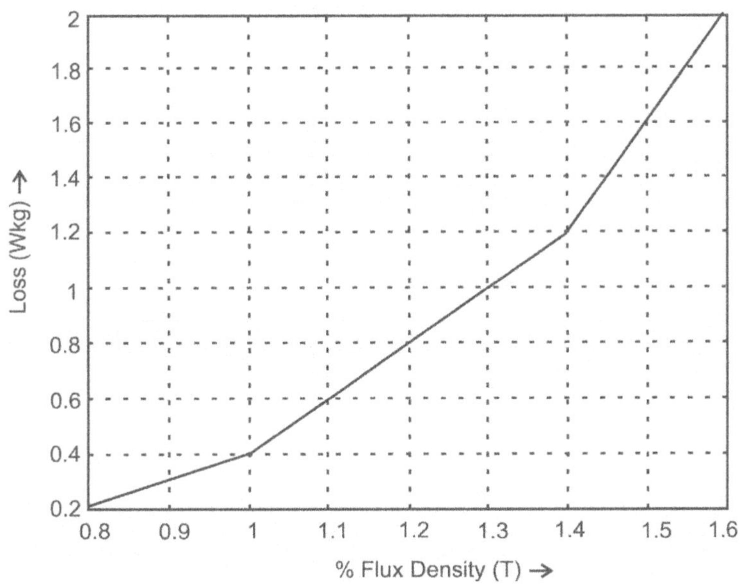

Fig. 5.1 Core loss for cold Rolled steels.

From Fig. 5.1,

Loss in Core corresponding to Bm = 1.5T (WpKgC) = 1.6 W/Kg

Weight of Core (KgC) = Ph × Ac × L × 7.55 × 1000

$$= 3 \times 0.0345 \times 0.59 \times 7.55 \times 1000 = 461.04 \text{ Kg}$$

Iron Loss in Core (PiC) = WpKgC × KgC = 1.6 × 461.04 = 737.67 W

Weight of Yoke (KgY) = 2 × Ay × W × 7.55 × 1000

$$= 2 \times 0.0397 \times 1.1 \times 7.55 \times 1000 = 659 \text{ W}$$

Flux density in Yoke (By) = Ac/Ay × Bm

$$= 0.0345/0.0397 \times 1.5 = 1.3043 \text{ T}$$

From Fig. 5.1,

Loss in Core corresponding to By = 1.3043T (WpKgY) = 1.009 W/Kg

Iron Loss in Yoke (PiY) = WpKgY × KgY

$$= 1.009 \times 659 = 664.73 \text{ W}$$

Considering 5% extra,

Total Iron Loss (Pi) = 1.05 × (PiC + PiY)/1000

$$= 1.05 \times (737.67 + 664.73)/1000 = 1.4725 \text{ KW};$$

5.2.2 *(a) Computer Program in "C" in MATLAB for Part-1*

```
% Design of Core Type Power Transformer
KVA = 800; HV = 6600; LV = 440; f = 50; Ph = 3; % Input Data
f2 = fopen ('Total_800KVA_Output.m','w');
fprintf (f2, '% 3d KVA, % 5d/ % 4d V, % 2d Hz, % 1d Phase,
'KVA, HV, LV, f, Ph);
fprintf (f2,' Delta/Star, Core Type, ON cooled Power
Transformer:\n', KVA, HV, LV, f, Ph);
fprintf (f2, '======================================\n');
fprintf (f2,'<-----1) Design of Magnetic Frame-----------------
->\n');
k = 0.6; ki = 0.92; K = 0.6; Bm = 1.5; cdav = 2.6; Lby D = 2.8;
% Assumptions
Et = K*sqrt (KVA/Ph); Ai1 = Et/(4.44*f*Bm); d1 = sqrt (Ai1/k);
d = ceil (d1*100) /100;
Ai = k*d^2; Et = 4.44*f*Bm*Ai; kw = 10/(30 + HV/1e3)*1.15;
Aw = KVA*1e3/(3.33*f*Bm*kw*cdav*1e6*Ai); L1 = sqrt(LbyD*Aw);
L = ceil(L1*100)/100;  D1 = Aw/L + d;  D = ceil(D1*100)/100;
LbyD = L/(D-d); W1 = 2*D + 0.9*d;   W = ceil(W1*10)/10;
Ac = Ai/ki; Ay = 1.15*Ac;  by = 0.9*d;hy = Ay/by;
BB = [0.8 1.0 1.2 1.4 1.6]; WpKg = [0.2 0.4 0.8 1.2 2];
% (for Cold Rolled)
Plot (BB, WpKg); grid; xlabel (% Flux density (T)-->');
ylabel ('Loss(W/Kg)-->');
title ('Core Loss for Cold rolled steels'); %
WpKgC = interp1(BB, WpKg, Bm, 'spline'); KgC = Ph*Ac*L*7.55e3;
PiC = WpKgC*KgC;  KgY = 2*Ay*W*7.55e3;   By = Ac/Ay*Bm;
WpKgY = interp1(BB, WpKg, By, 'spline');  PiY = WpKgY*KgY;
Pi = 1.07*(PiC + PiY)/1000;
```

5.2.3 No-Load Current (Part-2)

From Fig. 5.2,

AT/m for Core corresponding to Bm = 1.5T (atC) = 150 AT/m

From Fig. 5.2, AT/m for Yoke corresponding to By

= 1.3043 T (atY) = 109.6 AT/m

AT for Core (ATC) = Ph × at C × L = 3 × 150 × 0.59 = 265.5

AT for Yoke (ATY) = 2 × at Y × W = 2 × 109.6 × 1.1 = 241.12

Total AT/phase (ATpPh) = (ATC + ATY)/Ph

= (265.5 + 241.12)/3 = 168.87

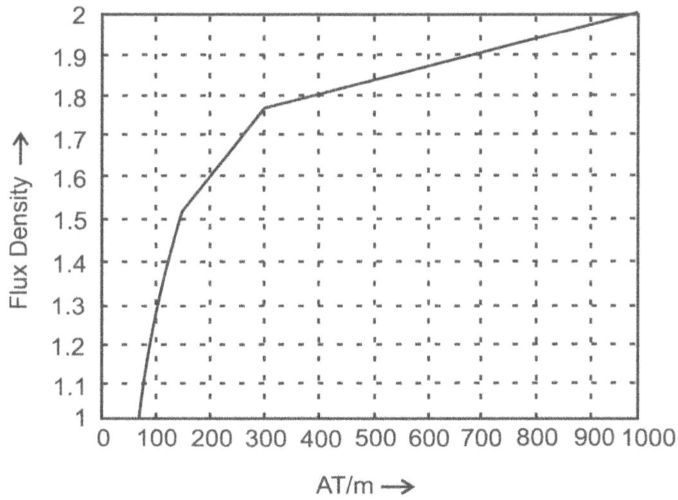

Fig. 5.2 Core loss for cold rolled steels.

$$\text{No. of Turns in LV wdg (T2)} = \frac{\text{LV}}{\sqrt{3} \times \text{Et}} = \frac{440}{\sqrt{3} \times 10.57} = 24.03 \approx 24 \text{ (Rounded off)}$$

$$\text{Phase Current in LV Wdg (I2)} = \frac{\text{KVA} \times 1000}{\sqrt{3} \times \text{LV}} = \frac{800 \times 1000}{\sqrt{3} \times 440} = 1049.7 \text{ A}$$

$$\text{Wattful Current (Iw)} = \frac{\text{Pi} \times 1000}{\sqrt{3} \times \text{LV}} = \frac{1.4725 \times 1000}{\sqrt{3} \times 440} = 1.9322 \text{ A}$$

$$\text{Magnetizing current (Im)} = \frac{1.15 \times \text{ATpPh}}{\sqrt{2} \times \text{T2}} = \frac{1.15 \times 168.87}{\sqrt{2} \times 24} = 5.7218 \text{ A}$$

$$\text{No load Current (I0)} = \sqrt{\text{Iw}^2 + \text{Im}^2} = \sqrt{1.9322^2 + 5.7218^2} = 6.0392 \text{ A}$$

Ratio of I0/I2 (I0byI2) = I0/I2 × 100 = 6.0392/1049.7 × 100

= 0.5753 (Lies between 0.5 to 1 and hence OK).

5.2.3 *(a) Computer Program in "C" in MATLAB for Part-2*

```
% <-----2) No-Load Current------------------->\n');
Bm = 1.5; By = 1.3043; W = 1.1; LV = 440; Et = 10.57;
KVA = 800; Pi = 1.4725; % Input Data
B1 = [1 1.25 1.5 1.75 2.0];    H1 = [70 100 150 300 1000];
% (Cold Rolled)
plot (H1,B1); grid; xlabel ('AT/m-->'); ylabel ('Flux
density(T)-->');
title ('Magnetization Curve for Cold rolled steels'); %
```

```
atC=interp1(B1,H1,Bm,'spline'); atY=interp1(B1,H1,By,'spline');
ATC  =  Ph*atC*L;      ATY  =  2*atY*W;    ATpPh  =  (ATC+ATY)/Ph;
T2  =  floor(LV/sqrt(3)/Et);      I2  =  KVA*1e3/(sqrt(3)*LV);
Iw  =  Pi*1000/(sqrt(3)*LV);   Im  =  1.15*  ATpPh/sqrt(2)/T2;
I0 = sqrt(Im^2 + Iw^2); I0byI2 = I0/I2*100;
% ---------------------------------------------------------------
```

5.2.4 Design of LV Winding (Part-3)

Assume:

1. Space available for turns (ALW) = 80% length
   $= 0.8 \times L = 0.8 \times 0.59 = 0.472$ m = 472 mm
2. 2 turns radially (T2r) and [T2a = T2/T2r
   = 24/2 = 12] Axially/Length-wise
3. Since 1049.7 Amps very high, 12 parallel strands (stP) in each conductor.

   Since 12 turns are accommodated in a length of 472 mm,

   space/turn (ALT) = 472/12 = 39.3 mm

   Since we can not have a single strand of 39.3 mm, let us select 3 strands (NstA = 3) axially

   No. of strands radially (NstR) = stP/NstA = 12/3 = 4

   Hence width of each strand (stW) = (ALT/NstA) – 0.5

   $= (39.3/3) - 0.5 = 12.6 \approx 12$ mm (bare)

   Considering 0.4 mm ins thickness for each strip,

   ins Strip width= 12.4mm

   Space occupied length-wise(ALWx)

   $= [\{(stW + 0.4) \times NstA\} + 2] \times T2a + 100$

   $= [\{(12 + 0.4) \times 3\} + 2] \times 12 + 100$

   = 570.4mm (Less than 590 and hence OK)

   (It is assumed 2mm as inter turn insulation and 100 mm as extra space)

   Assuming 0.98 factor due to rounding of edges of each strand,

   Slack in length (SlkLVax) = $L \times 1000 -$ ALWx

   $= 0.59 \times 1000 - 570.4 = 18.6$(Desirable > 7 and hence OK)

   CS area of conductor (a2) = stW $\times$ stT $\times$ stP $\times 0.98$

   $= 12 \times 3 \times 12 \times 0.98 = 423.36\ mm^2$;

   Current Density (cdLV) = I2/a2 = 1049.7/423.36 = 2.48 A/mm^2 (lies between 2.3 to 3.5 and hence OK)

   Radial Width of the winding (rwLV)

   $= NstR \times (stT + 0.4) \times T2r + 1.8 = 4 \times (3 + 0.4) \times 2 + 1.8 = 29$ mm

LV Winding of Core Type Transformer(dimensions are in mm)

Length/Axial-wise			
Cu.Strand	Bare + insulation	12 + 0.4 = 12.4	
Coil	3 Cu.Strands	3 × 12.4 = 37.2	
Coil + Insulation		37.2 + 2.0 = 39.2	
12 coils		12 × 39.2 =	470.4
Insulation			100.0
Axial Length			**570.4**

Radial-Wise			
Cu.Strand	Bare + insulation	3 + 0.4 = 3.4	
Each turn	4 Cu.Strands	4 × 3.4 = 13.6	
Turns	2 turns	2 × 13.6 = 27.2	27.2
Insulation			1.8
Radial-Width			**29.0**

Assuming Oil duct between Core and LV coil1 = 5mm,

Ins. cylinder between core and LV coil1 = 3 mm and

Oil duct between Insulating Cylider and LV coil2 = 5mm,

Inner dia of LV wdg (di2) = $d \times 1000 + 2 \times (5 + 3 + 5)$

$= 0.23 \times 1000 + 2 \times (13) = 256$ mm;

Outer dia of LV wdg (do2) = di2 + 2 × rwLV

$= 256 + 2 \times 29 = 314$ mm;

$$\text{Mean length of LV winding (Lmt2)} = \frac{\pi \times (\text{di2} + \text{do2})}{2 \times 1000}$$

$$= \frac{\pi \times (256 + 314)}{2 \times 1000} = 0.8954 \text{ m}$$

Res of LV wdg/ph (r2) = 0.02 × Lmt2 × T2/a2

$= 0.02 \times 0.8954 \times 24/423.36 = 1.0151$ mΩ

Copper loss in LV wdg(pcu2) = Ph × $I2^2$ × r2/1000

$= 3 \times 1049.7^2 \times 1.0151 \times 10^{-6} = 3.356$ KW

5.2.4 *(a) Computer Program in "C" in MATLAB for Part-3*

```
% <-----3) Design of LV Winding: ------->;
T2 = 24; L = 0.59; I2 = 1049.7; d = 0.23; Ph = 3; % Input Data
T2r = 2; stP = 12; NstA = 3; stT = 3; % Assumptions
T2a = T2/T2r; NstR = stP/NstA; ALW = 0.8*L*1000; ALT = ALW/T2a;
stW1= ALT/NstA-0.5; stW = floor(stW1);
ALWx=(((stW+ 0.4)*NstA) + 2)*T2a + 100;  SlkLVax = L*1000-ALWx;
a2   =   stW*stT*stP*0.98;                       cdLV   =   I2/a2;
rwLV= NstR*(stT + 0.4)*T2r + 1.8; di2 = d*1000 + 2*(5 + 3 + 5);
do2   =   di2   +   2*rwLV;   Lmt2   =   pi*(di2   +   do2)/2e3;
r2 = 0.02*Lmt2*T2/a2; pcu2 = Ph*I2^2*r2/1000;
% -------------------------------------------------------
```

5.2.5 Design of HV Winding (Part-4)

No. of turns/ph (T1) = T2 $\times$ HV/LV $\times$ $\sqrt{3}$

$= 24 \times 6600/440 \times \sqrt{3} = 623.5 = 624$ (rounded off)

Phase current in HV Wdg (I1) $= \dfrac{\text{KVA} \times 1000}{3 \times \text{HV}} = \dfrac{800 \times 1000}{3 \times 6600} = 40.404$ A

Disc type winding with rectangular cross section is used

Assuming "z-2" no. of coils will have "x" no. of turns and 2 coils (both are in either end) will have "0.65 $\times$ z",

Total no. of turns (T1) = x(z-2)+2 $\times$ 0.65 $\times$ x

from which x = T1/(z – 2 + 1.3);

Here in the program x1 = x and Axc = z;

Assuming total coils length-wise (AxC) = 14;

No. of turns in 12 coils (x1) $= \dfrac{\text{T1}}{\text{AxC} - 2 + 1.3} = \dfrac{624}{14 - 2 + 1.3} = 46.9 \approx 48$

Assuming each coil will have 4 (= cA) strands axially (length-wise), no. of strands in radial direction (x2)

= x1/cA = 48/4 = 12

No. of turns in each of the 2 end coils(x3)

= [T1– {x2 $\times$ cA $\times$ (AxC–2)}]/2

= [624 – {12 $\times$ 4 $\times$ (14 –2)}]/2 = 24

Check for Total turns (T1) = (12 $\times$ 48) + (2 $\times$ 24) = 624

Assuming 70% length is available for winding (ALW)

= 0.7 $\times$ L $\times$ 1000 = 0.7 $\times$ 590 = 413 mm

Space available per coil (ALPC) = ALW/A x C = 413/14 = 29.5 mm

Space for each strand (ALPC1) = ALPC/cA = 29.5/4 = 7.375 mm

Strand width (stW1) = ALPC1– 0.4 = 7.375– 0.4 = 6.975 mm;

Select 6 mm;

Assuming current density (cdHV) = cdav + 0.2

= 2.6 + 0.2 = 2.8 A/mm^2,

CS area of conductor in a turn (a1) = I1/cdHV

= 40.404/2.8 = 14.43 mm^2,

Thickness of strand in a turn (stT1) = a1/stW1

= 14.43/6 = 2.405, say 2.5 mm

Corrected CS area of strand in a turn (a1) = stT1 $\times$ stW1 $\times$ 0.98

= 2.5 $\times$ 6 $\times$ 0.98= 14.7 mm^2,

Current density (cdHV) = I1/a1 = 40.404/14.7 = 2.749 A/mm^2,

Axial length of 4 strands (aLc) = cA × (stW1 + 0.4)

$= 4 \times (6 + 0.4) = 25.6$ mm;

Radial width of Winding (rwHV) = cR × (stT1 + 0.4)

$= 12 \times (2.5 + 0.4) = 34.8$ mm;

Assuming 6mm insulation between coils,

Axial length occupied by all strands (AxLw)

$= AxC \times aLc + (AxC - 1) \times 6 = 14 \times 25.6 + (14 - 1) \times 6$

$= 436.6$ mm

Assuming 30mm gun metal end ring and 100mm end insulation, Axial Length (AxL)

$= AxLw + 30 + 100 = 436.6 + 130 = 566.4$ mm

Slack Available axially (SlkHVax) = L × 1000 – AxL

$= 0.59 \times 1000 - 566.4 = 23.6$ mm; (Desirable > 7 and hence OK)

HV Winding of Core Type Transformer(dimensions are in mm)

	Length/Axial-wise		
Cu.Strand	Bare + insulation	6 + 0.4 = 6.4	
Coil	4 Cu.Strands	4 × 6.4 = 25.6	
14 coils		14 × 25.6 =	358.4
Insulation bet coils		(14 – 1) × 6 = 78	78.0
End-Ring			30.0
End-Insulation			100.0
Axial Length			**566.4**

	Radial-Wise		
Cu.Strand	Bare + insulation	2.5 + 0.4 = 2.9	
Each turn	12 Cu.Strands	12 × 2.9 = 34.8	
Radial-Width			**34.8**

Assuming oil duct between LV and HV former of 5mm, Ins. cylinder of 6 mm, and Oil duct between HV former and HV winding as 5mm,

Inside dia oh HV wdg (di1) = do2 + 2 × (5 + 6 + 5)

$= 314 + 32 = 346$ mm

Outer dia of HV winding (do1) = di1 + (2 × rwHV)

$= 346 + 2 \times 34.8 = 415.6$ mm

Meal length of HV turns (Lmt1) = $\pi \times (di1 + do1)/(2 \times 1000)$

$= \pi \times (346 + 415.6)/2000 = 1.1963$ m

Res of HV wdg/ph(r1) = 0.02 × Lmt1 × T1/a1

$= 0.02 \times 1.1963 \times 624/14.7 = 1.0156\ \Omega$

Copper loss in HV Wdg (pcu1) = $3 \times I1^2 \times r1/1000$

$= 3 \times 40.404^2 \times r1/1000 = 1.0156/1000 = 4.9741$ KW

5.2.5 *(a) Computer Program in "C" in MATLAB for Part-4*

```
% <-----------4) Design of HV Winding:------->
HV = 6600; LV = 440; cdav = 2.6; do2 = 314; T2 = 24; KVA = 800;
L = 0.59;
T1a = T2*HV/LV*sqrt(3); T1 = ceil(T1a); I1 = KVA*1e3/(3*HV);
cA = 4; AxC = 14; cdHV = cdav + 0.2; %Assumptions
x1 = T1/(AxC-2+1.3); x2 = ceil(x1/cA);
x3 = (T1-(x2*cA*(AxC-2)))/2; T1x = (AxC-2)*cA*x2 + 2*x3;
cR = x2; ALW1 = floor(0.7*L*1000); ALPC = ALW1/AxC;
ALPC1 = ALPC/cA; stW11 = ALPC1-0.4;
stW1 = floor (stW11); a1a = I1/cdHV; stT11 = a1a/stW1;
stT1 = ceil (stT11*10)/10;a1 = stT1*stW1*0.98;
cdHV = I1/a1; aLc = cA*(stW1 + 0.4); rwHV = cR*(stT1 + 0.4);
AxLw = AxC*aLc + (AxC-1)*6; AxL = AxLw + 30 + 100;
SlkHVax = L*1000-AxL; di1 = do2 + 10 + 12 + 10;
do1 = di1 + 2*rwHV;
Lmt1 = pi*(di1 + do1)/2e3; r1 = 0.02*Lmt1*T1/a1;
pcu1 = 3*I1^2*r1/1000;
```

5.2.6 Performance Calculations (Part-5)

Assuming stray losses as 5%, Total Wdg Copper losses (pcuT)

$= 1.05 \times (\text{pcu1} + \text{pcu2}) = 1.05*(4.974 + 3.356) = 8.746$ KW

Total losses on Full Load (ptFL) = (pcuT + Pi)

$= 8.746 + 1.473 = 10.219$ KW;

Efficiency Calculation for 1st variant:

pf = 1.0; Ldpu = 1.0;

Total Loss (TL) = Pi + pcuT × Ldpu^2

$= 1.473 + (8.746 \times 1.0^2) = 10.219$ KW

Output (Opt) = Ldpu×KVA×pf = 1.0 ×800 ×1.0 = 800 KW

Input (Inp) = Opt + TL = 800 +10.219 = 810.219 KW

Efficiency(eff) = Opt/Inp ×100 = 800/810.219 ×100 = 98.7387 %;

Pf	Load(pu)	Losses (KW)	Output (KW)	Input (KW)	Efficiency = Output/Input × 100
1.0	1.0	10.219	800	810.219	98.7387 %
0.85	1.0	10.219	680	690.219	98.5195 %
0.85	0.75	6.3924	510	516.392	98.7621 %
0.85	0.5	3.6591	340	343.659	98.9352 %

Load for max. Efficiency (Ldmxef) = $\sqrt{\dfrac{\text{Pi}}{\text{pcuT}}}\times\text{KVA}$

$$=\sqrt{\frac{1.473}{8.746}}\times 800 = 328.25\ \text{KVA}$$

Max Efficiency (efmx)

$$=\frac{\text{Ldmxef}\times 0.85}{(\text{Ldmxef}\times 0.85)+(2\times\text{Pi})}=\frac{328.25\times 0.85}{(328.25\times 0.85)+(2\times 1.473)}$$

$$= 0.9895\text{pu} = 98.96\ \%$$

Mean Turn Length (Lmt) = (Lmt1 + Lmt2)/2

$$= (1.1963 + 0.8954)/2 = 1.0458\text{m}$$

Length of coil (Lc) = AxLw/1000 = 436.4/1000 = 0.4364 m;

AT/ph (AT) = $\text{I1}\times\text{T1} = 40.404\times 624 = 25212$

P.U.Reactance(Ex)

$$=\frac{2\times\pi\times f\times 4\times\pi\times 10^{-7}\times\text{Lmt}\times\text{AT}}{\text{Lc}\times\text{Et}}\times\left[0.016+\frac{\text{rwHV}+\text{rwLV}}{3\times 1000}\right]=0.0841\ \text{pu}$$

$$=\frac{2\times\pi\times 50\times 4\times\pi\times 10^{-7}\times 1.048\times 25212}{0.4364\times 10.57}\times\left[0.016+\frac{29+34.8}{3\times 1000}\right]=0.0841\ \text{pu}$$

P.U. Resistance (Er) = pcuT/KVA = 8.746/800 = 0.0109

P.U. Impedance (Ez) = $\sqrt{\text{Er}^2+\text{Ex}^2} = \sqrt{0.0109^2+0.0841^2} = 0.0848$ pu

Regulation at 0.85pf and FL (Reg85) = $\text{Er}\times 0.85+\text{Ex}\times\sqrt{1-0.85^2}$

$$= 0.0109\times 0.85 + 0.0849\times\sqrt{1-0.85^2} = 0.0536\text{pu}\ (5.36\%)$$

Regulation at UPF and FL (RegUPF) = Er = 0.0109 pu (1.09%)

5.2.6 *(a) Computer Program in "C" in MATLAB for Part-5*

```
% <-----------5) Performance Calculations :------>
pcu1 = 4.974; pcu2 = 3.356; Pi = 1.473; AxLw = 436.4;
I1 = 40.404; f = 50; Et = 10.57; % Input Data
rwHV = 29; rwLV = 34.8; KVA = 800; Lmt1 = 1.1963;
Lmt2 = 0.8954; T1 = 624; % Input Data
pcuT = 1.05*(pcu1 + pcu2); ptFL = (pcuT + Pi);
PF = [1 0.85 0.85 0.85]; LDPU = [1 1 0.75 0.5];
for I = 1:4;
pf = PF(i); Ldpu = LDPU(i);  TL(i) = Pi + pcuT*Ldpu^2;
Opt(i) = Ldpu*KVA*pf; Inp(i) = Opt(i) + TL(i);
eff(i) = Opt(i)/Inp(i)*100;end;
Ldmxef = sqrt(Pi/pcuT)*KVA;
```

```
efmx = Ldmxef*0.85/(Ldmxef*0.85 + 2*Pi)*100;
Lmt = (Lmt1 + Lmt2)/2; Lc = AxLw/1000; AT = I1*T1;
Ex= 2*pi*f*4*pi*1e-7*Lmt*AT/Lc/Et* (0.016 + (rwHV + rwLV)/3e3);
Er = pcuT/KVA; Ez = sqrt(Ex^2 + Er^2);
Reg 85 = Er*0.85 + Ex*sqrt(1-0.85^2); RegUPF = Er;
```

5.2.7 Tank Design and Weights (Part-6)

Assuming Length-wise clearance (dL) = 140 mm,

Length of Tank (Lt) = $2 \times D \times 1000 + (do1 + dL)$

$= 2 \times 0.44 \times 1000 + (415.6 + 140) = 1435.6$ mm

Assuming Width-wise clearance (dB) = 180mm,

Width of Tank (bt) = do1 + dB = 415.6 + 180 = 595.6 mm

Assuming Height-wise clearance (dH) = 500mm,

Height of Tank (ht) = $L \times 1000 + (2 \times hy \times 1000 + dH)$

$= 0.59 \times 1000 + (2 \times 0.1917 \times 1000 + 500) = 1473.3$ mm;

Volume of Tank (Vt) = $Lt \times bt \times ht/10^9 = 1435.6 \times 595.6 \times 1473.3/10^9 = 1.2598\ m^3$;

Cooling Surface area of tank (St) = $2 \times (bt + Lt) \times ht/10^6 = 2 \times (595.6 + 1435.6) \times 1473.3/10^6 = 5.9853\ m^2$;

Temperature-rise of Tank (Tr) = $\dfrac{ptFL \times 1000}{12.5 \times St} = \dfrac{10.219 \times 1000}{12.5 \times 5.9853} = 136.6\ ^\circ C$

Since Temp-rise is to be limited to (TRP) = 50 °C, Cooling tubes are to be provided

Selecting Tube dia (Dct) of 50mm and height (Hct) 0f 1000mm,

Area of each tube (At) = $\pi \times Dct \times Hct/10^6 = \pi \times 50 \times 1000/10^6 = 0.1571\ m^2$;

Required Tube Cooling Area (CAt) = $(ptFL \times 1000-12.5 \times St \times TRP)/(6.5 \times TRP \times 1.35)$

$= (10.219 \times 1000-12.5 \times 5.9853 \times 50)/(6.5 \times 50 \times 1.35) = 14.765\ m^2$;

No. of cooling tubes required (Nt) = 14.765/0.1571 = 94

Wt. of Copper in HV wdg (Wcu1) = $8.9 \times Lmt1 \times Tn\ LV\ w1 \times a1/1000 = 8.9 \times 1.1963 \times 624 \times 14.7/1000 = 97.665$ Kg

Wt. of Copper idg(Wcu2) = $8.9 \times Lmt2 \times T2 \times a2/1000 = 8.9 \times 0.8954 \times 24 \times 423.36/1000 = 80.967$ Kg

Wt. of Iron (Wiron) = KgC + KgY = 461 + 659 = 1120 Kg;

Assuming 1% Insulation Wt, Total Weight (Wtot) = $1.01 \times (Wcu1 + Wcu2 + Wiron)$

$= 1.01 \times (97.665 + 80.967 + 1120) = 1311.6$ Kg

Specific Wt (KgPkva) = Wtot/KVA = 1311.6/800 = 1.6395

5.2.7 *(a) Computer Program in "C" in MATLAB for Part-6*

```
% <-----------6) TANK Design andWeights------->
D = 0.44; do1 = 415.6; L = 0.59; hy = 0.1917; ptFL = 10.219;
Lmt1 = 1.1963; T1 = 624; % Input Data
Lmt2 = 0.8954; T2 = 24; a2 = 423.36; KgC = 461; KgY = 659;
KVA = 800; a1 = 14.7; % Input Data
dL = 140; dB = 180; dH = 500; Dct = 50; Hct = 1000; TRP = 50;
% Assumptions
Lt = 2*D*1e3 + (do1 + dL); bt = (do1 + dB);
ht = L*1e3 + (2*hy*1e3 + dH); Vt = Lt*bt*ht/1e9;
St=2*(bt+Lt)*ht/1e6; Tr = ptFL*1e3/12.5/St;At = pi*Dct*Hct/1e6;
CAt = (ptFL*1e3-12.5*St*TRP)/(6.5*TRP*1.35);
Nt = ceil(CAt/At); Wcu1 = 8.9*Lmt1*T1*a1/1e3;
Wcu2 = 8.9*1e3*Lmt2*T2*a2/1e6; Wiron = KgC+KgY;
Wtot = 1.01*(Wcu1 + Wcu2 + Wiron); KgPkva = Wtot/KVA;
% ---------------------------------------------------------------
```

5.2.8 Computer Program in "C" in MATLAB for Complete Design

```
% Design of Core Type Power Transformer
KVA = 800; HV = 6600; LV = 440; f = 50; Ph = 3; % Input Data
f2 = fopen ('Total_800KVA_Output.m','w');
fprintf (f2,' % 3d KVA, % 5d/ % 4d V, % 2d Hz, % 1d Phase, KVA,
HV, LV, f, Ph);
fprintf (f2, Delta/Star, Core Type, ON cooled Power
Transformer: n', KVA, HV, LV, f, Ph);
fprintf (f2,============================================== n');
fprintf (f2, <-----1) Design of Magnetic Frame----------->\n');
k = 0.6; ki = 0.92; K = 0.6; Bm = 1.5; cdav = 2.6; Lby D = 2.8;
% Assumptions
Et = K*sqrt (KVA/Ph); Ai1 = Et/(4.44*f*Bm); d1 = sqrt(Ai1/k);
d = ceil(d1*100)/100; Ai = k*d^2;
Et = 4.44*f*Bm*Ai; kw = 10/(30 + HV/1e3)*1.15;
Aw = KVA*1e3/ (3.33*f*Bm*kw*cdav*1e6*Ai);
L1 = sqrt(LbyD*Aw); L = ceil(L1*100)/100; D1 = Aw/L+d;
D = ceil(D1*100)/100; Lby D = L/(D-d);
W1 = 2*D + 0.9*d; W = ceil (W1*10)/10; Ac = Ai/ki; Ay =
1.15*Ac; by = 0.9*d; hy = Ay/by;
Fprintf (f2, Assumed: For 3stepped Core factor(k) = % 4.2f,
Value factor (K) = % 4.2f\n', k, K);
Fprintf (f2, Assumed: Window space factor (kw) = % 5.3f, Max
Core Flux Density (T) = % 3.1f\n, kw, Bm);
fprintf (f2, Core dia (d) (m) % 6.3f\n, d);
fprintf (f2, CS area of core (sq.m) % 6.4 f\n, Ai);
fprintf (f2, EMF/turn % 6.2f\n, Et);
fprintf (f2, Av-Current density (A/sq.mm) % 6.2f\n', cdav);
```

```
fprintf (f2,' Window Area (sq.m) % 6.4f\n', Aw);
fprintf (f2,' Length of core (L) (m) % 6.4f\n', L);
fprintf (f2, 'Dist betw Core-Centres (D) (m) % 6.4f\n', D);
fprintf (f2,' L/(D-d) ratio % 6.4f (Permissible: 2.5 to 4)\n',
LbyD);
fprintf (f2,' Length of Yoke (W) (m) % 6.4f\n', W);
fprintf (f2, 'Width X Ht of the yoke (m) % 6.4f and % 6.4f\n',
by, hy);
BB = [0.8 1.0 1.2 1.4 1.6];
WpKg = [0.2 0.4 0.8 1.2 2]; % for Cold Rolled
plot (BB, WpKg); grid; xlabel ('% Flux density (T) -->');
ylabel ('Loss(W/Kg)-->');
title ('Core Loss for Cold rolled steels');
WpKgC = interp1 (BB, WpKg, Bm, 'linear'); KgC = Ph*Ac*L*7.55e3;
PiC = WpKgC*KgC; KgY = 2*Ay*W*7.55e3;
By = Ac/Ay*Bm; WpKgY = interp1(BB, WpKg, By, 'linear');
PiY = WpKgY*KgY; Pi = 1.05*(PiC + PiY)/1000;
fprintf (f2,'Loss in Core = % 6.2fW/Kg * % 6.2f Kg = % 6.2f
W\n', WpKgC, KgC, PiC);
fprintf (f2, 'Flux Dens in Yoke (T)  % 6.4f\n', By);
fprintf (f2,' Loss in Yoke = % 6.2 fW/Kg * % 6.2f Kg = % 6.2f
W\n', WpKgY, KgY, PiY);
fprintf (f2, 'Total Iron Loss (KW) % 6.3f\n', Pi);
fprintf (f2,' <-----2) No-Load Current ------------------>\n');
B1 = [1 1.25 1.5 1.75 2.0];
H1 = [70 100 150 300 1000]; % for Cold Rolled
% plot (H1, B1); grid; xlabel ('AT/m-->'); ylabel ('Flux
density (T)-->');
% title ('Magnetization Curve for Cold rolled steels'); %
atC = interp1(B1, H1, Bm, 'spline'); at Y = interp1(B1, H1, By,
'spline'); ATC = Ph*atC*L;
ATY = 2*atY*W; ATpPh = (ATC + ATY)/Ph;
T2 = floor (LV/sqrt(3)/Et); I2 = KVA*1e3/(sqrt(3)*LV);
Iw = Pi*1000/(sqrt(3)*LV); Im = 1.15*ATpPh/sqrt(2)/T2;
I0 = sqrt(Im^2 + Iw^2);      I0 by I2 = I0/I2*100;
fprintf (f2,'AT/m: for Core = % 6.3 f and for Yoke = % 6.3f\n',
at C, at Y);
```

```
fprintf (f2, 'Total-AT: for Core = % 6.3 f and for Yoke = % 6.3f\n', ATC, ATY);

fprintf (f2,'AT/phase: % 6.2f\n', ATpPh);

fprintf(f2,'Turns in LV winding % 6.0f\n', T2);

fprintf (f2, 'Ph-Current in LV wdg  (Amp) % 6.1f\n',I2);

fprintf (f2, 'Magnetizing-Ct (I0 = Iw + jIm) % 6.3f = % 4.3f + j % 4.3f A\n', I0, Iw, Im);

fprintf (f2,' Ratio of I0/I2  (perc) % 6.3f (Permissible: 0.5 to 1)\n', I0 by I2);

fprintf (f2,'<-----3) Design of LV Winding:------->\n');

T2r = 2; stP = 12; NstA = 3; stT = 3;

T2a = T2/T2r; NstR = stP/NstA; ALW = 0.8*L*1000; ALT = ALW/T2a; stW1 = ALT/NstA-0.5; stW = floor(stW1);

ALWx=(((stW+ 0.4)* NstA) + 2)*T2a + 100; SlkLVax = L*1000-ALWx;

a2 = stW*stT*stP*0.98; cdLV = I2/a2;

rwLV = NstR*(stT+ 0.4)*T2r + 1.8; di2 = d*1000 + 2*(5 + 3 + 5);

do2 = di2 + 2*rwLV; Lmt2 = pi*(di2 + do2)/2e3; r2 = 0.02*Lmt2*T2/a2; pcu2 = Ph*I2^2*r2/1000;

fprintf (f2,'Turns (RadialXAxial) % 2.0f X % 2. 0f \n', T2r, T2a);

fprintf (f2,' Conductor: % 2d Strands in Parallel of W*t (mm*mm) % 2dX % 2d\n', stP, stW, stT);

fprintf (f2, Strands (RadialXAxial) % 2.0f X % 2.0f\n', NstR, NstA);

fprintf (f2, CS area of Conductor (sq.mm) % 6.2f\n', a2);

fprintf (f2, Current density (A/sq.mm) % 6.2f (Permissible:2.3 to 3.5)\n', cdLV);

fprintf (f2, Axial Slack in LV-Wdg (mm) % 6.2f (desirable>7)\n', SlkLVax);

fprintf (f2, Radial width of Wdg (mm) % 6.2f\n', rwLV);

fprintf (f2, Wdg-Dia (InnerXOuter) (mm) % 4.1f X % 4.1f\n', di2, do2);

fprintf (f2, Mean Length of Turn (m) % 6.2f\n', Lmt2);

fprintf (f2, Resistance/phase (m.ohm) % 6.4f\n',r2*1000);

fprintf (f2,' Cu.Loss in LV wdg (KW) % 6.3f\n', pcu2);

fprintf (f2,'<-----------4) Design of HV Winding:------->\n');
```

```
T1a = T2*HV/LV*sqrt(3); T1 = ceil(T1a); I1 = KVA*1e3/(3*HV);

cA = 4; AxC = 14; cdHV = cdav + 0.2; % Assumptions

x1 = T1/(AxC - 2 + 1.3); x2 = ceil(x1/cA);
x3 = (T1-(x2*cA*(AxC-2)))/2; T1x = (AxC-2)*cA*x2 + 2*x3;

cR = x2; ALW1 = floor (0.7*L*1000); ALPC = ALW1/AxC;
ALPC1 = ALPC/cA; stW11 = ALPC1-0.4;

stW1 = floor (stW11); a1a = I1/cdHV; stT11 = a1a/stW1;
stT1 = ceil(stT11*10)/10; a1 = stT1*stW1*0.98;

cdHV = I1/a1; aLc = cA*(stW1 + 0.4); rwHV = cR*(stT1 + 0.4);
AxLw = AxC*aLc + (AxC-1)*6;

AxL = AxLw + 30 + 100; SlkHVax = L*1000-AxL;
di1 = do2 + 10 + 12 + 10; do1 = di1 + 2*rwHV;

Lmt1 = pi*(di1 + do1)/2e3; r1 = 0.02*Lmt1*T1/a1;
pcu1 = 3*I1^2*r1/1000;

Fprintf (f2,'Turns in HV Wdg % 6.0f\n', T1);

Fprintf (f2, 'Ph-Current in HV wdg  (Amp) % 6.2f\n',I1);

Fprintf (f2,'Wdg: % 3d coils X % 3d turns + 2 coils X % 3d
turns = % 4. 0f\n', AxC-2, cA*cR, x3, T1x);

fprintf (f2,'Conductor(W*t) (mm*mm)  % 4.1f X % 4.1f\n', stW1,
stT1);

% fprintf (f2, Strands (RadialXAxial) % 2. 0f  X % 2.0f\n',
NstR, NstA);

fprintf (f2,'CS area of Conductor (sq.mm) % 6.2f\n', a1);

fprintf (f2, Current density (A/sq.mm) % 6.2f (Permissible:2.3
to 3.5)\n', cdHV);

fprintf (f2,' Axial Slack inHV-Wdg (mm) % 6.2f (desirable:
>7)\n', SlkHVax);

fprintf (f2,' Radial width of Wdg (mm) % 6.2f\n', rwHV);

fprintf (f2, Wdg-Dia (InnerXOuter) (mm) % 4.1f  X % 4.1f\n',
di1, do1);

fprintf (f2, Clearance bet HV Wdgs (mm) % 6.2f (desirable:
>15)\n', D*1000-do1);

fprintf (f2, Mean Length of Turn (m) % 6.2f\n', Lmt1);

fprintf (f2, 'Resistance/phase (ohm) % 6.4 f\n', r1);

fprintf (f2, Cu. Loss in LV wdg (KW) % 6.3f\n', pcu1);

fprintf (f2, <-----------5) Performance Calculations:------
>\n');
```

```
pcuT = 1.05*(pcu1 + pcu2); ptFL = (pcuT + Pi);
PF = [1 0.85 0.85 0.85]; LDPU = [1 1 0.75 0.5];
for i = 1:4;
pf = PF(i); Ldpu = LDPU(i);
TL(i ) = Pi + pcuT*Ldpu^2; Opt(i) = Ldpu*KVA*pf;
Inp(i) = Opt(i) + TL(i); eff(i) = Opt(i)/Inp(i)*100; end;
Ldmxef = sqrt(Pi/pcuT)*KVA;
efmx = Ldmxef*0.85/(Ldmxef*0.85 + 2*Pi)*100;
Lmt = (Lmt1 + Lmt2)/2; Lc = AxLw/1000; AT = I1*T1;
Ex= 2*pi*f*4*pi*1e-7*Lmt*AT/Lc/Et* (0.016 + (rwHV + rwLV)/3e3);
Er = pcuT/KVA;Ez = sqrt(Ex^2 + Er^2);
Reg85 = Er*0.85 + Ex*sqrt(1-0.85^2); RegUPF = Er;
Fprintf (f2, 'Efficiency at UPF and Full Load: % 5.2f\n',
eff(1));
fprintf (f2, 'Efficiency at 0.85 pf and at: 1pu Ld: % 5.2f,
0.75pu Ld: %5.2f and 0.5pu Ld: % 5.2f\n', eff(2), eff (3),
eff(4));
fprintf (f2, 'Max Efficiency = % 5.2f occuring at Load = %
5.2fKVA\n', efmx, Ldmxef);
fprintf (f2, 'Mean-turn Length = % 6.4fm, Axil Length of Wdg =
% 6.4fm, AT/ph = % 6.0f\n', Lmt, Lc, AT);
fprintf (f2,' Reactance = % 6.4fpu, Restance = % 6.4fpu,
Impedance = % 6.4fpu\n', Ex, Er, Ez);
fprintf (f2, Regulation: at FL and 0.85pf = % 5.4 fpu, at FL
and UPF = % 5.4 fpu\n', Reg85, RegUPF);
fprintf (f2,'<-----------6) TANK Design andWeights------->\n');
dL = 140; dB = 180; dH = 500; Dct = 50; Hct = 1000; TRP = 50;
% Assumptions;
Lt = 2*D*1e3 + (do1 + dL); bt = (do1 + dB); ht = L*1e3 +
(2*hy*1e3 + dH); Vt = Lt*bt*ht/1e9; St = 2*(bt + Lt)*ht/1e6;
Tr = ptFL*1e3/12.5/St; At = pi*Dct*Hct/1e6;
CAt = (ptFL*1e3-12.5*St*TRP) /(6.5*TRP*1.35);Nt = ceil(CAt/At);
Wcu1 = 8.9*Lmt1*T1*a1/1e3; Wcu2 = 8.9*1e3*Lmt2*T2*a2/1e6;
Wiron = KgC + KgY;
Wtot = 1.01* (Wcu1 + Wcu2 + Wiron); KgPkva = Wtot/KVA;
Fprintf (f2, 'Tank: Length = % 4.0fmm, Width = % 4.0 fmm,
Height = % 4.0 fmm and Volume = % 6.3fm3\n', Lt, bt, ht, Vt);
```

```
fprintf (f2,'Tank : Cooling area = % 6.4 fsq.m and Temp-Rise =
% 4.0 f is reduced to 50 deg-C\n', St, Tr);

fprintf(f2,' Cooling Tubes: Area/Tube = % 6.4 fsq.m, Required-
Area = % 6.4fsq.m and No of Tubes = % 3.0 f\n', At, CAt, Nt);

fprintf (f2,' Weight: HV-Wdg = % 6.2f + LV-Wdg : = % 6.2f +
(Core + Yoke): = % 6.2 f and Total = % 7. 1f Kg\n', Wcu1, Wcu2,
Wiron, Wtot);

fprintf (f2,' Kg/KVA = % 6.3f\n', KgPkva);

fprintf (f2, <----------- End of Program------->\n');

fclose (f2);
```

5.2.8 *(a) Computer Output Results for Complete Design*

800 KVA, 6600/ 440 V, 50 Hz, 3 Phase, Delta/Star, Core Type, ON cooled Power Transformer:

==

<-----1) Design of Magnetic Frame----------------->

Assumed: For 3stepped Core factor (k) = 0.60, Value factor (K) = 0.60

Assumed: Window space factor (kw) = 0.314, Max Core Flux Density (T) = 1.5

Core dia (d) (m) 0.230

CS area of core (sq.m) 0.0317

EMF/turn 10.57

Av-Current density (A/sq.mm) 2.60

Window Area (sq.m) 0.1235

Length of core (L) (m) 0.5900

Dist betw Core-Centres (D) (m) 0.4400

L/(D-d) ratio 2.8095 (Permissible:2.5to4)

Length of Yoke (W) (m) 1.1000

Width X Ht of the yoke (m) 0.2070 and 0.1917

Loss in Core = 1.60W/Kg × 461.04 Kg = 737.67 W

Flux Dens in Yoke (T) 1.3043

Loss in Yoke = 1.01W/Kg × 659.00 Kg = 664.73 W

Total Iron Loss (KW) 1.473

<-----2) No-Load Current----------------->

AT/m: for Core = 150.000 and for Yoke = 109.600

Total-AT: for Core = 265.500 and for Yoke = 241.119

AT/phase: 168.87

Turns in LV winding 24

Ph-Current in LV wdg (Amp) 1049.7

Magnetizing-Ct (I0 = Iw + jIm) 6.039 = 1.932 + j5.722 A

Ratio of I0/I2 (perc) 0.575(Permissible: 0.5 to 1)

<-----3) Design of LV Winding: ------->

Turns (Radial × Axial) 2 × 12

Conductor:-12 Strands in Parallel of W × t (mm × mm) 12 × 3

Strands (Radial × Axial) 4× 3

CS area of Conductor (sq.mm) 423.36

Current density (A/sq.mm) 2.48 (Permissible: 2.3 to 3.5)

Axial Slack in LV-Wdg (mm) 19.60 (desirable>7)

Radial width of Wdg (mm) 29.00

Wdg-Dia (Inner × Outer) (mm) 256.0 × 314.0

Mean Length of Turn (m) 0.90

Resistance/phase (m.ohm) 1.0151

Cu.Loss in LV wdg (KW) 3.356

<-----------4) Design of HV Winding:------->

Turns in HV Wdg 624

Ph-Current in HV wdg (Amp) 40.40

Wdg: 12 coils × 48 turns + 2 coils × 24 turns = 624

Conductor(W × t) (mm × mm) 6.0 × 2.5

CS area of Conductor (sq.mm) 14.70

Current density (A/sq.mm) 2.75 (Permissible: 2.3 to 3.5)

Axial Slack in HV-Wdg (mm) 23.60 (desirable: >7)

Radial width of Wdg (mm) 34.80

Wdg-Dia(InnerXOuter) (mm) 346.0 × 415.6

Clearance bet HV Wdgs (mm) 24.40 (desirable: > 15)

Mean Length of Turn (m) 1.20

Resistance/phase (ohm) 1.0156

Cu. Loss in LV wdg (KW) 4.974

<-----------5) Performance Calculations:------>

Efficiency at UPF and Full Load: 98.74

Efficiency at 0.85 pf and at: 1pu Ld: 98.52, 0.75 pu Ld: 98.76 and 0.5pu Ld: 98.94

Max Efficiency = 98.96 occuring at Load = 328.25KVA

Mean-turn Length = 1.0458m, Axil Length of Wdg = 0.4364m, AT/ph = 25212

Reactance = 0.0841pu, Restance = 0.0109pu, Impedance = 0.0848pu

Regulation: at FL and 0.85pf = 0.0536pu , at FL and UPF = 0.0109pu

<-----------6) TANK Design andWeights------->

Tank: Length = 1436mm, Width = 596 mm, Height = 1473 mm and

Volume = 1.260 m^3

Tank: Cooling area = 5.9853 sq.m and Temp-Rise = 137 is reduced to 50 deg-C

Cooling Tubes: Area/Tube = 0.1571 sq.m, Required-Area = 14.7650 sq.m and No. of Tubes = 94

Weight: HV-Wdg: = 97.66 + LV-Wdg: = 80.97 + (Core + Yoke): = 1120.04 and Total = 1311.6 Kg

Kg/KVA = 1.6395

<-----------End of Program------->

5.2.9 Modifications to be done in the above Program to get Optimal Design

1. Insert "for" Loops for the following parameters to iterate the total program between min and max permissible limits for selecting the feasible design variants:
   (a) No. of values of "K" (Factor of ratio of core to copper)
   (b) No. of Values of "Bm", (Core flux density)
   (c) No. of Average current density values
   (d) No. of "L/D" ratios.
2. Insert also minimum or maximum range of required objective functional values as constraint values, for example, (a) Efficiency (b) Kg/KVA (c) I0/I2 ratio (d) Volume of Tank etc.
3. Run the program to get various possible design variants.

Note: From the feasible design variants printed in the output, select that particular design fulfilling the objective of optimal parameter for the design

5.2.10 Computer Program in "C" in MATLAB for Optimal Design

```
% Design of Core Type Power Transformer,

KVA = 800; HV = 6600; LV = 440; f = 50; Ph = 3; % Input Data

f2 = fopen ('Optimal_800KVA_Output.m','w');

fprintf (f2,' % 3d KVA, % 5d/ % 4d V, % 2d Hz, % 1d Phase,
KVA, HV, LV, f, Ph);

fprintf (f2, Delta/Star, Core Type, ON cooled Power
Transformer:\n', KVA, HV, LV, f, Ph);

fprintf                                              (f2,'
==============================================\n');

fprintf (f2,' Sn   K   Bm   d   L   D    W  L/(D-d) cdav
cdLV CDHV I0/I2 eff Kg/KVA Reg Tank Nt\n');

fprintf (f2,'--- ---- --- ---- ---- ----- ----- -----  ----
----- ----- ----- ----- ---- ---- ----\n');

k = 0.6; ki = 0.92; sn = 0; M1 = 0; M2 = 0; M3 = 0; M4 = 0;
EFFmax = 98; min I0 by I2 = 1; min KgPkva = 2; minVt = 2;
% Assumptions

kw = 10/(30 + HV/1e3)*1.15;

for K = 0.6:0.01:0.65; % K--> 0.6 to 0.65

for Bm = 1.5: 0.1: 1.6;  % Bm --> 1.5 to 1.7 for CRGO and PT

for cdav = 2.5: 0.1: 3.5; % cdav --> 2.5 to 3.5

for Lby D = 2.6:0.2:4; % Lbyd --> 2.5 to 4

% fprintf (f2,'<-----1) Design of Magnetic Frame-----------
----->\n');

Et   =   K*sqrt   (KVA/Ph);   Ai1   =   Et/(4.44*f*Bm);
d1 = sqrt(Ai1/k); d = ceil(d1*100)/100; Ai = k*d^2;

Et   =   4.44*f*Bm*Ai;   kw   =   10/(30   +   HV/1e3)*1.15;
Aw = KVA*1e3/(3.33*f*Bm*kw*cdav*1e6*Ai);

L1 = sqrt(LbyD*Aw); L = ceil(L1*100)/100;

D1  =  Aw/L  +  d;  D  =  ceil(D1*100)/100;  LbyD1  =  L/(D-d);
if LbyD <= 2.5 || LbyD > 4 continue; end;
```

```
W1 = 2*D + 0.9*d; W = ceil(W1*10)/10; Ac = Ai/ki;
Ay = 1.15*Ac; by = 0.9*d; hy = Ay/by;

% fprintf (f2,'<-----2) Performance of Magnetic Frame-------
------>\n');

BB = [0.8 1.0 1.2 1.4 1.6]; WpKg = [0.2 0.4 0.8 1.2 2]; %
Cold Rolled

WpKgC = interp1(BB,WpKg, Bm, 'linear');
KgC = Ph*Ac*L*7.55e3; PiC = WpKgC*KgC; KgY = 2*Ay*W*7.55e3;

By = Ac/Ay*Bm; WpKgY = interp1(BB, WpKg, By, 'linear');
PiY = WpKgY*KgY; Pi = 1.05*(PiC+PiY)/1000;

% fprintf (f2,'<-----3) No-Load Current------------------
>\n');

B1 = [1 1.25 1.5 1.75 2.0]; H1 = [70 100 150 300 1000]; %
Cold Rolled

atC = interp1(B1, H1, Bm, 'spline');
atY = interp1(B1,H1,By,'spline'); ATC = Ph*atC*L;

ATY = 2*atY*W; ATpPh = (ATC + ATY)/Ph;
T2 = floor(LV/sqrt(3)/Et); I2 = KVA*1e3/(sqrt(3)*LV);

Iw = Pi*1000/(sqrt(3)*LV); Im = 1.15*ATpPh/sqrt(2)/T2;
I0 = sqrt(Im^2 + Iw^2); I0 by I2 = I0/I2*100;

if I0 by I2 > = 0.66 continue; end;

% fprintf (f2,'<-----3) Design of LV Winding:------->\n');

T2r = 2; stP = 12; NstA = 3; stT = 3;

T2a = T2/T2r; NstR = stP/NstA; ALW = 0.8*L*1000;
ALT = ALW/T2a; stW1 = ALT/NstA-0.5; stW = floor(stW1);

ALWx=(((stW + 0.4)*NstA)+ 2)*T2a+100; SlkLVax = L*1000-ALWx;
if SlkLVax < = 7 continue; end;

a2 = stW*stT*stP*0.98; cdLV = I2/a2;

rwLV=NstR*(stT+0.4)*T2r + 1.8; di2 = d*1000 + 2*(5 + 3 + 5);

do2 = di2 + 2*rwLV; Lmt2 = pi*(di2 + do2)/2e3;
r2 = 0.02*Lmt2*T2/a2; pcu2 = Ph*I2^2*r2/1000;

% fprintf (f2,'<------4) Design of HV Winding: ------->\n');

T1 = ceil (T2*HV/LV*sqrt(3)); I1 = KVA*1e3/(3*HV);
```

```
cdHV = cdav + 0.2; cA = 4; AxC = 14;  % Assumptions

x1 = T1/(AxC-2 + 1.3); x2 = ceil(x1/cA);
x3 = (T1-(x2*cA*(AxC-2)))/2; T1x = (AxC-2)*cA*x2 + 2*x3;

cR = x2; a1a = I1/cdHV; ALW1 = floor (0.7*L*1000);
ALPC = ALW1/AxC; ALPC1 = ALPC/cA; stW11 = ALPC1-0.4;

stW1 = floor (stW11); stT11=a1a/stW1;
stT1 = ceil (stT11*10)/10; a1 = stT1*stW1*0.98;

cdHV = I1/a1; aLc = cA*(stW1 + 0.4);
rwHV = cR*(stT1 + 0.4); AxLw = AxC*aLc + (AxC-1)*6;

AxL = AxLw + 30 + 100; SlkHVax = L*1000-AxL;
if SlkHVax < = 7 continue; end;

di1 = do2 + 10 + 12 + 10; do1 = di1 + 2*rwHV;
if(D*1000-do1) < = 15 continue; end;

Lmt1 = pi*(di1 + do1)/2e3; r1 = 0.02*Lmt1*T1/a1;
pcu1 = 3*I1^2*r1/1000;

% fprintf (f2,'<---5) Performance Calculations:------>\n');

pcuT = 1.05*(pcu1 + pcu2); ptFL = (pcuT + Pi);

PF = [1 0.85 0.85 0.85]; LDPU = [1 1 0.75 0.5];

for i = 1:4;

pf = PF(i); Ldpu = LDPU(i);

TL(i ) = Pi + pcuT*Ldpu^2; Opt(i) = Ldpu*KVA*pf; Inp(i) =
Opt(i) + TL(i); eff(i) = Opt(i)/Inp(i)*100; end;

if eff(2) < = 98.50 continue; end;

Ldmxef = sqrt(Pi/pcuT)*KVA;

efmx = Ldmxef*0.85/(Ldmxef*0.85 + 2*Pi)*100;

Lmt = (Lmt1 + Lmt2)/2; Lc = AxLw/1000; AT = I1*T1;

Ex=2*pi*f*4*pi*1e-7*Lmt*AT/Lc/Et*(0.016+ (rwHV + rwLV)/3e3);
Er = pcuT/KVA; Ez = sqrt (Ex^2 + Er^2);

Reg85 = Er*0.85 + Ex*sqrt(1-0.85^2); Reg UPF = Er;

dL = 140; dB = 180; dH = 500; Dct = 50; Hct = 1000;
TRP = 50;  % Assumptions

Lt = 2*D*1e3 + (do1 + dL); bt = (do1 + dB);
```

```
St = 2*(bt + Lt)*ht/1e6; Vt = Lt*bt*ht/1e9;
ht = L*1e3 + (2*hy*1e3 + dH);

Tr = ptFL*1e3/12.5/St; At = pi*Dct*Hct/1e6;
CAt = (ptFL*1e3-12.5*St*TRP)/(6.5*TRP*1.35);
Nt = ceil(CAt/At); Wcu1 = 8.9*Lmt1*T1*a1/1e3;
Wcu2 = 8.9*1e3*Lmt2*T2*a2/1e6; Wiron = KgC + KgY;

Wtot = 1.01*(Wcu1 + Wcu2 + Wiron); KgPkva = Wtot/KVA;
if KgPkva > = 1.67 continue; end;

sn = sn + 1; if eff(2) > = EFFmax EFFmax = eff(2); end;
if abs (eff(2)-EFFmax) < = 2e-3 M2 = sn; end;

if KgPkva < = minKgPkva  minKgPkva = KgPkva; end;

if abs (KgPkva-minKgPkva) < = 0.001 M1 = sn; end;

if I0 by I2 < = min I0 by I2 min I0 by I2 = I0by I2; end;
if abs (I0 by I2-min I0 by I2) < = 0.0001 M3 = sn; end;

if Vt < = minVt minVt = Vt; end;
if abs (Vt-minVt) < = 0.001 M4 = sn; end;

fprintf (f2,' % 3d % 5.2f % 4.1f % 5.2f % 5.2f % 6.3f % 6.3f
% 6.2f % 6.2f % 6.3f % 6.3f % 6.3f % 6.2f % 5.2f % 5.2f %
5.2f % 3d\n', sn, K, Bm, d, L, D, W, Lby D1, cdav, cdLV,
cdHV, I0byI2, eff(2), KgPkva, Reg85*100, Vt, Nt);

end; end; end; end;

fprintf (f2,'--- ---- --- ---- ---- ----- ----- ----- ----
----- ----- ----- ----- ---- ---- ---- \n');

fprintf (f2, Selection of Design Variant based on
Optimization Criteria:');

fprintf (f2,\nIf Maximum Efficiency is Required, Select
Variant (Sn) = % 3d          (% 5.2f perc)', M2, EFFmax);

fprintf (f2,\nIf Minimum Kg/KVA is Required, Select Variant
(Sn) = % 3d                  (% 5.2f)', M1, minKgPkva);

fprintf (f2,\nIf Minimum ratio of I0/I2 is Reqd, Select
Variant (Sn) = % 3d  (% 4.3f)', M3, min I0 by I2);

fprintf (f2,\nIf Minimum Volume of Tank is Reqd, Select
Variant (Sn) = % 3d  (% 5.2f)', M4, minVt);

fclose (f2);
```

5.2.10 *(a) Computer Output Results for Optimal Design*

800 KVA, 6600/ 440 V, 50 Hz, 3 Phase, Delta/Star, Core Type, ON cooled Power Transformer:

Sn	K	Bm	d	L	D	W	L/(D-d)	cdav	cdLV	DHV	I0/I2	eff	Kg/KVA	Reg	Tank	Nt
1	0.60	1.5	0.23	0.59	0.440	1.100	2.81	2.60	2.480	2.749	0.575	98.52	1.64	5.36	1.26	94
2	0.60	1.5	0.23	0.58	0.440	1.100	2.76	2.70	2.480	2.863	0.570	98.49	1.62	5.33	1.24	98
3	0.60	1.5	0.23	0.63	0.420	1.100	3.32	2.80	2.289	2.945	0.596	98.51	1.68	4.61	1.22	96
4	0.61	1.5	0.23	0.59	0.440	1.100	2.81	2.60	2.480	2.749	0.575	98.52	1.64	5.36	1.26	94
5	0.61	1.5	0.23	0.58	0.440	1.100	2.76	2.70	2.480	2.863	0.570	98.49	1.62	5.33	1.24	98
6	0.61	1.5	0.23	0.63	0.420	1.100	3.32	2.80	2.289	2.945	0.596	98.51	1.68	4.61	1.22	96
7	0.62	1.5	0.23	0.59	0.440	1.100	2.81	2.60	2.480	2.749	0.575	98.52	1.64	5.36	1.26	94
8	0.62	1.5	0.23	0.58	0.440	1.100	2.76	2.70	2.480	2.863	0.570	98.49	1.62	5.33	1.24	98
9	0.62	1.5	0.23	0.63	0.420	1.100	3.32	2.80	2.289	2.945	0.596	98.51	1.68	4.61	1.22	96
10	0.63	1.5	0.23	0.59	0.440	1.100	2.81	2.60	2.480	2.749	0.575	98.52	1.64	5.36	1.26	94
11	0.63	1.5	0.23	0.58	0.440	1.100	2.76	2.70	2.480	2.863	0.570	98.49	1.62	5.33	1.24	98
12	0.63	1.5	0.23	0.63	0.420	1.100	3.32	2.80	2.289	2.945	0.596	98.51	1.68	4.61	1.22	96
13	0.64	1.5	0.23	0.59	0.440	1.100	2.81	2.60	2.480	2.749	0.575	98.52	1.64	5.36	1.26	94
14	0.64	1.5	0.23	0.58	0.440	1.100	2.76	2.70	2.480	2.863	0.570	98.49	1.62	5.33	1.24	98
15	0.64	1.5	0.23	0.63	0.420	1.100	3.32	2.80	2.289	2.945	0.596	98.51	1.68	4.61	1.22	96

Selection of Design Variant based on Optimization Criteria :

If Maximum Efficiency is Required, Select Variant (Sn) = 13 (98.52 perc)

If Minimum Kg/KVA is Required, Select Variant (Sn) = 14 (1.62)

If Minimum ratio of I0/I2 is Required, Select Variant (Sn) = 14 (0.570)

If Minimum Volume of Tank if Required, Select Variant (Sn) = 15

5.3 Shell Type Power Transformer

Total design is split into six parts in a proper sequence. Design calculations are given for a given Rating of a Transformer, followed by Computer Program written in "C" language using MATLAB software for each part. Finally all the Programs are added together to get the total Program by running which we get the total design. Computer Out put of total design is given.

This design may not be the optimum one. Now optimization objective and Design Constraints are inserted into this total Program. When this program is run we will get various alternative feasible designs from which the selected variant based on the Optimization Criteria can be picked up. Computer output showing the important design parameters for various feasible alternatives is given in the end of this chapter along with a logic diagram.

Problem: Design a 125KVA, 3300/ 440 V, 50 Hz, 1-Phase, Shell Type, oil cooled Power-class Transformer: (Limitations: Temp-rise to 40 deg-C, No-load current to 1.5%, Regulation to 4 % and Total losses to 2%).

5.3.1 Sequential Steps for Design of Each Part and Programming Simultaneously

(a) Calculate the dimensions of Magnetic frame consisting of Core, window andYoke. Calculate Flux densities in those parts and Iron losses.

(b) Calculate Amp-Turns and No load current.

(c) Calculate No. of turns, size of Copper and final dimensions of LV winding.

(d) Calculate No. of turns, size of Copper and final dimensions of HV winding.

(e) Check weather clearances between Magnetic frame and windings are OK.

(f) Calculate Windings copper Losses, total losses, Efficiency, Reactances and % Regulation.

(g) Calculate Dimensions of cooling tank, number and size of cooling tubes, temperature rise, total weight and Kg/KVA.

Note: By adding programs established for each part sequentially we get the Program for complete design.

5.3.2 Design of Magnetic Frame (Part-1)

Assuming Magnetic frame (both core and yoke) are made with Hot rolled Silicon Steel with Rectangular Cross section and Value factor (K) = 1.1,

Volts/turn (Et) = $K \times \sqrt{KVA} = 1.1 \times \sqrt{125} = 12.2984$ V/t

Assuming Max. Flux density in Core and Yoke (Bm) = 1.1 T,

$$\text{Net CS area of Core (Ain)} = \frac{\text{Et}}{4.44 \times \text{f} \times \text{Bm}} = \frac{12.2984}{4.44 \times 50 \times 1.1} = 0.0504 \text{ m}^2;$$

Assuming Iron factor (ki) = 0.9,

gross CS area of Core (Aig) = Ain/ki = 0.0504/0.92 = 0.056 m^2;

Assuming Length / width of core (R1) = 2.5,

$$\text{Width of main Core (bmc)} = \sqrt{\text{Aig}/\text{R1}}$$

$$= \sqrt{0.056/2.5} = 0.1496\text{m} \approx 150 \text{ mm (Rounded off)}$$

Length of main core (Lmc) = R1 × bmc = 2.5 × 150 = 375 mm

Assuming Window Space factor (kw) = 0.27 and

Average Current density (cdav) = 2.5 A/mm^2;

$$\text{Window Area (Aw)} = \frac{\text{KVA} \times 1000}{2.22 \times \text{f} \times \text{Bm} \times \text{kw} \times \text{cdav} \times 10^6 \times \text{Ain}}$$

$$= \frac{125 \times 1000}{2.22 \times 50 \times 1.1 \times 0.27 \times 2.5 \times 10^6 \times 0.0504} = 0.0301 \text{ m}^2;$$

Assuming Height / width of window (R2) = 2.1,

$$\text{Width of window (bw)} = \sqrt{\text{Aw}/\text{R2}} = \sqrt{0.0301/2.1}$$

$$= 0.1198\text{m} \approx 120 \text{ mm (Rounded off)}$$

Height of Window (hw) = R2 × bw = 2.1 × 120 = 252 mm

Since in a 1-ph shell type Transformer, side core and yoke carry half the total flux (for same Bm)

Area of yoke (Ay) = Ain/2 = 0.0504 / 2 = 0.0252 m^2;

Height of Yoke (hy) = Width of side core (bsc) = bmc/2 = 150/2 = 75 mm

Width of yoke (by) = Ay/hy = 0.0252/75 × 10^6 = 335.75 mm

Length of Yoke (Ly) = 2 × (bsc + bw) + bmc = 2 × (75 + 120) + 150 = 540 mm

Volume of central core (Vcc) = Ain × hw/1000 = 0.0504 × 252/1000 = 0.0127 m^3;

Volume of 2 side cores (V2sc) = Vcc = 0.0127 m^3;

Total Volume of core (Vc) = Vcc + V2sc = 0.0127 + 0.0127 = 0.0254 m^3;

Volume of both yokes (Vy) = 2 × Ly × Ay/1000 = 2 × 540 × 0.0252/1000 = 0.0272 m^3;

Volume of Iron (Viron) = Vy + Vc = 0.0272 + 0.0245 = 0.0526 m^3;

Weight of Iron (Wiron) = Viron × 7.55 × 1000 = 0.0526 × 7.55 × 1000 = 397 Kg;

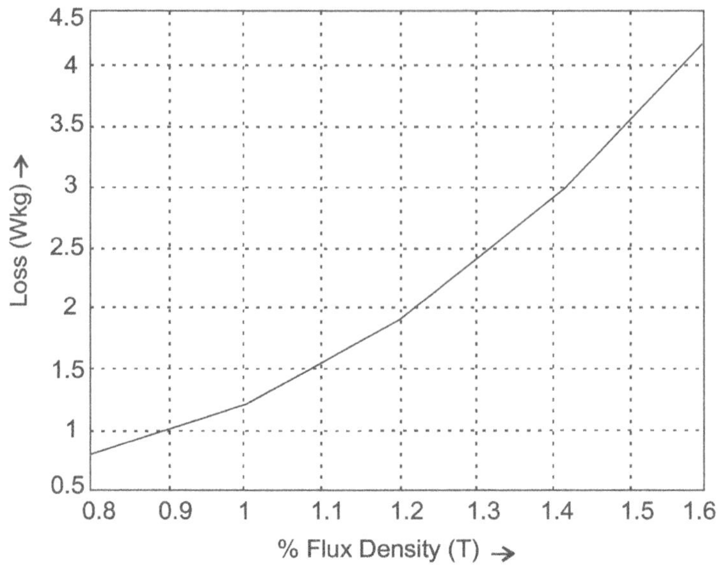

Fig. 5.3 Core loss for hot rolled steels.

From Fig. 5.3, Loss in Core corresponding to Bm = 1.1T (scL) = 1.55 W/Kg

Taking 5% more for laminations, Total Iron Loss (Pi) = 397 × 1.55 × 1.05 = 646 W;

$$\text{Percentage Iron Loss (pcPi)} = \frac{\text{Pi} \times 100}{\text{KVA} \times 1000} = \frac{646 \times 100}{125 \times 1000} = 0.5168\%$$

5.3.2 *(a) Computer Program in "C" in MATLAB for Part-1*

```
% Design of Shell Type (1-ph) Transformer
KVA = 125; HV = 3300; LV = 440; f = 50; pf = 0.85; % Input Data
% <-----1) Design of Magnetic Frame------------------->
K = 1.1; Bm = 1.1; cdav = 2.5; kw = 0.27; ki = 0.9; R1 = 2.5;
R2 = 2.1; % Assumptions
Cdlv = cdav-0.1; cdhv = cdav + 0.1;     Et = K*sqrt (KVA);
Ain = Et/(4.44*f*Bm); Aig = Ain/ki;
bmc1 = sqrt(Aig/R1); bmc = ceil(bmc1*100)*10; Lmc = R1*bmc;
Aw = KVA*1e3/(2.22*f*Bm*kw*cdav*1e6*Ain); bw1 = sqrt (Aw/R2);
bw = ceil (bw1*100)*10;
Hw = R2*bw; Ay = Ain/2; sc = bmc/2; hy = bsc; by = Ay/hy*1e6;
Ly = 2*(bsc + bw) + bmc; Vcc = Ain*hw/1000; V2sc = Vcc;
Vc = Vcc + V2sc; Vy = 2*Ly*Ay/1e3; Viron = Vy + Vc;
Wiron = Viron*7.55e3;
BB = [0.8 1.0 1.2 1.4 1.6];
WpKg = [0.8 1.2 1.9 2.9 4.2]; % (Hot Rolled)
% plot (BB, WpKg); grid; xlabel ('% Flux density (T)-->');
ylabel ('Loss(W/Kg)-->');
```

```
% title ('Core Loss for Hot rolled steels');
scL = interp1(BB, WpKg, Bm, 'linear');    Pi = Wiron*scL*1.05;
pcPi = Pi/KVA/10;
```

5.3.3 No-Load Current (Part-2)

Wattful Current (Iw) = $\dfrac{\text{Pi}}{\text{LV}} = \dfrac{646.06}{440} = 1.4683$ A

Length of flux path in Main and side cores (Lfpc) = 2 × hw/1000

= 2 × 252/100 = 0.504 m

Length of flux path in Yoke (Lfpy) = Ly/1000 = 540/1000 = 0.54 m

Total length of Flux path in magnetic frame (LfpT) = Lfpc + Lfpy

= 0.504 + 0.54 = 1.044m

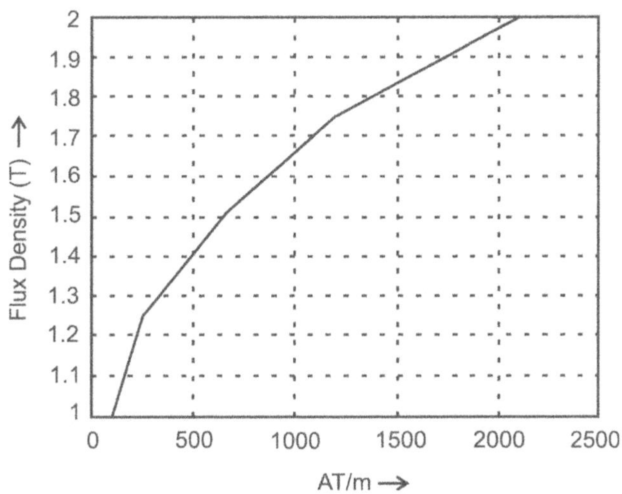

Fig. 5.4 Magnetization curve for hot rolled steels.

From Fig. 5.4, AT/m for total frame corresponding to Bm

= 1.1T (atpm) = 118.8 AT/m

Total Ampere-turns (AT) = atpm × LfpT = 118.8 × 1.044 = 124.03 ;

No of Turns in LV wdg (T2) = $\dfrac{\text{LV}}{\text{Et}} = \dfrac{440}{12.2984} = 35.8 \approx 36$ (Rounded off)

Magnetizing current (Im) = $\dfrac{1.1 \times \text{AT}}{\sqrt{2} \times \text{T2}} = \dfrac{1.1 \times 124}{\sqrt{2} \times 36} = 2.68$ A

No load Current (I0) = $\sqrt{\text{Iw}^2 + \text{Im}^2} = \sqrt{1.4683^2 + 2.68^2} = 3.0556$ A

Phase Current in LV Wdg (I2) $= \frac{KVA \times 000}{LV} = \frac{125 \times 1000}{440} = 284.1$ A

Ratio of I0/I2 (I0 by I2) = I0/I2*100 = 3.0556/284.1 × 100

= 1.0756 (< 2 and hence OK)

Assuming average current density (cdav) = 2.5,

Current desity in LV wdg (cdlv) = cdav – 0.1 = 2.5-0.1 = 2.4 A/mm^2,

Current desity in HV wdg (cdhv) = cdav + 0.1 = 2.5 + 0.1 = 2.6 A/mm^2,

CS area of LV conductor (a2a) $= \frac{I2}{cdlv} = \frac{284.1}{2.4} = 118.37 mm^2$,

No. of Turns on HV side (T1) $= \frac{HV}{LV} \times T2 = \frac{3300}{440} \times 36 = 270$

Phase Current in HV Wdg (I1) $= \frac{KVA \times 1000}{HV} = \frac{125 \times 1000}{3300} = 37.879$ A

CS area of HV conductor (a1a) $= \frac{I1}{cdhv} = \frac{37.879}{2.6} = 14.57 mm^2$,

5.3.3 *(a) Computer Program in "C" in MATLAB for Part-2*

```
% <-----2) No-Load Current------------------>

Pi = 646.06; LV = 440; hw = 252; Ly = 540; Bm = 1.1;
Et = 12.2984; KVA = 125; % Input Data

cdlv = 2.4; HV = 3300; cdhv = 2.6; % Input Data

Iw = Pi/LV; Lfpc= 2*hw/1000;Lfpy = Ly/1000; LfpT = Lfpc + Lfpy;

B1 = [1 1.25 1.5 1.75 2.0]; H1 = [100 250 650 1200 2100];
% (Hot Rolled)

% plot (H1, B1); grid; xlabel ('AT/m-->'); ylabel ('Flux
density (T) -->');

% title ('Magnetization Curve for Hot rolled steels'); %

Atpm = interp1 (B1, H1, Bm, 'spline'); AT = atpm*LfpT;
T2a = LV/Et;   T2 = ceil (T2a);

Im = 1.1*AT/(sqrt(2)*T2);      I0 = sqrt(Iw^2 + Im^2);
I2 = KVA*1000/LV;   I0 by I2 = I0/I2*100;   a2a = I2/cdlv;
T1 = T2*HV/LV;  I1 = KVA*1000/HV;  a1a = I1/cdhv;

% -----------------------------------------------------------
```

5.3.4 Design of LV Winding (Part-3)

Assuming axial Space occupied by insulation is 85 mm (axTins),

pace available for LV and HV coils (axspc) = (hw-axTins) = 252–85 = 167mm

Assuming each winding is split in to 4 coils,

Axil space available/coil (axspc) = 167/(4 + 4) = 20.875mm

Since I2 = 284.1 A is very high, the conductor is split into 4 parallel strips (nps2) and area of CS of each strip (as2a) = a2a/nps2 = 118.37/4 = 29.59 mm^2;

Assuming thickness of each strip (tst2) = 3.5mm,

width (wst2a) = 29.59/3.5 = 8.4551 mm (say 8 mm)

Corrected strip area (as2) = 0.97 × wst2 × tst2 = 0.97 × 8 × 3.5 = 27.16 mm^2,

(Factor 0.97 is for strip edges rounding)

Assuming strip insulation of 0.5 thk and coil insulation is 1mm, Axial space required for LV coil (axsp2a) = (nps2)/ 2 × (wst2 + 0.5) + (2 × 1) = 4/2 × (8 + 0.5) + 2 = 19 mm

Assuming Ins of 1mm between Coils,

Axial length of LV wdg with 4 coils (axL2) = 4 × axsp 2a + (4-1) × 1

= (4 × 19) + 3 = 79 mm

Radial width of LV coils (rsp2) = [2 × (tst2 + 0.5) + 1] × T2/nc2

= [2 × (3.5 + 0.5) + 1] × 36/4 = 81 mm

Assuming outer Insulation as 20 mm,

Radial width of complete LV winding (rw2) = rsp2 + 20 = 81 + 20 = 101 mm;

LV Winding of Shell Type Transformer(dimensions are in mm)

	Length/Axial-wise		
Cu.Strand	Bare + insulation	8 + 0.5 = 8.5	
Coil	2 Cu.Strands	2 × 8.5 = 17	
Coil + Insulation		17 + 2 = 19	
4 Coils		4 × 19 = 76	76.0
Insulation	(bewteen coils)	(4-1) × 1 = 3	3.0
Axial Length			**79.0**

	Radial-Wise		
Cu.Strand	Bare + insulation	3.5 + 0.5 = 4	
Coil	2 Cu.Strands	2 × 4 = 8	
Coil + Insulation		8 + 1 = 9	
9-Coils		9 × 9 = 81	81.0
Insulation	Outer		20.0
Radial-Width			**101.0**

Clearance between winding and side core (cbwsc2) = bw-(rw2)

= 120 – (101) = 19 mm (OK);

Mean Length of turn in LV wdg (Lmt2) = $(2 \times \text{Lmc} + 2 \times \text{bmc} + \pi \times \text{rw2})/1000$

$= (2 \times 375 + 2 \times 150 + \pi \times 101)/1000 = 1.3673$ m

Res of LV wdg (r2) = $0.02 \times \text{Lmt2} \times \text{T2/a2} = 0.02 \times 1.3673 \times 36/108.64 = 9.062\text{m}\Omega$

Assuming Stray losses as 5%, Copper loss in LV wdg (pcu2) = $1.05 \times I2^2 \times r2$

$= 3 \times 284.1^2 \times 9.062 \times 10^{-3} = 767.91\text{W}$

5.3.4 *(a) Computer Program in "C" in MATLAB for Part-3*

```
% <-----3) Design of LV Windings:------->

hw = 252; a2a = 118.37; T2 = 36; bw = 120; Lmc = 375; bmc
= 150; I2 = 284.1; % Input Data

axTins= 85; nc1 = 4;nc2 = 4; nps2 = 4; tst2 = 3.5; % assumption

axspc = (hw-axTins)/(nc1+nc2); as2a = a2a/nps2;
wst2a = as2a/tst2; wst2 = floor(wst2a);

as2 = 0.97*wst2*tst2; a2 = as2*nps2;
axsp2a = nps2/2*(wst2 + 0.5) + 2; axslk2 = axspc-axsp2a;

% if axslk2 < = 0.2 continue; end; axL2 = 2*axsp2a + 3;
rsp2 = (2*(tst2 + 0.5) + 1)*T2/nc2;

cbwsc2 = bw-(rsp2 + 20); % if cbwsc2 < = 10 continue; end;
rw2 = rsp2 + 20; Lmt2 = (2*Lmc + 2*bmc + pi*rw2)/1000;
r2 = 0.02*Lmt2*T2/a2; pcu2 = 1.05*I2^2*r2;
```

5.3.5 Design of HV Winding (Part-4)

Disc type winding with rectangular cross-section is used with 4 coils

Axial space per turn (axspt) = axspc/2 = 20.875/2 = 10.4375mm

Assuming thickness of strip (tst1) = 1.6mm,

Width of strip (wst1) = a1a/tst1 =14.5688/1.6 = 9.1(say 10mm);

Corrected CS area of HV conductor (a1) = wst1 × tst1 × 0.98

$= 10 \times 1.6 \times 0.98 = 15.68\ \text{mm}^2$;

Axial space reqd by HV coil (axsp1a) = 2 × (wst1 + 0.5) + 2

= 2 × (10 + 0.5) + 2 = 23mm;

Axial length of HV Wdg with 4 Coils (axL1) = 4 × axsp1a + (4-1) × 5

= (4 × 23) + 15 = 107 mm;

Since LV and HV coils are mounted one after the other Axially,

Total Axial Space Occupied = 79 + 107 = 186 mm (out of 252 mm) → OK

Radial turns (Tr1) = T1/8 = 270 / 8 = 33.75, Say 40

[Total turns = 2 (40 + 40) + 2(30 + 25) = 270] so that

max no of turns radially = 40 turns

Radial space required (rsp1a) = Tr1 × (tst1 + 0.5) + 20

= 40 × (1.6 + 0.5) + 20 = 104 mm

HV Winding of Core Shell Transformer (dimensions are in mm)

	Length/Axial-wise		
Cu.Strand	Bare + insulation	10 + 0.5 = 10.5	
Coil	2 Cu.Strands	2 × 10.5 = 21	
Coil + Insulation		21 + 2 = 23	
4 Coils		4 × 23 = 92	92.0
Insulation	(bewteen coils)	(4-1) × 5 = 15	15.0
Axial Length			**107.0**

	Radial-Wise		
Cu.Strand	Bare + insulation	1.6 + 0.5 = 2.1	
40-turns		40 × 2.1 = 84	84.0
Insulation	Outer		20.0
Radial-Width			**104.0**

Since LV and HV coils are mounted one after the other axially,

Total Axial Space Occupied =79 + 107 = 186 mm (out of 252mm)-->OK

Clearance between HV wdg and side core (cbwsc1)

= bw-(rsp1a) = 120 – (104) = 16 mm (OK);

Meal length of HV turns (Lmt1) = Lmt2 = 1.367 m

Res of HV wdg/ph(r1) = 0.02 × Lmt1 × T1/a1 = 0.02 × 1.367 × 270/15.68 = 0.4709 Ω

Assuming Stray losses as 5%, Copper loss in HV Wdg (pcu1)

$= 1.05 \times I1^2 \times r1/1000 = 1.05 \times 37.88^2 \times 0.4709 = 709.4$ W

5.3.5 *(a) Computer Program in "C" in MATLAB for Part-4*

```
% <-----4) Design of HV Windings:------->

axspc  =  20.875;  a1a  =  14.5688;  T1  =  270;  bw  =  120;
Lmt2 = 1.367; I1 = 37.88; % Input data

tst1 = 1.6; % Assumptions
```

```
axspt = axspc/2; wst1a = a1a/tst1; wst1 = ceil(wst1a);
a1 = wst1*tst1*0.98;   axsp1a = 2*(wst1 + 0.5) + 2;
axL1 = 2*axsp1a + 5;        Tr1 = ceil(T1/8/10)*10;
rsp1a = Tr1*(tst1 + 0.5);   cbwsc1 = bw-(rsp1a + 20);
% if cbwsc1 < = 10 continue; end; Lmt1 = Lmt2;
r1 = 0.02*Lmt2*T1/a1; pcu1 = 1.05*I1^2*r1;
```

5.3.6 Performance Calculations (Part-5)

Total Winding Copper losses (pcuT) = (pcu1 + pcu2) = (709.4 + 767.9) = 1477.3W

Copper Loss Percentage (pcPcu) = $\dfrac{\text{pcuT}}{\text{KVA} \times 1000} \times 100 = 1.182\%$

Total losses on Full Load (ptFL) = (pcuT + Pi) = 1477.3 + 646.1 = 2123.4 KW;

Total Losses Percentage (pcTL) = $\dfrac{\text{ptFL}}{\text{KVA} \times 1000} \times 100 = \dfrac{2123.4}{125 \times 1000} \times 100 = 1.699\%$

Efficiency Calculation at pf = 0.85 and FL

Output (Opt) = KVA × pf = 125 × 0.85 = 106.25 KW

Input (Inp) = OPt + ptFL/1000 = 106.25 + (2123.4/1000) = 108.37 KW

Efficiency (eff) = Opt/Inp × 100 = 106.25/108.37 × 100 = 98.04 %;

Load for max. Efficiency (Ldmxef) = $\sqrt{\dfrac{\text{Pi}}{\text{pcuT}}} \times \text{KVA} = \sqrt{\dfrac{646.1}{1477.3} \times 125} = 82.66 \text{ KVA}$

Max Efficiency (efmx)

$$= \frac{\text{Ldmxef} \times \text{pf}}{(\text{Ldmxef} \times \text{pf}) + (2 \times \text{Pi})} = \frac{82.66 \times 0.85}{(82.66 \times 0.85) + (2 \times 0.6461)} = 0.9819\text{pu} = 98.19\ \%$$

Mean Turn Length (Lmt) = (Lmt1 + Lmt2)/2 = (1.3673 + 1.3673)/2 = 1.3673m

AT (AT) = I1 × T1 = 37.9 × 270 = 10227

Assuming Axial Clearance between LV and HV coils (ac12) = 12 mm = 0.0012 m,

$$\text{P.U.Reactance (Ex)} = \frac{\pi \times f \times 4 \times \pi \times 10^{-7} \times \text{Lmt} \times \text{AT}}{\text{Et} \times \text{rsp2}/1000 \times 2} \times \left[\text{ac12} + \frac{\text{axL1} + \text{axL2}}{6 \times 1000}\right]$$

$$= \frac{\pi \times 50 \times 4 \times \pi \times 10^{-7} \times 1.3673 \times 10227}{12.2984 \times 81/1000 \times 2} \times \left[0.012 + \frac{51 + 41}{6 \times 1000}\right] = 0.0379 \text{ pu}$$

P.U. Resistance (Er) = pcuT/KVA = 1477.3 /(125 × 1000) = 0.0118

P.U.Impedance (Ez) = $\sqrt{\text{Er}^2 + \text{Ex}^2} = \sqrt{0.0118^2 + 0.0379^2} = 0.0397\text{pu}$

Regulation at pf = 0.85 and FL (Reg) = $\text{Er} \times \text{pf} + \text{Ex} \times \sqrt{1 - \text{pf}^2}$

$$= 0.0118 \times 0.85 + 0.0379 \times \sqrt{1 - 0.85^2} = 0.03\text{pu (3\%)}$$

5.3.6 *(a) Computer Program in "C" in MATLAB for Part-5*

```
% <--------5) Performance Calculations:------>

ac12 = 0.012; % Specification

pcu1 = 709.4; pcu2 = 767.9; KVA = 125; Pi = 646.1; pf = 0.85;
Lmt1 = 1.3673; Lmt2 = 1.3673; % Input Data

I1 = 37.9; T1 = 270; f = 50; Et = 12.2984; rsp2 = 81;
axL1 = 51; axL2 = 41;                  % Input Data

pcuT=pcu1+ pcu2; pcPcu=pcuT/(KVA*1000)*100; ptFL = (pcuT + Pi);
pcTL=ptFL/(KVA*1000)*100; Opt=KVA*pf; Inp= Opt + ptFL/1000;
eff = Opt/Inp*100; Ldmxef = sqrt(Pi/pcuT)*KVA;
efmx = Ldmxef*pf/(Ldmxef*pf + 2*Pi/1000)*100;
Lmt = (Lmt1 + Lmt2)/2; AT = I1*T1;

Ex=pi*f*4*pi*1e-7*Lmt*AT/(Et*rsp2/1e3*2)*(ac12+(axL1+
axL2)/6e3);

Er = pcuT/(KVA*1000); Ez = sqrt(Ex^2 + Er^2);
Reg = (Er*pf + Ex*sqrt(1-pf^2))*100;
```

5.3.7 Tank Design and Weights (Part-6)

Assuming Length-wise clearance (dL) = 80 mm,

Length of Tank (Lt) = (Ly + dL) = (540 + 80) = 620 mm

Assuming Width-wise clearance (dB) = 100 mm,

Width of Tank (bt) = Lmc + 2 × rw2 + dB; = 375 + 2 × 101 + 100 = 677 mm

AssumingHeight -wise clearance (dH) = 450mm,

Height of Tank(ht) = hw + 2 × (hy + dH) = 252 + 2 × (75 + 450) = 852 mm ;

Volume of Tank (Vt) = Lt × bt × ht/10^9 = 620 × 677 × 852/10^9 = 0.3576 m^3;

Cooling Surface area of tank (St) = 2 × (bt + Lt) × ht/10^6 = 2 × (677 + 620) × 852/10^6 = 2.2101 m^2

$$\text{Temperature-rise of Tank (Tr)} = \frac{\text{ptFL}}{12.5\times \text{St}} = \frac{2123.4}{12.5\times 2.2101} = 76.86\ ^\circ\text{C}$$

Since Temp-rise is to be limited to (TRP) = 40 °C, Cooling tubes are to be provided

Selecting Tube dia (Dct) of 50 mm and height (Hct) 0f 550 mm,

Area of each tube (At) = π × Dct × Hct/10^6 = π × 50 × 550/10^6 = 0.0864 m^2;

Required Tube Cooling Area (CAt)

$$= \frac{\text{ptFL} - 12.5\times \text{St}\times \text{TRP}}{6.5\times \text{TRP}\times 1.35} = \frac{2123.4 - 12.5\times 2.2101\times 40}{6.5\times 40\times 1.35} = 2.9012\ \text{m}^2$$

No. of cooling tubes required (Nt) = CAt/At = 2.9012/0.0864 = 33

Wt. of Copper in HV wdg (Wcu1)

$= 8.9 \times \text{Lmt1} \times \text{T1} \times \text{a1}/1000 = 8.9 \times 1.3673 \times 270 \times 15.68/1000 = 51.519$ Kg

Wt. of Copper in LV wdg (Wcu2)

$= 8.9 \times \text{Lmt2} \times \text{T2} \times \text{a2}/1000 = 8.9 \times 1.3673 \times 36 \times 108.64/1000 = 47.59$ Kg

Total Weight (Wtot) = $1.01 \times$ (Wcu1 + Wcu2 + Wiron)

$= 51.519 + 47.59 + 396.96 = 501.03$ Kg

Specific Wt (KgPkva) = Wtot/KVA = 501.03/125 = 4.008

5.3.7 *(a) Computer Program in "C" in MATLAB for Part-6*

```
% <-----------6) TANK Design andWeights------->

Ly = 540; Lmc = 375; rw2 = 101; hw = 252; hy = 75;
ptFL = 2123.4; Lmt1 = 1.3673; T1 = 270; % Input Data

a1 = 15.68; Lmt2 = 1.3673; T2 = 36; a2 = 108.64;
Wiron = 396.96; KVA = 125; % Input Data

dL = 80; dB = 100; dH = 450; Dct = 50; Hct = 550; TRP = 40;
% Assumptions

Lt = Ly + dL; bt = Lmc + 2*rw2 + dB; ht = hw + (2*hy + dH);
Vt = Lt*bt*ht/1e9;  St = 2*(bt + Lt)*ht/1e6;

Tr = ptFL/12.5/St; At = pi*Dct*Hct/1e6;
CAt = (ptFL-12.5*St*TRP)/ (6.5*TRP*1.35); Nt = floor(CAt/At);
Wcu1 = 8.9*Lmt1*T1*a1/1e3; Wcu2 = 8.9*1e3*Lmt2*T2*a2/1e6;
Wtot = 1.01*(Wcu1 + Wcu2 + Wiron); KgPkva = Wtot/KVA;

% --------------- END OF PROGRAM ------------>
```

5.3.8 Computer Program in "C" in MATLAB for Complete Design

```
% Design of Shell Type(1-ph) Power Transformer

KVA= 125; HV = 3300; LV = 440; f = 50; pf = 0.85;  % Input Data

f2 = fopen ('Total_125KVA_Output.m','w');

fprintf(f2,'% 3d KVA, % 5d/ % 4d V, % 2d Hz,1-Phase, KVA, HV,
LV, f);

fprintf (f2, Shell Type, Oil cooled Power Class 1-Ph
Transformer:\n');

fprintf(f2,'=================================================
=================================\n');
```

```
fprintf (f2,'<-----1) Design of Magnetic Frame----------->\n');

K = 1.1; Bm = 1.1; cdav = 2.5; kw = 0.27; ki = 0.9; R1 = 2.5;
R2 = 2.1; % Assumptions

cdlv = cdav-0.1; cdhv = cdav + 0.1; Et = K*sqrt(KVA);
Ain = Et/(4.44*f*Bm); Aig = Ain/ki;

bmc1 = sqrt(Aig/R1); bmc = ceil(bmc1*100)*10; Lmc = R1*bmc;

Aw = KVA*1e3/(2.22*f*Bm*kw*cdav*1e6*Ain); bw1 = sqrt(Aw/R2);
bw = ceil(bw1*100)*10;

Hw = R2*bw; Ay = Ain/2; bsc = bmc/2; hy = bsc; by = Ay/hy*1e6;
Ly = 2*(bsc + bw) + bmc; Vcc = Ain*hw/1000;

V2sc = Vcc; Vc = Vcc + V2sc; Vy = 2*Ly*Ay/1e3; Viron = Vy + Vc;
Wiron = Viron*7.55e3;

BB = [0.8 1.0 1.2 1.4 1.6]; WpKg = [0.8 1.2 1.9 2.9 4.2];
% Data for Hot Rolled

% plot (BB, WpKg); grid; xlabel (% Flux density (T)-->');
ylabel ('Loss(W/Kg)-->');

% title (Core Loss for Hot rolled steels');

scL = interp1(BB, WpKg, Bm, linear'); Pi = Wiron*scL*1.05;
pcPi = Pi/KVA/10;

% Et, Ain, Aig, bmc1, bmc, Lmc, Aw, bw1, bw, hw, Ay, bsc, hy,
by, Ly,

% Vcc, V2sc, Vc, Vy, Viron, Wiron, scL, Pi, pcPi,

fprintf (f2, Assumed: Value factor (K) % 6.2f\n', K);

fprintf (f2,'Volts/turn % 6.2f\n', Et);

fprintf (f2, Assumed: Window space factor (kw) = % 5.3f, Max
Core Flux Density (T) = % 3.1 f\n', kw, Bm);

fprintf (f2, Core-CS-area (Net and Gross) (sq.m) % 6.4f and %
6.4 f\n', Ain, Aig);

fprintf (f2, Main Core Length × Width (mm) % 6.1f × % 6.1f\n',
Lmc, bmc);

fprintf (f2, Window Area (sq.m) % 6.4f\n', Aw);

fprintf (f2, Window Height × Width (mm) % 6.1f × % 6.1f\n', hw,
bw);

fprintf (f2, 'Width × Ht of the yoke (m) % 6.0 f and % 3.0
f\n', by, hy);

fprintf (f2, Length of Yoke (Ly) (mm) % 6.0f\n', Ly);
```

```
fprintf (f2, <-----2) No-Load Current------------------>\n');

Iw=Pi/LV; Lfpc = 2*hw/1000; Lfpy = Ly/1000; LfpT = Lfpc + Lfpy;

B1= [1 1.25 1.5 1.75 2.0];

H1 = [100 250 650 1200 2100]; % Data for Hot Rolled

% plot (H1, B1); grid; xlabel ('AT/m-->'); ylabel ('Flux
density(T)-->');

% title ('Magnetization Curve for Hot rolled steels'); %

atpm = interp1(B1, H1, Bm, 'spline'); AT = atpm*LfpT;
T2a = LV/Et; T2 = ceil(T2a);

Im = 1.1*AT/(sqrt(2)*T2); I0 = sqrt(Iw^2 + Im^2);
I2 = KVA*1000/LV; I0byI2 = I0/I2*100; a2a = I2/cdlv;

T1 = T2*HV/LV; I1 = KVA*1000/HV; a1a = I1/cdhv;

% Iw, Lfpc, Lfpy, LfpT, atpm, AT, T2a, T2, Im, I0, I2, I0 by
I2, a2a, T1, I1, a1a,

fprintf (f2,'Magnetizing-Ct(I0 = Iw + jIm) % 6.3f = % 4.3f + j
% 4.3f A\n', I0, Iw, Im);

fprintf (f2, Ratio of I0/I2 (perc) % 6.3f (Permissible: <
2)\n', I0 by I2);

fprintf (f2,'<-----3) Design of  LV Windings:------->\n');

axTins = 85; nc1 = 4; nc2 = 4; nps2 = 4; tst2 = 3.5;
% assumption

axspc = (hw - axTins)/(nc1 + nc2); as2a = a2a/nps2;
wst2a = as2a/tst2; wst2 = floor(wst2a);

as2 = 0.97*wst2*tst2; a2 = as2*nps2;
axsp2a = nps2/2*(wst2 + 0.5) + 2; axslk2 = axspc-axsp2a;

axL2 = 2*axsp2a + 3; rsp2 = (2*(tst2 + 0.5) + 1)*T2/nc2;
cbwsc2 = bw - (rsp2 + 20); % if cbwsc2 < = 10 continue; end;

rw2 = rsp2 + 20; Lmt2 = (2*Lmc + 2*bmc + pi*rw2)/1000;
r2 = 0.02*Lmt2*T2/a2; pcu2 = 1.05*I2^2*r2;

% axspc, as2a, wst2a, wst2, as2, axsp2a, axslk2, axL2, rsp2,
cbwsc2, rw2, Lmt2, r2, pcu2,

fprintf (f2,'Turns in LV winding % 6.0f\n', T2);

fprintf (f2, 'Ph-Current in LV wdg (Amp) % 6.1f\n', I2);

fprintf (f2,'Width × Thick of Cond (mm) % 6.1f and % 3.1 f\n',
wst2, tst2); I2,

fprintf (f2, CS area of Conductor (sq.mm) % 6.3f\n', a2);
```

```
fprintf (f2, Current density (A/sq.mm) % 6.2f (Permissible:2.3 to 3.5)\n', cdlv);

fprintf (f2, Axial Slack in LV-Wdg (mm) % 6.2f (desirable > 0.2)\n', axslk2);

fprintf (f2, Clearance bet LV Wdg and core(mm) % 6.2f (desirable: > 10)\n', cbwsc2);

fprintf (f2, Radial width of Wdg (mm)  % 6.2f\n', rw2);

fprintf (f2, Mean Length of Turn (m) % 6.3f\n', Lmt2);

fprintf (f2, Resistance/phase (m.ohm)  % 6.4f\n', r2*1000);

fprintf (f2, <-----4) Design of HV Windings:------->\n');

tst1 = 1.6; % Assumptions

axspt = axspc/2; wst1a = a1a/tst1; wst1 = ceil (wst1a); a1 = wst1*tst1*0.98; axsp1a = 2*(wst1 + 0.5) + 2;

axL1 = 2*axsp1a + 5; Tr1 = ceil (T1/8/10)*10; rsp1a = Tr1*(tst1 + 0.5); cbwsc1 = bw-(rsp1a + 20);

% if cbwsc1 < = 10 continue; end;

Lmt1 = Lmt2; r1 = 0.02*Lmt2*T1/a1; pcu1 = 1.05*I1^2*r1;

% axspt, a1a, wst1a, wst1, a1, axsp1a, axL1, Tr1, rsp1a, cbwsc1, r1, pcu1,

fprintf (f2,'Turns in HV winding % 6.0f\n', T1);

fprintf (f2, Ph-Current in HV wdg (Amp) % 6.1f\n', I1);

fprintf (f2, Width × Thick of Cond (mm) % 6.1f and % 3.1f\n', wst1, tst1);

fprintf (f2, CS area of Conductor (sq.mm) % 6.3f\n', a1);

fprintf (f2, Current density (A/sq.mm) % 6.2f (Permissible: 2.3 to 3.5)\n', cdhv);

fprintf (f2, Clearance bet HV Wdg and core (mm) % 6.2f (desirable: >10)\n', cbwsc1);

fprintf (f2, Mean Length of Turn (m) % 6.3 f\n', Lmt1);

fprintf (f2, Resistance/phase (ohm) % 6.4f\n', r1);

fprintf (f2, <--------5) Performance Calculations:------>\n');

ac12 = 0.012; % Specification

pcuT = pcu1 + pcu2; pcPcu = pcuT/(KVA*1000)*100; ptFL = (pcuT + Pi); pcTL = ptFL/(KVA*1000)*100;
```

```
Opt = KVA*pf; Inp = Opt + ptFL/1000; eff = Opt/Inp*100;
Ldmxef = sqrt(Pi/pcuT)*KVA;

efmx = Ldmxef*pf/(Ldmxef*pf + 2*Pi/1000)*100;
Lmt = (Lmt1 + Lmt2)/2; AT = I1*T1;

Ex = pi*f*4*pi*1e-7*Lmt*AT/(Et*rsp2/1000*2)*(ac12 + (axL1 +
axL2)/6e3);

Er = pcuT/(KVA*1000); Ez = sqrt (Ex^2 + Er^2);
Reg = (Er*pf + Ex*sqrt(1-pf^2))*100;

% pcuT, pcPcu, ptFL, pcTL, Opt, Inp, eff, Ldmxef, efmx, Lmt,
AT, Ex, Er, Ez, Reg,

Fprintf (f2, Cu. Losses of: HV-Wdg: = % 4.1f, LV-Wdg: = % 4.1 f
and Iron Loss: = % 4.1f and Total = % 6.1 f W\n', pcu1, pcu2,
Pi, ptFL);

fprintf (f2, Percentage Loss: Iron Loss: = % 6.3f, Copper =
% 6.2 f and Total = % 6.3 f perc\n', pcPi, pcPcu, pcTL);

fprintf (f2, Efficiency at pf = % 4.2 f and Full Load = % 5.2f
perc\n', pf, eff);

fprintf (f2, Max Efficiency = % 5.2f occuring at Load = % 5.2 f
KVA\n', efmx, Ldmxef);

fprintf (f2, Mean-turn Length = % 6.4fm, AT = % 6.0f\n', Lmt,
AT);

fprintf (f2, Reactance = % 6.4 fpu, Restance = % 6.4fpu,
Impedance = % 6.4 fpu\n', Ex, Er, Ez);

fprintf (f2, Regulation: at pf = % 4.2 f and Full Load = %
6.2f\n', pf, Reg);

fprintf (f2,'<-----------6) TANK Design and Weights-------
>\n');

dL = 80; dB = 100; dH = 450; Dct = 50; Hct = 550; TRP = 40; %
Assumptions

Lt = Ly + dL; bt = Lmc + 2*rw2 + dB; ht = hw + (2*hy + dH);
Vt = Lt*bt*ht/1e9; St = 2*(bt + Lt)*ht/1e6;

Tr = ptFL/12.5/St; At = pi*Dct*Hct/1e6;
CAt = (ptFL-12.5*St*TRP)/ (6.5*TRP*1.35); Nt = floor(CAt/At);

Wcu1 = 8.9*Lmt1*T1*a1/1e3; Wcu2 = 8.9*1e3*Lmt2*T2*a2/1e6;
Wtot = 1.01*(Wcu1 + Wcu2 + Wiron); KgPkva = Wtot/KVA;

Lt, bt, ht, Vt, St, Tr, At, CAt, Nt, Wcu1,Wcu2, Wtot, KgPkva,
Lmt1, a1, a2, T1, T2,
```

```
Fprintf (f2, 'Tank: Length = % 4.0fmm, Width = %4. 0fmm, Height
= % 4.0 fmm and Volume = % 6.3fm3\n', Lt, bt, ht, Vt);
fprintf (f2,'Tank: Cooling area = % 6.4 fsq.m and Temp-Rise = %
4.1f is reduced to % 3d deg-C\n', St, Tr, TRP);
fprintf(f2, Cooling Tubes:Area/Tube = % 6.4 fsq.m, Required-
Area = % 6.4 fsq.m and No of Tubes = % 3.0 f\n', At, CAt, Nt);
fprintf (f2, Weight: HV-Wdg: = % 6.2f + LV-Wdg: = % 6.2f +
(Core + Yoke): = % 6.2f and Total = % 7.1 f Kg\n', Wcu1,Wcu2,
Wiron, Wtot);
fprintf (f2,'Kg/KVA = % 6.3f\n', KgPkva); axL1, axL2,
fprintf (f2,<-----------End of Program------->\n');
fclose (f2);
```

5.3.8 *(a) Computer Output Results for Complete Design*

125 KVA, 3300/ 440 V, 50 Hz, 1-Phase, Shell Type, Oil cooled Power Class 1-Ph Transformer:

==

<-----1) Design of Magnetic Frame----------------->

Assumed: Value factor (K) 1.10

Volts/turn 12.30

Assumed: Window space factor (kw) = 0.270, Max Core Flux Density (T) = 1.1

Core-CS-area (Net and Gross) (sq.m) 0.0504 and 0.0560

Main Core Length × Width (mm) 375.0 × 150.0

Window Area (sq.m) 0.0301

Window Height × Width (mm) 252.0 × 120.0

Width × Ht of the yoke (m) 336 and 75

Length of Yoke (Ly) (mm) 540

<-----2) No-Load Current ----------------->

Magnetizing-Ct (I0 = Iw + jIm) 3.056 = 1.468 + j2.680 A

Ratio of I0/I2 (perc) 1.076 (Permissible: < 2)

<-----3) Design of LV Windings:------->

Turns in LV winding 36

Ph-Current in LV wdg (Amp) 284.1

Width × Thick of Cond (mm) 8.0 and 3.5

CS area of Conductor (sq.mm) 108.640

Current density (A/sq.mm) 2.40 (Permissible: 2.3 to 3.5)

Axial Slack in LV-Wdg (mm) 1.88 (desirable > 0.2)

Clearance bet LV Wdg and core (mm) 19.00(desirable: >10)

Radial width of Wdg (mm) 101.00

Mean Length of Turn (m) 1.367

Resistance/phase (m.ohm) 9.0616

<-----4) Design of HV Windings:------->

Turns in HV winding 270

Ph-Current in HV wdg (Amp) 37.9

Width × Thick of Cond (mm) 10.0 and 1.6

CS area of Conductor (sq.mm) 15.680

Current density (A/sq.mm) 2.60 (Permissible: 2.3 to 3.5)

Clearance bet HV Wdg and core (mm) 16.00 (desirable: > 10)

Mean Length of Turn (m) 1.367

Resistance/phase (ohm) 0.4709

<--------5) Performance Calculations:------>

Cu. Losses of: HV-Wdg: = 709.4, LV-Wdg: = 767.9 and Iron Loss: = 646.1 and Total = 2123.4 W

Percentage Loss:- Iron Loss: = 0.517, Copper = 1.18 and Total = 1.699 perc

Efficiency at pf = 0.85 and Full Load = 98.04 perc

Max Efficiency = 98.19 occuring at Load = 82.66KVA

Mean-turn Length = 1.3673m, AT = 10227

Reactance = 0.0379pu, Restance = 0.0118pu, Impedance = 0.0397pu

Regulation: at pf = 0.85 and Full Load = 3.00

<-----------6) TANK Design and Weights------->

Tank: Length = 620mm, Width = 677mm, Height = 852mm and Volume = 0.358m3

Tank : Cooling area = 2.2101sq.m and Temp-Rise = 76.9 is reduced to 40 deg-C

Cooling Tubes: Area/Tube = 0.0864 sq.m, Required-Area = 2.9012 sq.m and No of Tubes = 33

Weight: HV-Wdg: = 51.52 + LV-Wdg: = 47.59 + (Core + Yoke): = 396.96 and Total = 501 Kg

Kg/KVA = 4.008

<----------- End of Program------->

5.3.9 Modifications to be done in the above Program to get Optimal Design

1. Insert "for" Loops for the following parameters to iterate the total program between min and max permissible limits for selecting the feasible design variants:
   (a) No. of values of "K" (Factor of ratio of core to copper)
   (b) No. of Values of "Bm", (Core flux density)
   (c) No. of Average current density values
   (d) No. of "L/D" ratios
   (e) No. of values of "R1" (Ratio of Length to width of main core)
   (f) No. of values of "R2" (Ratio of Length to width Window).
2. Insert also minimum or maximum range of required objective functional values as constraint values, for example (a) Efficiency (b) Kg/KVA (c) I0/I2 ratio (d) Volume of Tank etc.
3. Run the program to get various possible design variants.

Note: From the feasible design variants printed in the output, select that particular design that fulfills the objective of optimal parameter for the design.

5.3.10 Computer Program in "C" in MATLAB for Optimal Design

```
% Design of Shell Type (1-ph) Power Transformer

KVA = 125; HV = 3300; LV = 440; f = 50; pf = 0.85; % Input
Data

f2 = fopen ('Optimal_125KVA_Output.m','w');

fprintf (f2, % 3d KVA, % 5d/ % 4d V, % 2d Hz, 1-Phase, KVA,
HV, LV, f);

fprintf (f2, Shell Type, Oil cooled Power Class 1-Ph
Transformer:\n');

fprintf(f2,=================================================
============================================\n');

fprintf (f2, Sn K Bm R1 R2 cdav T1 T2 Lmc bmc hw
bw hy Ly I0byI2 eff Reg Vt KgPkva\n')

K = 1.1; Bm = 1.1; cdav = 2.3; kw = 0.27; ki = 0.9;
r1 = 2.5; r2 = 2; scL = 1.65; atpm = 150; cdlv = 2.2;
cdhv = 2.4; % Assumptions

Sn = 0; M1 = 0; M2 = 0; M3 = 0; M4 = 0; EFFmax = 97;
minI0by I2 = 2; minKgPkva = 5; min Vt = 2; % Assumptions

for K = 1:0. 1:1. 1; % (Range: 1 to 1.1)
```

```
for Bm = 1.1: 0.1: 1.3; % (Range: 1.1 to 1.3)

for cdav = 2.3: 0.1: 3.5 ; % (Range: 2.3 to 3.5)

for R1 = 2:0. 1:3; % (Range: 2 to 3)

for R2 = 2:0. 1:3; % (Range: 2 to 3)

cdlv = cdav-0.1; cdhv = cdav + 0.1;

Et = K*sqrt (KVA); Ain = Et/(4.44*f*Bm); Aig = Ain/ki;
bmc1 = sqrt(Aig/R1); bmc = ceil(bmc1*100)*10;

Lmc = R1*bmc; Aw = KVA*1e3/(2.22*f*Bm*kw*cdav*1e6*Ain);

bw1 = sqrt(Aw/R2); bw = ceil(bw1*100)*10; hw = R2*bw;
Ay = Ain/2; bsc = bmc/2; hy = bsc; by = Ay/hy*1e6;

Ly = 2*(bsc + bw) + bmc; Vcc = Ain*hw/1000; V2sc = Vcc;
Vc = Vcc + V2sc; Vy = 2*Ly*Ay/1e3; Viron = Vy + Vc;

Wiron = Viron*7.55e3;

BB = [0.8 1.0 1.2 1.4 1.6]; WpKg = [0.8 1.2 1.9 2.9 4.2];
% Data for Hot Rolled

scL = interp1 (BB, WpKg, Bm, linear'); Pi = Wiron*scL*1.05;
pcPi = Pi/KVA/10;

% fprintf (f2,'<-----2) No-Load Current----------->\n');

Iw = Pi/LV; Lfpc = 2*hw/1000; Lfpy = Ly/1000; LfpT = Lfpc +
Lfpy;

B1 = [1 1.25 1.5 1.75 2.0]; H1 = [100 250 650 1200 2100];
% Data for Hot Rolled

atpm = interp1 (B1, H1, Bm, spline'); AT = atpm*LfpT;
T2a = LV/Et; T2 = ceil(T2a);

Im = 1.1*AT/(sqrt(2)*T2); I0 = sqrt(Iw^2 + Im^2);
I2 = KVA*1000/LV; I0 by I2 = I0/I2*100;

if I0 by I2 > = 1.09 continue; end;

a2a = I2/cdlv; T1 = T2*HV/LV; I1 = KVA*1000/HV;
a1a = I1/cdhv;

% fprintf (f2,'<-----3) Design of LV Windings:------->\n');

axTins = 85; nc1 = 4; nc2 = 4; nps2 = 4; tst2 = 3.5;
% assumption

axspc = (hw - axTins)/(nc1 + nc2); as2a = a2a/nps2;
wst2a = as2a/tst2; wst2 = floor(wst2a);
```

```
as2 = 0.97*wst2*tst2; a2 = as2*nps2;
axsp2a = nps2/2*(wst2 + 0.5) + 2; axslk2 = axspc - axsp2a;

if axslk2 < = 0.2 continue; end;

axL2 = 2*axsp2a + 3; rsp2 = (2*(tst2 + 0.5) + 1)*T2/nc2;
cbwsc2 = bw - (rsp2 + 20); if cbwsc2 < = 10 continue; end;

rw2 = rsp2 + 20; Lmt2 = (2*Lmc + 2*bmc + pi*rw2)/1000;
r2 = 0.02*Lmt2*T2/a2; pcu2 = 1.05*I2^2*r2;

tst1 = 1.6; % Assumptions

axspt = axspc/2; wst1a = a1a/tst1; wst1 = ceil(wst1a);
a1 = wst1*tst1*0.98; axsp1a = 2*(wst1 + 0.5) + 2;

axL1 = 2*axsp1a + 5; Tr1 = ceil (T1/8/10)*10;
rsp1a = Tr1*(tst1 + 0.5);cbwsc1 = bw - (rsp1a + 20);

% if cbwsc1 < = 10 continue; end;

Lmt1 = Lmt2; r1 = 0.02*Lmt2*T1/a1; pcu1 = 1.05*I1^2*r1;

% fprintf (f2,'<--------5) Performance Calculations:------
>\n');

ac12 = 0.012; % Specification

pcuT = pcu1 +pcu2; pcPcu = pcuT/(KVA*1000)*100;
ptFL = (pcuT+Pi); pcTL = ptFL/(KVA*1000)*100;

Opt = KVA*1000*pf; Inp = Opt + ptFL; eff = Opt/Inp*100;
if eff < = 97.95 continue; end;

Ldmxef = sqrt(Pi/pcuT)*KVA;

efmx = Ldmxef*pf/(Ldmxef*pf + 2*Pi/1000)*100;
Lmt = (Lmt1 + Lmt2)/2; AT = I1*T1;

Ex = pi*f*4*pi*1e-7*Lmt*AT/(Et*rsp2/1000*2)*(ac12 + (axL1 +
axL2)/6e3);

Er = pcuT/(KVA*1000); Ez = sqrt(Ex^2 + Er^2);
Reg = (Er*pf + Ex*sqrt(1 -pf^2))*100;

% fprintf (f2,'<-------6) TANK Design and Weights-------
>\n');

dL = 80; dB = 100; dH = 450; Dct = 50; Hct = 550; TRP = 40;
% Assumptions

Lt = Ly + dL; bt = Lmc + 2*rw2 + dB; ht = hw + (2*hy + dH);
Vt = Lt*bt*ht/1e9; St = 2*(bt + Lt)*ht/1e6;
```

```
Tr = ptFL/12.5/St; At = pi*Dct*Hct/1e6;
CAt = (ptFL -12.5*St*TRP)/(6.5*TRP*1.35); Nt=floor (CAt/At);

Wcu1 = 8.9*Lmt1*T1*a1/1e3; Wcu2 = 8.9*1e3*Lmt2*T2*a2/1e6;
Wtot = 1.01*(Wcu1 + Wcu2 + Wiron); KgPkva = Wtot/KVA;

if KgPkva > = 4.0 continue; end; Sn = Sn + 1;  Sn,

if eff > = EFFmax  EFFmax = eff; end; if abs (eff-EFFmax) <
= 2e-3 M2 = Sn; end;

if KgPkva < = minKgPkva minKgPkva = KgPkva; end; if abs
(KgPkva - minKgPkva) < = 0.001 M1 = Sn; end;

if I0byI2 < = min I0 by I2 min I0 by I2 = I0 by I2; end;
if abs (I0 by I2-min I0 by I2) < = 0.0001 M3 = Sn; end;

if Vt < = minVt minVt = Vt; end;
if abs(Vt - minVt) < = 0.001 M4 = Sn; end;

fprintf (f2, '% 3d % 4.1 f % 4.1 f % 4.1 f % 4.1 f % 5.1 f %
4.0 f % 4.0 f % 5.0 f % 5.0 f % 5.0 f % 5.0 f % 5.0 f % 5.0
f % 6.2 f % 7.2 f % 6.2 f % 6.2 f % 6.2 f\n', Sn, K, Bm, R1,
R2, cdav, T1, T2, Lmc, bmc, hw, bw, hy, Ly, I0 by I2, eff,
Reg, Vt, KgPkva);

% Sn, K, Bm, R1, R2, cdav T1 T2 Lmc, bmc, hw, bw, hy,  Ly,
I0 by I2, eff, Reg, Vt, KgPkva);

end; end; end; end; end;

fprintf(f2,=================================================
==================================================\n');

fprintf (f2, Selection of Design Variant based on
Optimization Criteria:');

fprintf (f2,\nIf Maximum Efficiency is Required, Select
Variant(Sn) = % 3d (% 5.2 f perc)', M2, EFFmax);

fprintf (f2,\nIf Minimum Kg/KVA is Required, Select
Variant(Sn) = % 3d  (% 5.2 f)', M1, minKgPkva);

fprintf (f2,\nIf Minimum ratio of I0/I2 is Reqd, Select
Variant(Sn) = % 3d  (% 4.3f)', M3, min I0 by I2);

fprintf (f2,\nIf Minimum Volume of Tank is Reqd, Select
Variant (Sn) = % 3d  (% 5.2fm3), M4, min Vt);

fclose (f2);
```

5.3.10 (a) Computer Output Results for Optimal Design

125 KVA, 3300/ 440 V, 50 Hz, 1-Phase, Shell Type, Oil cooled Power Class 1-Ph Transformer:

Sn	K	Bm	R1	R2	cdav	T1	T2	Lmc	bmc	hw	bw	hy	Ly	I0 by I2	eff	Reg	Vt	KgPkva
1	1.0	1.1	2.3	2.2	2.3	300	40	345	150	286	130	75	560	1.05	98.02	3.22	0.38	4.08
2	1.0	1.1	2.6	2.0	2.3	300	40	364	140	280	140	70	560	1.04	98.01	3.27	0.38	4.06
3	1.0	1.1	2.6	2.2	2.3	300	40	364	140	286	130	70	540	1.03	98.01	3.27	0.37	4.04
4	1.0	1.1	2.7	2.0	2.3	300	40	378	140	280	140	70	560	1.04	97.98	3.33	0.39	4.08
5	1.0	1.1	2.7	2.2	2.3	300	40	378	140	286	130	70	540	1.03	97.98	3.33	0.38	4.06
6	1.0	1.1	2.8	2.2	2.3	300	40	392	140	286	130	70	540	1.03	97.95	3.40	0.39	4.08
7	1.0	1.1	2.0	2.0	2.4	300	40	320	160	280	140	80	600	1.08	97.95	3.16	0.39	4.09
8	1.0	1.1	2.0	2.1	2.4	300	40	320	160	273	130	80	580	1.05	97.97	3.16	0.37	3.99
9	1.0	1.1	2.0	2.2	2.4	300	40	320	160	286	130	80	580	1.07	97.96	3.16	0.38	4.07
10	1.0	1.1	2.0	2.0	2.5	300	40	320	160	260	130	80	580	1.02	97.98	3.16	0.37	3.92
11	1.0	1.1	2.0	2.1	2.5	300	40	320	160	273	130	80	580	1.05	97.97	3.16	0.37	3.99
12	1.0	1.1	2.0	2.2	2.5	300	40	320	160	286	130	80	580	1.07	97.96	3.16	0.38	4.07
13	1.0	1.1	2.3	2.0	2.5	300	40	345	150	260	130	75	560	1.00	97.96	3.23	0.37	3.89
14	1.0	1.1	2.6	2.0	2.5	300	40	364	140	260	130	70	540	0.99	97.95	3.28	0.36	3.84
15	1.1	1.1	2.9	2.2	2.4	270	36	406	140	264	120	70	520	1.08	98.00	3.09	0.36	4.05
16	1.1	1.1	3.0	2.2	2.4	270	36	420	140	264	120	70	520	1.08	97.97	3.15	0.37	4.06
17	1.1	1.1	2.5	2.1	2.5	270	36	375	150	252	120	75	540	1.08	98.04	3.00	0.36	4.01
18	1.1	1.1	2.6	2.1	2.5	270	36	390	150	252	120	75	540	1.08	98.01	3.07	0.37	4.03
19	1.1	1.1	2.7	2.1	2.5	270	36	405	150	252	120	75	540	1.08	97.98	3.13	0.37	4.04
20	1.1	1.1	2.8	2.1	2.5	270	36	420	150	252	120	75	540	1.08	97.95	3.20	0.38	4.06
21	1.1	1.1	2.9	2.1	2.5	270	36	406	140	252	120	70	520	1.05	98.01	3.09	0.36	3.97
22	1.1	1.1	2.9	2.2	2.5	270	36	406	140	264	120	70	520	1.08	98.00	3.09	0.36	4.05
23	1.1	1.1	3.0	2.1	2.5	270	36	420	140	252	120	70	520	1.05	97.98	3.15	0.36	3.99
24	1.1	1.1	3.0	2.2	2.5	270	36	420	140	264	120	70	520	1.08	97.97	3.15	0.37	4.06
25	1.1	1.1	2.5	2.1	2.6	270	36	375	150	252	120	75	540	1.08	97.97	3.00	0.36	3.97

Selection of Design Variant based on Optimization Criteria:

If Maximum Efficiency is Required, Select Variant (Sn) = 17 (98.04 perc)

If Minimum Kg/KVA is Required, Select Variant (Sn) = 14 (3.84)

If Minimum ratio of I0/I2 is Required, Select Variant (Sn) = 14 (0.985)

If Minimum Volume of Tank if Required, Select Variant (Sn) = 25 (036m3).

CHAPTER 6

Synchronous Machines

6.1 Introduction

Synchronous machines are of two types (viz) Salient Pole type and Round Rotor type. Theory portion of Design is not given in this book, but necessary formulae, curves and tables given in standard books are made use of. Since the machine is the same for both operations of Generator and Motor, the same design is applicable for both.

6.2 Salient Pole Type

Total design is split into eight parts in a proper sequence. Design Calculations are given for a given Rating of Generator, followed by Computer Program written in "C" language using MATLAB software for each part. Finally all the programs are added together to get the total program by running which we get the total design. Computer output of total design is given.

This design may not be the optimum one. Now optimization objective and Design Constraints are inserted into this total program. When this program is run we will get various alternative feasible designs from which the selected variant based on the Optimization Criteria can be picked up. Computer output showing the important design parameters for various feasible alternatives is given in the end of this chapter along with a logic diagram.

6.2.1 Sequential Steps for Design of Each Part and Programming Simultaneously

(a) Calculate Output Coefficient, Main dimensions of Stator Core (viz) D, L and Flux/Pole, Turns/phase, No. of slots, checking the Peripheral velocity and Slot Loading

(b) Calculate size of slot, conductor size, checking current density, slot balance. Calculate tooth flux density, Depth of core, Wt. of Copper, Copper losses and Leakage reactance

(c) Calculate Air gap length, Rotor dia, Dimensions of Poles and Field Coils.

(d) Calculate Carter Coefft and Ampere-turns for Air gap, Stator tooth, core, Poles, rotor core and Total No-load AT

(e) Calculation and plotting of Open circuit characteristic (OCC)

(f) Calculation of Field AT at rated Load and PF

(g) Calculate Copper size, No. of turns in Field winding. Calculation of Iron loss and total losses and Efficiency

(h) Calculation of Temp-rises, Total Weight and KG/KVA.

Note: By adding programs established for each part sequentially we get the Program for complete design.

Problem: Design a 2000KVA, 3300V, 375 RPM, 50 Hz, 0.85 PF, 3 ph, Star connected Salient-pole Generator.

6.2.2 Calculation of the Stator Main Dimensions and Flux/Pole (Part - I)

SPECIFIC MAGNETIC LOADING and SPECIFIC ELECTRIC LOADING

KVA	100	200	500	1000	5000	10000	20000
Bav (T)	0.52	0.54	0.56	0.58	0.61	0.63	0.65
Amp-Cond/m	20000	23000	26000	29000	34000	37000	40000

Following Data is assumed:

No. of ventilating ducts (nv) = 9; Width of ventilating ducts (bv) = 10 mm

Winding factor (Kw) = 0.955; Length/Pole pitch = 2.3; Pole Arc = Core Length (L),

Slots/pole/ph (spp) = 3; Iron factor (k_i) = 0.9;

Values of Specific magnetic loading (B_{av}) = 0.6089 T and

Specific electric loading (q) = 33400 ac/m → are read from the table given above.

Calculations

$$\text{No. of Poles} = P = \frac{120 \times f}{N} = \frac{120 \times 50}{375} = 16$$

Output Coefft (K) = 11 × Bav × q × Kw × 1e-3

= 11 × 0.6089 × 33400 × 0.955 × 1e-3 = 213.6465

$$\text{Speed (n)} = \frac{2 \times f}{P} = \frac{2 \times 50}{16} = 6.25\text{rps}$$

$$D^2L = \frac{KVA}{K \times n} = \frac{2000}{213.6465 \times 6.25} = 1.4978\,m^3$$

$$D = \sqrt[3]{\frac{D^2L \times P}{L/PP \times \pi}} = \sqrt[3]{\frac{1.4978^2 L \times 16}{2.3 \times \pi}} = 1.491\text{ m} = 1491\text{mm} \approx 1500\text{ mm (Rounded off)}$$

$$L = \frac{D^2L}{D^2} = \frac{1.4978}{1.5^2} = 0.6657\text{m} = 665.7 \text{ mm} \approx 670 \text{ mm (Rounded off)}$$

$$L_s = (\text{L-nv} \times \text{bv}) = [670-(9 \times 10)] = 580 \text{ mm}$$

$$L_i = (\text{L-nv} \times \text{bv}) \times k_i = 580 \times 0.91 = 522 \text{ mm}$$

$$\text{Pole Pitch (PP)} = \frac{\pi \times D}{P} = \frac{\pi \times 1.5}{16} = 294.52\text{mm}$$

$$\text{Peripheral speed (Vr)} = \pi \times D \times n = \pi \times 1500/1000 \times 6.25 = 29.45 \text{ m/s}$$

$$\text{Flux/Pole } (\Phi) = \frac{B_{av} \times \pi \times D \times L \times 10^{-6}}{P} = \frac{0.6089 \times \pi \times 1500 \times 670 \times 10^{-6}}{16} = 0.1202\text{T}$$

$$\text{Turns/ph (Tph)} = \frac{V}{\sqrt{3}} \times \frac{1}{4.44 \times f \times Kw \times \Phi}$$

$$= \frac{3300}{\sqrt{3}} \times \frac{1}{4.44 \times 50 \times 0.955 \times 0.1202} = 74.79$$

Conductors/ph (Zph) = Tph × 2 = 74.79 × 2 = 149.58

No. of Stator slots(S) = spp × P × 3 = 3 × 16 × 3 = 144

$$\text{Slot pitch (sp)} = \frac{\pi \times D}{S} = \frac{\pi \times 1500}{144} = 32.7249 \text{ mm}$$

$$\text{Cond/slot(Zs)} = \frac{Z_{ph} \times 3}{S} = \frac{149.58 \times 3}{144} = 3.12 \rightarrow \text{Rounded to 4 (equal to multiples of 2)}$$

$$\text{Revised Tph} = \frac{Z_s \times S}{6} = \frac{4 \times 144}{6} = 96$$

$$\text{Phase Current (Iph)} = \frac{KVA \times 1000}{V \times \sqrt{3}} = \frac{2000 \times 1000}{3300 \times \sqrt{3}} = 349.9 \text{ amps}$$

Slot Loading (SLoad) = Iph × Zs = 349.9 × 4 = 1400 amps/slot;

$$\text{Corrected } \Phi = \frac{V}{\sqrt{3}} \times \frac{1}{4.44 \times f \times Kw \times Tph}$$

$$= \frac{3300}{\sqrt{3}} \times \frac{1}{4.44 \times 50 \times 0.955 \times 96} = 0.0936\text{Wb}$$

6.2.2 *(a) Computer Program in "C" in MATLAB for Part-1*

```
KVA   =   2000;   V   =   3300;   N   =   375;   f   =   50;   pf   =   0.85;
% Star connected Salient Pole Gen

KVAS = [100 200 500 1000 5000 10000 20000];

BavS = [0.52 0.54 0.56 0.58 0.61  0.63  0.65];

qS    =    [20e3    23e3    26e3    29e3    34e3       37e3       40e3];
% plot (KVAS,BavS); grid;
```

```
Bav      =      interp1(KVAS,      BavS,      KVA,      'spline');
q = interp1 (KVAS, qS, KVA, 'spline');
Kw = 0.955; LbyPP = 2.3; nv = 9; bv = 10; ki = 0.9; spp = 3;
% Assumptions Pole Arc = L;
P = 120*f/N; K = 11*Bav*q*Kw*1e-3; n = 2*f/P; DsqL = KVA/(K*n);
D1  =  (DsqL*P/(LbyPP*pi))^(1/3)*1000;  D  =  ceil(D1/10)/100;
L1 = DsqL/D^2; L = ceil(L1*100)*10; D = D*1000; Ls = (L-nv*bv);
Li   =   Ls*ki;   PP   =   pi*D/P;        Vr   =   pi*D*n/1000;
flux1 = Bav*pi*D*L/P/1e6;  Tphi = V/sqrt(3)/(4.44*f*Kw*flux1);
Zph  =  Tphi*2;  S  =  spp*P*3;  sp  =  pi*D/S;  Zsi  =  Zph*3/S;
Zs = ceil  (Zsi/2)*2; Tph = Zs*S/6; Iph = KVA*1e3/(1.7321*V);
SLoad = Iph*Zs;  FI = V/sqrt(3)/ (4.44*f*Kw*Tph);
```

6.2.3 Design of Armature Winding and Core (Part-2)

Assuming current density (cds) = 4 A/mm^2;

Cond Area of CS (As) = $\dfrac{\text{Iph}}{\text{cds}} = \dfrac{349.9}{4} = 87.475\text{mm}^2$

No of teeth under pole arc (TpP) = $\dfrac{\text{PoleArc}}{\text{SlotPitch}}$

$= \dfrac{0.7 \times 294.52}{32.7249} = 6.3 \cong 7$ (Rounded to higher Integer)

Tooth width at air gap (bt0) = $\dfrac{\Phi}{B_{t0} \times \text{Li} \times \text{TpP}} = \dfrac{0.0936}{1.8 \times 522 \times 7} = 14.23\text{ mm}$

Slot Width (Ws) = Slot Pitch – bt0 = 32.72 – 14.23 = 18.49 mm

Cond Width (Wc) = Slot width – width of Ins = 18.49 – 6

= 12.49 m ≅ 13 mm (Rounded off)

Cond Thk (Hc) = $\dfrac{\text{As}}{\text{Wc}} = \dfrac{87.475}{13} = 6.73 \cong 7\text{mm}$ (Rounded off)

Corrected Width of Slot (Ws) = 1 × Wc + Wins = (1 × 13) + 6 = 19 mm;

Corrected current density (cds) = $\dfrac{\text{Iph}}{\text{Wc} \times \text{Hc} \times 0.98} = \dfrac{349.9}{13 \times 7 \times 0.98} = 3.9235\text{A}/\text{mm}^2$

(assuming a factor of 0.98 for rounding the conductor edges)

Height of Slot (Hs) = Zs × Hc + Zs × 2 × 0.5 + Hins + Hw + HL

= [(4 × 7) + 4 × (2 × 0.5) + 8 + 4 + 1] = 45mm (Refer Table given below)

Mean Length of Turn (Lmt) = (2 × L + 2.5 × PP + 50 × KV + 150)/1000

= (2 × 670) + (2.5 × 294.52) + (50 × 3300/1000) + 150 = 2.3913 m

$$\text{Stator Wdg Res/ph (Rph)} = 0.021 \times \text{Lmt} \times \frac{\text{Tph}}{\text{As}} = 0.021 \times 2.3913 \times \frac{96}{89.18} = 0.0541\,\Omega$$

Cross-Section of Stator Slot (dimensions in mm)

Height-Wise			
Lip			1.0
Wedge			4.0
Ins.Under Wedge			1.0
Cu.Coductor	bare + insulation	7 + 1 = 8	
Top-Layer	for 2 conductors	2 × 8 = 16	16.0
Ins.between Layers			4.0
Bottom- Layer	for 2 conductors	2 × 8 = 16	16.0
Bottom Insulation			1.0
Slack			2.0
Total			**45.0**

Width-Wise			
Cu.Conductor	bare + insulation	13 + 1 = 14	14.0
Layer Insulation			2.0
Insulation on sides	2 × 1 = 2		2.0
Slack			1.0
Total			**19.0**

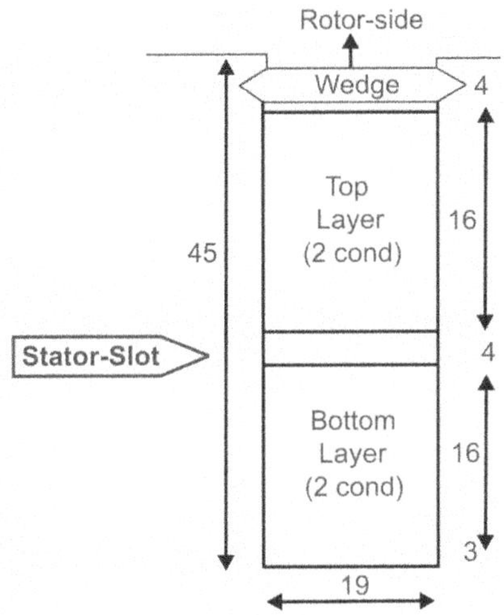

DC Copper Loss (Pcus) = 3 × Iph^2 × Rph × 10^{-3}

= 3 × 349.9^2 × 0.0541 × 10^{-3} = 19.8548 KW

$$\alpha = \sqrt{Wc/Ws} = \sqrt{13/19} = 0.8272$$

$$\text{Av Loss Factor (Kdav)} = 1 + (\alpha Hc)^4 \times \frac{Zs^2}{9} = 1 + (0.8272 \times 0.007)^2 \times \frac{4^2}{9} = 1.1998$$

Eddy Current Loss (Peddy) = (Kdav-1) × Pcus = 0.1998 × 19.8548 = 3.9675 KW

Assuming Stray Losses = 15%,

Total Copper Losses in Stator Wdg (Pts) = 1.15 × (19.8548 + 3.9675) = 27.396 KW

Effective PU Resistance (Reff) = Iph × Rph × Kdav × V/ √3

= 349.9 × 0.0541 × 1.1998/ (3300/√3) = 0.0119 pu

Wt of Stator Copper (Wcus) = 3 × Lmt × As × Tph × 8.9/1000

= 3 × 2.3913 × 89. 18 × 96 × 8.9/1000 = 546.6Kg

For Leakage Reactance Calculation

h1 = Zs × Hc + (4 + 2 + 2) = (4 × 7) + 8 = 36 mm; h2 = 2 + 2 = 4 mm;

h3 = Hw = 4mm; h4 = HL = 1mm; Width at opening of slot (Ws_0) = 9 mm;

$$\text{Specific Permeance of Slot } (\lambda_s) = \frac{h_1}{3 \times W_s} + \frac{h_2}{W_s} + \frac{2h_3}{W_s + W_{s0}} + \frac{h_4}{W_{s0}}$$

$$= \frac{36}{3 \times 19} + \frac{4}{19} + \frac{2 \times 4}{19 + 9} + \frac{1}{9} = 1.2389$$

Slot Leakage Flux (ϕ_s) = $2 \times \sqrt{2} \times \mu0 \times$ Iph $\times$ Zs $\times$ Ls $\times \lambda_s$

$$= 2 \times \sqrt{2} \times 4 \times \pi \times 10^{-7} \times 349.9 \times 4 \times 580/1000 \times 1.2389 = 0.0036 \text{ Wb}$$

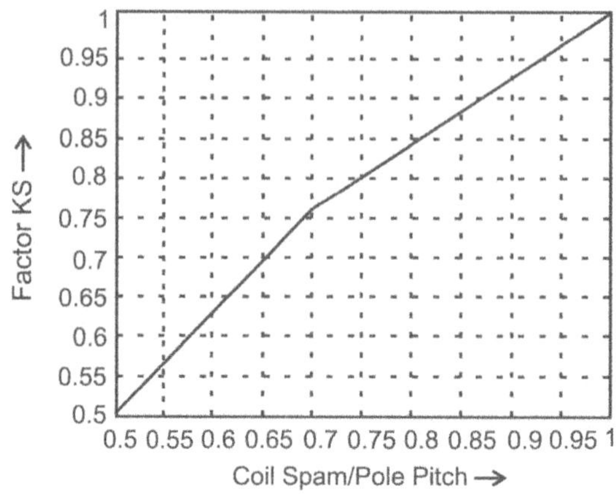

Fig. 6.1 Leakage factor.

$$L_0\lambda_0 = \frac{K_s \times \tau_p^{\ 2}}{\pi\tau_s} = \frac{1 \times (294.52/1000)^2}{\pi \times 32.725/1000} = 0.8437$$

($\tau_p \rightarrow$ Pole pitch and $\tau_s \rightarrow$ Slot Pitch and $K_s = 1$ from above graph (Fig. 6.1))

Overhang Leakage Flux (ϕ_0) $= 2 \times \sqrt{2} \times \mu_0 \times I_{Ph} \times Z_s \times L_0\lambda_0$

$$= 2 \times \sqrt{2} \times 4 \times \pi \times 10^{-7} \times 349.9 \times 4 \times 0.8437 = 0.0042$$

Total Leakage Flux $= \phi_L = \phi_S + \phi_0 = 0.0036 + 0.0042 = 0.0078$

Leakage Rectance $(X_L) = \dfrac{\phi_L}{\phi} = \dfrac{0.0078}{0.0936} = 0.0833$pu

Flux in core (ϕ_c) $= \dfrac{\phi}{2} = \dfrac{0.0936}{2} = 0.0468$ Wb

Depth of Core (dc) $= \dfrac{\phi_c}{B_c \times L_i} = \dfrac{0.468}{1.1 \times 522} = 81.5$ mm

Outer diameter of Core (D_0) = D + 2 × (Hs + dc)

$$= 1500 + 2 \times (45 + 81.5) = 1753 \approx 1760 \text{ mm (Ronded off)}$$

Coorected Depth of Core(dc) = (1760 – 1500)/2 – 45 = 85 mm

6.2.3 *(a) Computer Program in "C" in MATLAB for Part-2*

```
% (2) Design of Stator Winding, Reactance, Core
D = 1500; L = 670; FI = 0.0936; Iph = 349.9; PP = 294.52;
Zs = 4; Tph = 96; sp = 32.7249; % Specification
```

```
V = 3300; Ls = 580; Li = 522; nv = 9; bv = 10; Wins = 6;
Hins = 8; Hw = 4; HL = 1; % Specification
Ki = 0.9; cds = 4; PAtoPP = 0.7; Bt0 = 1.8; Bc = 1.1; Ks = 1;
% Assumptions
% Wc = 0.8; tins = 0.5; Hw = 5; HL = 2; InsIL = 5; Hins = 6;
InsM = 2; % Assumptions
As = Iph/cds; TpP = ceil (PAtoPP*PP/sp);
bt0 = FI*1e6/(Bt0*Li*TpP); Ws1 = sp-bt0;
Wc = ceil((Ws1)-Wins); Hc = ceil(As/Wc*2)/2; Ws = 1*Wc + Wins;
As=Wc*Hc*0.98; cds=Iph/As; Hs=Zs*Hc+ Zs*2*0.5+Hins+Hw+HL;
rat1 = Hs/Ws; Lmt = (2*L + 2.5*PP + 50*KV + 150)/1000;
Rph = 0.021*Lmt*Tph/As; Pcus = 3*Iph^2*Rph/1e3;
Wcus = 3*Lmt*As*Tph*8.9e-3; alfa = sqrt(Wc/Ws);
Kdav = 1 + (alfa*Hc/10)^4*Zs^2/9; Peddy = (Kdav-1)*Pcus;
Pts= (Pcus + Peddy)*1.15; Reff = Iph*Rph*Kdav/(KV/sqrt(3)*1e3);
<------------ Leakage Reactance --------->
h1 = Zs*Hc + 4 + 2 + 2; h2 = 2 + 2; h3 = Hw; h4 = HL; Ws0 = 9;
Lmdas = h1/(3*Ws) + h2/Ws + 2*h3/(Ws + Ws0) + h4/Ws0;
Fis = 2*sqrt(2)*4*pi*1e-7*Iph*Ls/1000*Lmdas*Zs;
CSbPP = [0.5 0.6 0.7 0.8 0.9 1];
ks = [0.5 0.63 0.76 0.84 0.92 1];
plot (CSbPP, ks);grid; xlabel ('Coil Span/Pole Pitch-->');
Ylabel ('Factor KS-->');
title ('Leakage Factor'); y1 = 1; KS = interp1 (CSbPP, ks, y1,
'spline');
L0Lmd0 = Ks*PP^2/(pi*sp*1000); T0 = Zs; FIo =
2*sqrt(2)*4*pi*1e-7*Iph*T0*L0Lmd0; FIL = FIs + FIo; XL =
FIL/FI; FIc = FI/2; dc1 = FIc*1e6/ (Bc*Li); Do1 = D + 2*(Hs +
dc1); Do = ceil(Do1/10)*10; dc = (Do - D)/2 - Hs;
```

6.2.4 Design of Rotor and Calculation of AT (Part – 3)

Armature reaction:

$$\text{AT/Pole}\,(\text{AT}_a) = \frac{1.35 \times \text{Tph} \times \text{Iph} \times \text{kw}}{P/2} = \frac{1.35 \times 96 \times 349.9 \times 0.955}{16/2} = 5413.3$$

No Load AT/Pole (ATf0) = scr × Ata = 0.9 × 5413.3 = 4872

ATg = 0.75 × ATf0 = 0.75 × 4872 = 3654

$$\text{AirGapFluxDensity}(B_g) = \frac{B_{av}}{\text{PoleArc/Polepitch}} = \frac{0.6089}{0.7} = 0.8699\text{T}$$

Assuming Air Gap Coefficient (Kg) = 1.17 (Varies from 1.12 to 1.18);

$$\text{Air Gap Length}\,(L_g) = \frac{\text{AT}_g}{0.796 \times B_g \times K_g \times 10^6} = 0.0045\,\text{m} = 4.5\text{mm}$$

$$\frac{\text{Air gap}}{\text{Pole pitch}} = \frac{0.0045}{0.2928} = 0.0154 \text{ (Usual value} = 0.012 \text{ to } 0.016)$$

Rotor dia (Dr) = D – 2 × Lg = 1500 – 2 × 4.5 = 1491mm

Axial Length of Pole (Lp) = L – 15 = 670 – 15 = 655 mm

Assuming Stacking factor = 0.9, Iron Length of Pole (Lpi) = 0.90 × 655 = 589.5 mm

Assuming Pole Leakage factor = 1.1,

Flux in the Pole body (Φ_p)

$$= 1.1 \times \Phi = 1.1 \times 0.0936 = 0.103\text{Wb}$$

Assuming flux density in Pole body (Bp) = 1.5T,

Area of Pole (Ap) = Φ_p / Bp = 0.103/1.5 = 0.0686 m^2

$$\text{Width of Pole (Wp)} = \frac{A_p}{L_{pi}} = \frac{0.0686 \times 10^6}{589.5} = 116.44 \text{ mm}$$

W poles = Ap × Lp × P × 7800 = 0.0686 × 655 × 16 × 7.8 = 5610.9 Kg

Depth of Field Winding (df)

Pole-Pitch (m)	0.1	0.2	0.3	0.4	0.5
Windg-Depth (m)	0.03	0.036	0.04	0.045	0.05

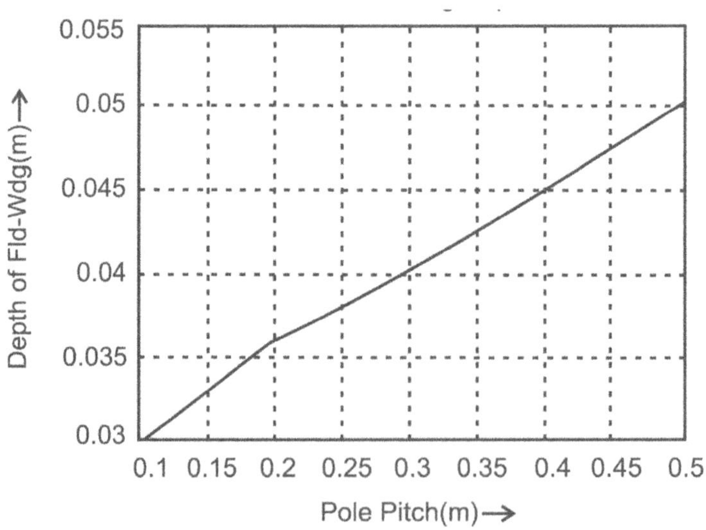

Fig. 6.2 Field winding depth.

From the Fig. 6.2, Depth of Field Winding (df) corresponding to pole pitch of 0.2928m → 0.0398

Assuming Full Load AT (IfTf) = 1.8 times ATa, IfTf = 1.8 × 5413.3 = 9743.9

Height of the Field Coil (hf)

$$= \frac{\text{IfTf}}{10^4\sqrt{\text{df}\times\text{Sf}\times\text{Pf}}} = \frac{9743.9}{10^4\sqrt{0.0398\times 0.75\times 725}} = 0.20955\text{ m} = 209.5\text{ mm}$$

Where Sf = Space Factor = 0.75 (assumed) and

Pf = Permissible Loss/m^2 = 725 (assumed)

Assuming Height of Pole shoe at the center as 4 mm and the space for insulation as 12% of Pole pitch,

Height of the Pole (Hp) = hf + 40 + 0.12 × Pole Pitch

$$= 209.5 + 40 + (0.12 \times 294.52) = 284.9\text{ mm}$$

Wrot = $\pi \times (\text{Dr} - 2 \times \text{Hp})^2/4 \times \text{L} \times 7800$

$$= \pi \times (1500 - 2 \times 284.9)^2/4 \times 670 \times 7.8 \times 10^{-6} = 3483.3\text{Kg}$$

Flux in Rotor Core (Φ_{cr}) = Flux in Pole Body /2 = 0.1030/2 = 0.0515 wb

Assuming Flux density in Rotor Core (Bcr) = 1.1 T,

$$\text{Depth of Rotor Core (dcr)} = \frac{\Phi_{cr}}{\text{Bcr}\times\text{L}} = \frac{0.0515\times 10^6}{1.1\times 670} = 69.851\text{ mm}$$

6.2.4 *(a) Computer Program in "C" in MATLAB for Part-3*

```
% (3) Pole Dimensions

KVA = 2000; V = 3300; N = 375; f = 50; P = 16; D = 1500;
L = 670; % Star connected Salient Pole Gen

Tph = 96; Iph = 349.9; Bav = 0.6089; FI = 0.09361;
sp = 32.7249; Ws = 19; % Specification

nv = 9; bv = 10; Hs = 45; S = 144; TpP = 7; PP = 294.52;
Li = 522; % Specification

kw = 0.955; scr = 0.9; Kg = 1.17; PAtoPP = 0.7; ki = 0.9;
Bp = 1.5; Sf = 0.75; Pf = 725; % Assumptions

Bcr = 1.1; CCs = 0.47; CCv = 0.4; Bc = 1.1; dc = 85; %
Assumptions

Ata = 1.35*Tph*Iph*kw/(P/2); ATf0 = scr*ATa; ATg = 0.75*ATf0;
Bg = Bav/PAtoPP; Lg1 = ATg/(0.796*Bg*Kg*1e3);
Lg = floor(Lg1*10)/10; rat2 = Lg/PP; Dr = D - 2*Lg;

Lp = L - 15; Lpi = Lp*ki; FIp = 1.1*FI; Ap = FIp/Bp;
Wp = Ap/Lpi*1e6; Wpoles = Ap*Lp*P*7.8;
```

```
% <--Table for Depth of Field Winding------------>
PPS = [0.1 0.2 0.3 0.4 0.5]; DF = [0.03 0.036 0.04 0.045 0.05];
% plot (PPS,DF); grid; xlabel ('Pole Pitch(m)-->');
ylabel('Depth of Fld-Wdg (m)-->');
% title ('Field Winding Depth');
df = interp1 (PPS, DF, PP/1e3, 'spline'); IfTf = 1.8*ATa;
hf = IfTf/(1e4*sqrt(df*Sf*Pf))*1000; Hp = hf + 40 + 0.12*PP;
dcr = FIp*1e6/2/Bcr/L;Wrot = pi*(Dr - 2*Hp)^2/4*L*7.8e-6;
```

6.2.5 Carter Coefficients and Ampere Turns for Various Parts of Magnetic Circuit (Part-4)

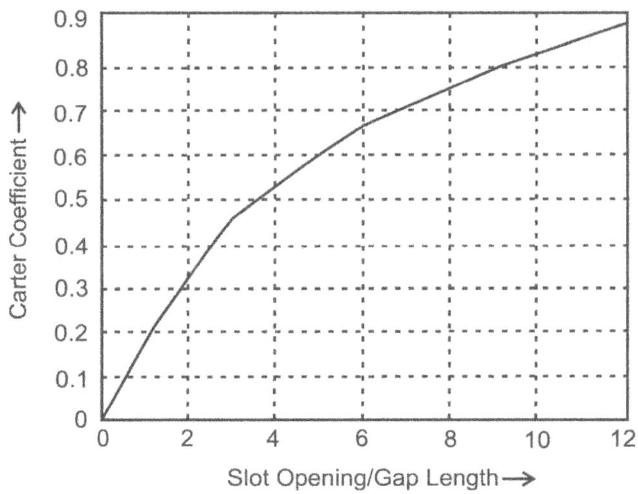

Fig. 6.3 Carters coefft for semi-closed slots.

$$\frac{\text{SlotWidth}}{\text{Airgap}} = \frac{19}{4.5} = 4.2222$$

Carters Coefficient read from graph (Fig. 6.3) corresponding to 4.2222 is 0.5459

$$\text{Slot Coefficient (Kgs)} = \frac{\text{Slot Pitch}}{\text{Slot Pitch} - \text{Slot width} \times \text{Carter Coefft}}$$

$$= \frac{32.725}{32.725 - 19 \times 0.5459} = 1.4641$$

$$\frac{\text{DuctWidth}}{\text{GapLength}} = \frac{10}{4.5} = 2.2222$$

Carters Coefficient (CCv) Read from Graph (Fig. 6.3) corresponding to 2.2222 is 0.3598

$$\text{Duct Coefft (Kgv)} = \frac{L}{L - nv \times bv \times CCv} = \frac{.670}{670 - 9 \times 10 \times 0.3598} = 1.0508$$

Air gap coefficient (kg) = Kgs × Kgv = 1.4641 × 1.0508 = 1.5384

(a) **Air gap AT** (ATg) = $0.796 \times Bg \times kg \times Lg \times 10^3$

$= 0.796 \times 0.8699 \times 1.5384 \times 4.5 \times 10^3 =$ **4793.4**

(b) **Stator Teeth**

Dia of stator teeth at ⅓ height from narrow end (D13)

= D + 2 × Hs/3 = 1500 + 2 × 45 / 3 = 1530 mm

$$\text{Slot Pitch at above dia (sp13)} = \frac{\pi \times D13}{S} = \frac{\pi \times 530}{144} = 33.38\text{ mm}$$

Width of tooth at above dia (bt13) = (sp13 – Ws) = (33.38 – 19) = 14.38 mm

$$\text{Flux density in the tooth (B13)} = \frac{\Phi}{Li \times bt13 \times TpP} = \frac{0.0936 \times 10^6}{522 \times 14.38 \times 7} = 1.7814\text{ T}$$

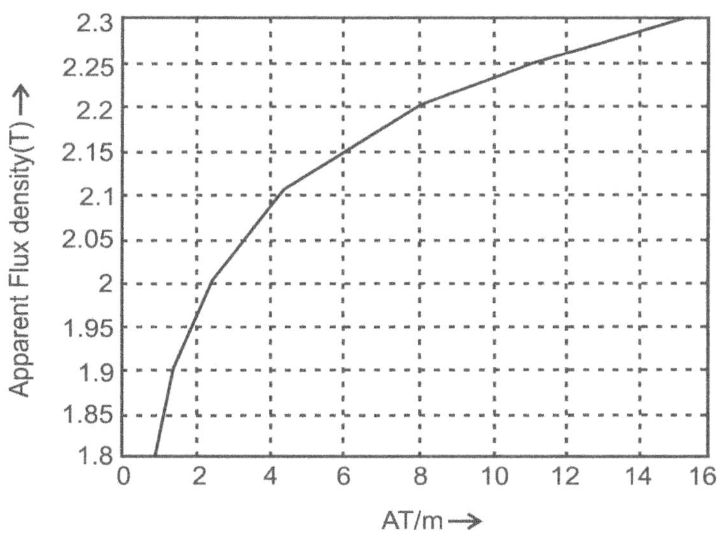

Fig. 6.4 Tooth Magnetization curve $K_s = 1$

AT/m for tooth corresponding to B13 = 1.7814T as read from magnetization curve (Fig. 6.4) = 8611.5

AT required for the **Stator tooth** (for tooth height of 45mm)

= ATt = 8611.5 × 45/1000 = **387.52**

(c) Stator Core

Length of Magnetic Path in the Stator Core (Lc)

$$= \frac{\pi(D + 2Hs + dc)}{2P} = \frac{\pi(1500 + 2\times 45 + 85)}{2\times 16} = 164.44 \text{ mm}$$

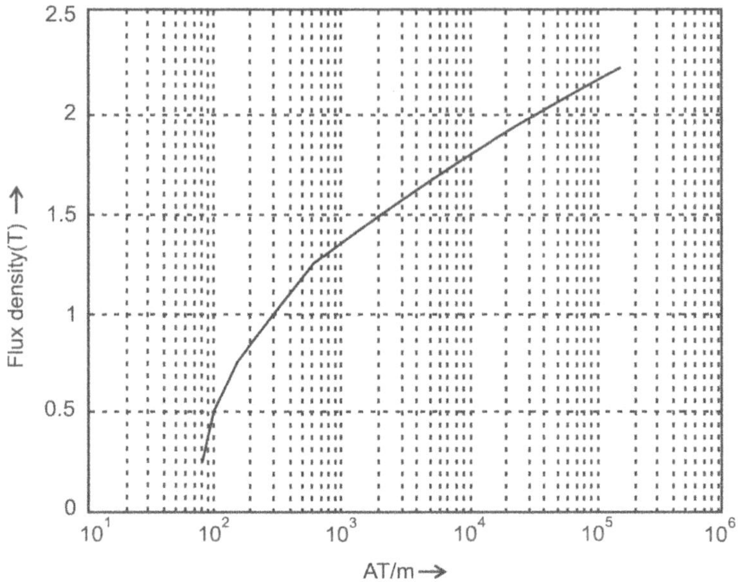

Fig. 6.5 Magnetization curve for dynamo stamping steel.

Assuming Flux density in the Core (Bc) = 1.1T, AT/m for Core corresponding to Bc = 1.1T as read from magnetization curve (Fig. 6.5) = 417.4

AT required for **Stator Core** = ATc = 417.4 × 164.44/1000 = **68.64**

(d) Pole

AT for Airgap + Stator(AT_L) = ATg + ATt + ATc = 4893.4 + 387.5 + 68.64 = 5249.6

Assuming Leakage between Pole Shoes/pole (Φ_{LS})

$= 19 \times 10^{-8} \times AT_L$ (Varies from17 × 10^{-8} to 20 × 10^{-8} times)

$\Phi_{LS} = 19 \times 10^{-8} \times 5249.6 = 0.001$ Wb

Assuming Leakage between Poles/pole (Φ_{Lp})

$= 72 \times 10^{-8} \times AT_L$ (Varies from70 × 10^{-8} to 80 × 10^{-8} times)

$\Phi_{LP} = 72 \times 10^{-8} \times 5249.6 = 0.0038$ Wb

Flux at top of pole (Φ_{pmin}) = $\Phi + \Phi_{LS}$ = 0.0936 + 0.001 = 0.0946 Wb

Flux at bottom of pole (Φ_{pmax}) = $\Phi_{pmin} + \Phi_{Lp}$ = 0.0946 + 0.0038 = 0.0984 Wb

Min Flux Density in the pole (B_{pmin}) = Φ_{pmin}/Ap = 0.0946/0.0686 = 1.3782 T

Max Flux Density in the pole (B_{pmax}) = Φ_{pmax}/Ap = 0.0984/0.0686 = 1.4332 T

AT/m corresponding to 1.3782 T from magnetization curve (Fig.6.5)

$$= \text{atp1} = 815.2$$

AT/m corresponding to 1.4332 T from magnetization curve (Fig.6.5)

$$= \text{atp2} = 1172$$

$$\text{AT for Pole} = \left[\frac{2}{3} \times \text{atp1} \times \text{Hp}\right] + \left[\frac{1}{3} \times \text{atp2} \times \text{Hp}\right]$$

$$= \left[\frac{2}{3} \times 815.2 \times \frac{284.9}{1000}\right] + \left[\frac{1}{3} \times 1172 \times \frac{284.9}{1000}\right] = 266.1$$

(e) Rotor Core

Length of Magnetic Path in the Rotor Core (Lcr)

$$= \frac{\pi(\text{Dr} - 2\text{Hp} - \text{dcr})}{2\text{P}} = \frac{\pi(1491 - 2 \times 284.9 - 69.85)}{2 \times 16} = 83.583 \text{ mm}$$

Assuming Flux density in the Rotor Core (Bcr) = 1.1T, AT/m for Core corresponding to Bc = 1.1T as read from magnetization curve (Fig.6.5) = 417.45

AT required for **Rotor Core** = ATrc = 417.45 × 83.583/1000 = **34.89**

Total No-Load AT for the Magnetic Circuit at 100% rated Voltage (ATf0)

$$= \text{ATg} + \text{ATt} + \text{ATc} + \text{ATp} + \text{ATrc}$$

$$= 4793.4 + 387.5 + 68.64 + 266.1 + 34.89 = \mathbf{5550.6}$$

6.2.5 *(a) Computer Program in "C" in MATLAB for Part-4*

```
%-------------------Ampere-Turns at 100% V-------------------->

%<---B-H Curve for St Tooth for ks=1 ---->

BtS = [1.8  1.9   2   2.1  2.15   2b    2.25   2.3];

attS  =  [9e3  14e3  24e3  43e3    60e3    80e3  110e3  152e3];
% ks = 1.0

% plot (attS, BtS); grid; xlabel ('AT/m-->'); ylabel ('Apparent
Flux density(T)-->');

% title ('Tooth Magnetization Curve for Ks = 1');

% ---- Carters Coefft for Semi-closed slots--->

W0byLgS = [0  1   2   3   4   5  6   7   8   9   10  11  12];
% Slot opening/Gap

CC = [0 .18 .33 .45 .53 .6 .66 .71 .75 .79 .82 .86 .89];

% plot (W0byLgS, CC); grid; xlabel ('Slot Opening/Gap Length--
>'); Ylabel ('Carter Coefficient-->');

% title ('Carters Coefft for Semi-closed slots');

% <---B-H Curve common for St Core, Rt core and Poles---->
```

```
BS = [0 0.25 0.5 0.75  1.0 1.25 1.5 1.75 2.0 2.25];

atS = [0  80  100  150  300   580  2e3  8e3 3e4 16e4];

%   semilogx   (atS,   BS);   grid;   xlabel   ('AT/m-->');
ylabel ('Flux density(T)-->');

% title ('Magnetization Curve for DynamoStamping Steel');

WsbyLg = Ws/Lg; CCg = interp1 (W0byLgS, CC, WsbyLg, 'spline');
kgs = sp/(sp-Ws*CCg);

DbyG = bv/Lg; CCv = interp1 (W0byLgS, CC, DbyG, 'spline');

kgv = L/(L-nv*bv*CCv); kg = kgs*kgv;

%-----------------------------------------------------------

ATg = 0.796*Bg*kg*Lg*1e3; D13 = D + 2*Hs/3; sp13 = pi*D13/S;
bt13    =    sp13-Ws;    B13    =    FI/(Li*bt13*TpP)*1e6;
att = interp1(BtS, attS, B13, 'spline');

ATt = att*Hs/1000; Lc = pi*(D + 2*Hs + dc)/P/2;

tc  =  interp1(BS,  atS,  Bc,  'spline');  ATc  =  atc*Lc/1000;
ATL = ATg + ATt + ATc; FILs = 19e-8*ATL; FILp = 72e-8*ATL;
FIpmin = FI + FILs; FIpmax = FIpmin + FILp; Bpmin = FIpmin/Ap;
atp1 = interp1(BS, atS, Bpmin, 'spline');   Bpmax = FIpmax/Ap;

atp2 = interp1 (BS, atS, Bpmax, 'spline');

ATp=(2/3*atp1+1/3*atp2)*Hp/1e3; Lcr = pi*(Dr - 2*Hp - dcr)/P/2;

atrc = interp1 (BS, atS, Bcr, 'spline');

ATrc = atrc*Lcr/1000;     ATf0 = ATg + ATt + ATc + ATp + ATrc;

<-----------------------------------------------------
```

6.2.6 Open Circuit Characteristic Curve (OCC) (Part-5)

A plot of Ampere Turns on X-Axis and Various Terminal voltages (0 % to 130% of rated Voltage).

Procedure for Plotting Open Circuit Characteristic Curve (OCC)

1. Consider Terminal Voltages of 100%, 110%, 120% and 130% of rated voltage.
2. Fill up all the details pertaining to Air-Gap, Stator tooth, Stator Core, Poles and Rotor Core calculated above under the 100% Voltage Column in the table.
3. Calculate the Flux densities in various parts at other 3 Voltages which are linearly proportional to Voltage (Ex:-If Flux density = 1.1T at 100% Volts, then at 110%Volts it will be (110/100) × 1.1 = 1.21T).
4. At these values of Flux densities measure the "**AT/m**" from appropriate curves as was referred for calculation for 100%V).
5. Calculate Values of "**AT**" for each part by directly multiplying the "**AT/m**" value with corresponding "**length**" of each part.

6. Add values of "AT" of all the parts under each % Voltage Colum to get the **Total AT** required for each voltage.
7. Plot Values of AT on X-Axis and % Rated Voltages on Y-Axis to get **OCC.**

Open Circuit Characteristic Curve (OCC) of Salient Pole Generator

Part	Length	100%(	3300	Volts)	110%(	3630	Volts)	120%(	3960	Volts)	130%(	4290	Volts)
	m	B(tesla)	AT/m	AT	B(tesla)	AT/m	AT	B(tesla)	AT/m	AT	B(tesla)	AT/m	AT
Air-Gap	0.0045	0.8699		4793.4	0.95689		5272.7	1.04388		5752.1	1.13087		6231.42
Stator Teeth	0.045	1.7816	8610	387	1.95976	19370	871.7	2.13792	55650	2504	2.31608	168080	7564
Stator-Core	0.16444	1.1	417.4	69	1.21	545.2	89.7	1.32	647.5	106.5	1.43	1144.3	188.2
Pole(min)		1.3782	815		1.51602	2278.2		1.65384	5563.4		1.79166	8963.4	
	0.28489			266.1			772.4			1748.1			2823.5
Pole(max)		1.4332	1172		1.57652	3577		1.71984	7282		1.86316	11805	
Rot-Core	0.08358	1.1	417.4	35	1.21	545.2	45.6	1.32	647.5	54.1	1.43	1144.3	95.6
Total-AT=				**5550.5**			**7052**			**10165**			**16903**
Field-Amps	**(OCC)**			**58.4**			**74.2**			**107**			**177.9**

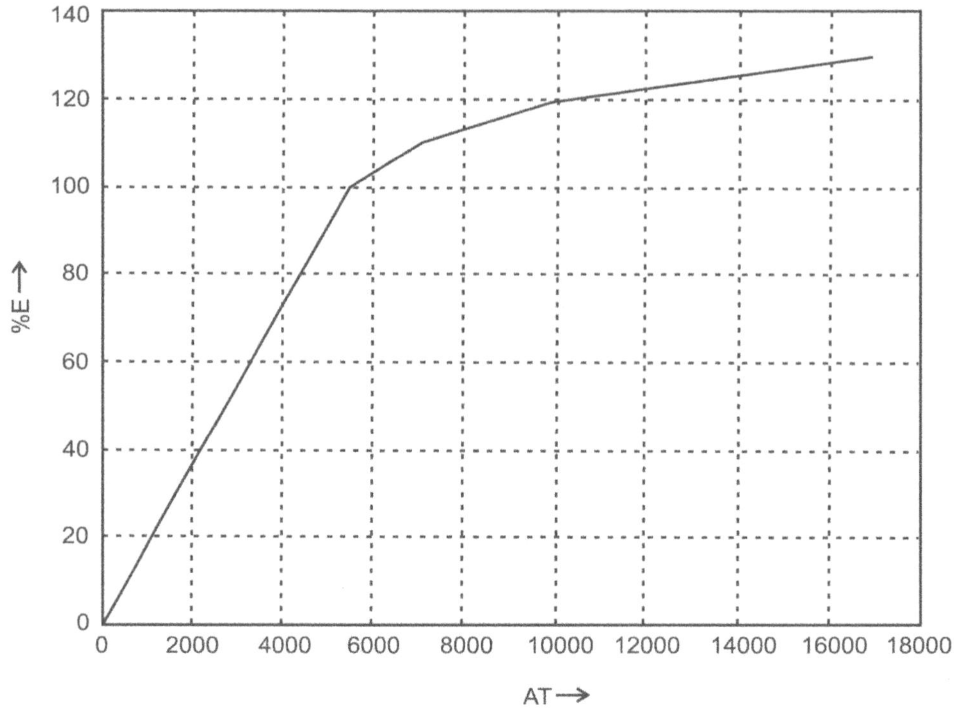

Fig. 6.6 OCC of salient pole generator.

6.2.6 *(a) Computer Program in "C" in MATLAB for Part-5*

```
% (4) OCC-> Magnetization Curve of Salient Pole TG
f2 = fopen ('OCC Output.m', 'w');
fprintf (f2,'OCC Table:-\n');
fprintf (f2,' E A-Gap  St-T  St-C  Pole Rt-C TotAT\n');
fprintf (f2,'---  ----  ----  ----  ----  ----   ----- \n');
Ws = 19; Lg = 4.5; Bg = 0.8699; Hs = 45; B13 = 1.7816;
Bc = 1.1; Lc = 164.44; % Specification
sp = 32.725; L = 670; nv = 9; bv = 10; % Specification
Bpmin = 1.3782; Bpmax = 1.4332; Hp = 284.9; Lcr = 83.58;
Bcr = 1.1; % Specification
E = [100 110 120 130]; EE (1) = 0; ATT (1) = 0;
%<---B-H Curve for St Tooth for ks=1)----->
BtS = [1.8  1.9   2   2.1  2.15   2.2  2.25   2.3];
attS = [9e3  14e3  24e3  43e3   60e3   80e3  110e3  152e3];
% ks = 1.0
%----Carters Coefft for Semi-closed slots--->
```

```
W0byLgS = [0  1   2   3   4   5  6   7   8   9   10  11  12];
% Slot opening/Gap

CC =  [0 .18 .33 .45 .53 .6 .66 .71 .75 .79 .82 .86 .89];

% <--- B-H Curve common for St Core, Rt core and Poles----->

BS = [0 0.25 0.5 0.75  1.0  1.25  1.5 1.75 2.0 2.25];

atS = [0  80  100  150  300   580  2e3  8e3 3e4 16e4];

WsbyLg = Ws/Lg; CCg = interp1 (W0byLgS, CC, WsbyLg, 'spline');
kgs = sp/(sp - Ws*CCg);

DbyG  =  bv/Lg;  CCv  =  interp1  (W0byLgS,  CC,  DbyG,  'spline');
kgv = L/(L -nv*bv*CCv); kg = kgs*kgv;

for i = 1:4;

Bgx(i) = Bg*E(i)/100; ATgx(i) = 0.796*Bgx(i)*kg*Lg*1e3;

Bstx(i) = B13*E(i)/100;

atST(i) = interp1(BtS, attS, Bstx(i), 'spline');

ATstx(i) = atST(i)*Hs/1000;

Bscx(i) = Bc*E(i)/100;

atSC(i)=interp1(BS,atS,Bscx(i),'spline');
ATscx(i) = atSC(i)*Lc/1000;

Bpminx(i) = Bpmin*E(i)/100;

atP1(i) = interp1(BS, atS, Bpminx(i), 'spline');

Bpmaxx(i) = Bpmax*E(i)/100;

atP2(i) = interp1(BS, atS, Bpmaxx(i), 'spline');

ATpx(i) = (2/3*atP1(i) + 1/3*atP2(i))*Hp/1000;

Brcx(i) = Bcr*E(i)/100;

atRC(i) = interp1(BS, atS, Brcx(i), 'spline');

ATrcx(i) = atRC(i)*Lcr/1000;

ATTx(i) = ATgx(i) + ATstx(i) + ATscx(i) + ATpx(i) + ATrcx(i);
ATT(i + 1) = ATTx(i); EE(i + 1) = E(i);

Fprintf  (f2,  '%3.0  f%  6.0  f%  6.0  f%  6.0  f%  6.0  f%6.0  f%  8.0
f\n',  E(i),  ATgx(i),  ATstx(i),  ATscx(i),  ATpx(i),  ATrcx(i),
ATTx(i)); end; ATTx,

Fprintf (f2,'---  ----  ----  ----  ----  ----   ----- \n');

plot  (ATT,EE);  grid;  xlabel  ('AT-->');  ylabel  ('%E-->');
title ('OCC of Salient Pole Generator');

fclose (f2);
```

6.2.7 Field Current at Rated Load and PF (Part-6)

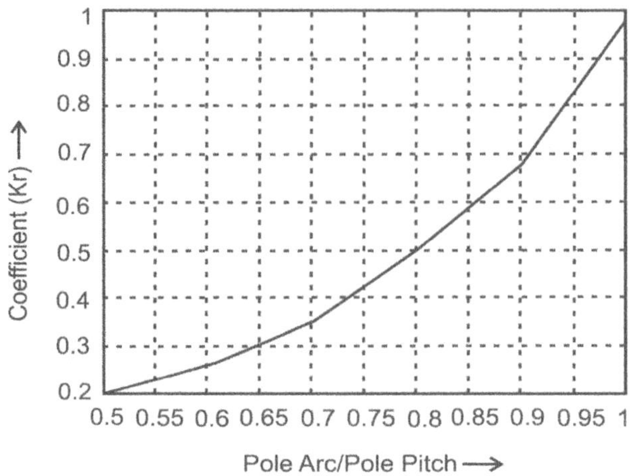

Fig. 6.7 Relative permeances.

From Fig. 6.7, Coefficient corresponding to Pole Arc/Pole Pitch = 0.7, Kr = 0.35

$$\text{PFangle} = \phi = \cos^{-1}(0.85) = 0.5548 \text{ rad}$$

$$V_{ph} = \frac{V}{\sqrt{3}} = \frac{3300}{\sqrt{3}} = 1905.3 \text{ V}$$

$$Z_{pu} = \frac{V_{ph}}{I_{ph}} = 5.4451\,\Omega$$

$$X = Z_{pu} \times X_L = 5.4451 \times 0.083 = 0.4519\,\Omega$$

$$E_g = \sqrt{(\text{Vph} \times \text{PF} + \text{Iph} \times \text{Rph})^2 + (\text{Vph} \times \sin[\phi] + \text{Iph} \times \text{X})^2}$$

$$E_g = \sqrt{(1905.3 \times 0.85 + 349.9 \times 0.0541)^2 + (1905.3 \times \sin[0.5548] + 349.9 \times 0.4519)^2} = 2008.5\text{V}$$

$$\text{Egpu} = \frac{\text{Eg}}{\text{Vph}} = \frac{2008.5}{1905.3} = 1.0542\text{pu} = 105.42\%$$

AmpereTurns corresponding to 105.42% Volts read from OCC(ATx)=6224 AT

$$AT_1 = \sqrt{(\text{ATx} \times \text{PF})^2 + (\text{ATx} \times \sin[\phi] + \text{Kr} \times \text{ATa})^2}$$

$$AT_1 = \sqrt{(6224 \times 0.85)^2 + (6224 \times \sin[0.5548] + 0.35 \times 5413.3)^2} = 7400 \text{ AT}$$

$$\text{AA} = \sin^{-1}(\text{ATx} \times \sin[\pi/2 + \phi]/AT_1) = \sin^{-1}(6224 \times \sin[\pi/2 + 0.5548]/7400) = 0.7966\text{rad}$$

$$\text{AT2} = (1 - \text{Kr}) \times \text{ATa} \times \cos(\text{AA}) = (1 - 0.35) \times 5413.3 \times \cos(0.7966) = 2460 \text{ AT}$$

AT at Full Load and Rated PF (ATFL) = AT1 + AT2 = 7400 + 2460 = 9860 AT

Voltage corresponding to AT = 9860 as read from OCC (E_0) = 119.46%

$$\text{Voltage Regulation}=\frac{E_0 - V}{V}=\frac{119.46-100}{100}=19.46\%.$$

6.2.7 *(a) Computer Program in "C" in MATLAB for Part-6*

```
E = [0 100 110 120 130]; ATT = [0 5551 7052 10165 16902];
V = 3300; pf = 0.85; % Specification

Iph = 349.9; Rph = 0.0541; XL = 0.083; ATa = 5413.3;
PAbyPP = 0.7; % Specification

% <---Kr)---->

PAbyPPS = [0.5 0.6 0.7 0.8 0.9 1.0]; % Pole Arc by PP Ratio

KrS = [0.2 0.26 0.35 0.5 0.68 0.98];

Kr = interp1(PAbyPPS, KrS, PAbyPP, 'spline');

A = acos(pf); Eph = V/sqrt(3); Zpu = Eph/Iph; X = Zpu*XL;

Eg = sqrt((Eph*cos(A) + Iph*Rph)^2 + (Eph*sin(A) + Iph*X)^2);

Egpu = Eg/Eph; ATx = interp1(E, ATT, Egpu*100, 'spline');
AT1 = sqrt ((ATx*cos(A))^2 + (ATx*sin(A) + Kr*ATa)^2);
AA = asin(ATx*sin(pi/2 + A)/AT1);

AT2 = (1 - Kr)*ATa*cos(AA); ATFL = AT1 + AT2;
E0 = interp1(ATT,E,ATFL,'spline'); reg=(E0-E(2))/E0*100;
plot(PAbyPPS,KrS); grid; xlabel('PoleArc/PolePitch-->');
ylabel ('Coefficient(Kr)-->'); title('Relative Permeances');
```

6.2.8 Field Winding Design + Losses + Efficiency (Part-7)

Assuming Exciter Voltage (Vex) = 220 V, Voltage/Field Coil = 0.8 × 220/16 = 11V

Assuming thickness of insulation on Pole (ti) = 0.01m,

Mean Length of turn of field coil (Lmtf) = 2 × (Lp + Wp) + pi × (df + 2 × ti)

$$= 2 \times (655 + 116.45)/1000 + \pi \times (0.0397 + 2 \times 0.01) = 1.7308 \text{ m}$$

Area of CS of field conductor (af)

$$=\frac{\rho \times \text{Lmtf} \times \text{IfTf}}{\text{Vc}}=\frac{0.02\times 1.7308\times 9860}{11}=31.03\text{ mm}^2$$

Selecting Conductor Thk (Tcf) = 1.9mm, Width of Conductor (Wcf)

$$=\frac{\text{af}}{\text{Tcf}}=\frac{31.03}{1.9}=16.45\cong 17\text{ mm}$$

Corrected Area of CS of field conductor (af) = Wcf × Tcf × 0.98

$= 17 \times 1.9 \times 0.98 = 31.654\ \text{mm}^2$

(assuming a factor of 0.98 for conductor edges rounding)

Assuming current density in the field winding (cdfw) as 3.3 A/ mm^2

Field Current (If) = af × cdfw = 31.654 × 3.3 = 104.46 A

No. of Field Winding Turns (Tf) = $\dfrac{\text{IfTf}}{\text{If}} = \dfrac{9860}{104.46} = 94.39 \cong 95$

Resistance of Field Winding (Rf) = $\dfrac{\rho \times \text{Lmtf} \times \text{Tf}}{\text{af}} = \dfrac{0.02 \times 1.7308 \times 95}{31.654} = 0.1039\ \Omega$

Field Winding Copper Loss (Pcuf) = $I_f^2 \times \text{Rf} \times \text{P} = 104.46^2 \times 0.1039 \times 16 = 18137$ W

Brush loss at slip rings (Pbr) = Vb × If = 2 × 104.46 = 208.9 W

(assuming Voltage drop across a pair of brushes = 2 × 1 = 2 V)

Total field copper losses (PcufT) = (Pcuf + Pbr)/1000

= (18137 + 208.9)/1000 = 18.346 KW

Weight of Field Copper (Wcuf) = Lmtf × af × Tf × P × 8.9/1000

= 1.7308 × 31.654 × 95 × 16 × 8.9/1000 = 741.14 Kg

Assuming Exciter Efficiency = 98%, Input to Exciter = 18.346/0.98 = 20.85 KW

Exciter Losses (Pex) = 20.85 – 18.346 = 2.5 KW

Assuming friction and windage Loss (Pm) as 1% of output

= 0.01 × KVA × pf = 0.01 × 2000 × 0.85 = 17 KW

Dia at centre of St tooth (D12) = D + 2 × Hs/2 = 1500 + 45 = 1545 mm

Slot pitch at D12(sp12) = π × D12/S = π × 1545/144 = 33.707 m

With of tooth at D12 (bt12) = 33.707 – 19 = 14.707 mm

Tooth (KgT) = 7800 × S × bt12 × Li × Hs

= 7800 × 144 × 14.707 × 522 × 45 = 375.5 Kg

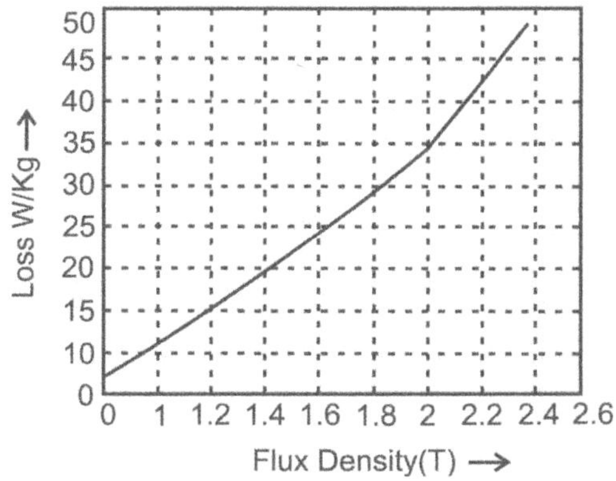

Fig. 6.8 Core loss for 0.5 mm stampings.

From Fig. 6.8, Iron loss in tooth / Kg at B13 = 1.7942T (PitpKg) = 28.2 W/Kg

Iron Losses in teeth (Pit) = PitpKg × KgT = 28.2 × 388/1000 = 10.94 Kw

Mean Dia of Stator Core (Dmc) = D + (2 × Hs + dc)

= 1500 + (2 × 45) + 85 = 1675 mm

Weight of Core(KgC) = 7800 × 10^{-6} × (dc × Li) × (pi × Dmc)

= 7800 × 10^{-6} × (85 × 522) × (π × 1675) = 1821.2 Kg

From Fig. 6.8, Iron loss in Core / Kgat Bc = 1.1T (PicpKg) = 12.8574 W/Kg

Iron Losses in Core (Pic) = 12.8574 × 1821.2/1000 = 23.42 Kw

Total Iron Loss (Pi) = Pit + Pic = 10.94 + 23.42 = 34.36 Kw

Total Losses ar Rated Load (PT) = Pts + Pi + PcufT + Pex + Pm

= 27.427 + 34.36 + 18.346 + 2.5 + 17 = 99.6 KW

Rated Output (KW) = KVA × pf = 2000 × 0.85 = 1700 KW

Input = Output + Losses = 1700 + 99.6 = 1799.6 KW

$$\textbf{Efficiency} = \frac{\text{Output}}{\text{Input}} \times 100 = \frac{1700}{1799.6} \times 100 = 94.47\%$$

6.2.8 *(a) Computer Program in "C" in MATLAB for (Part-7)*

```
% (6) Field Winding Design + Losses + Efficiency

KVA = 2000; pf = 0.85; P = 16; Lp = 655; Wp = 116.45;
df = 0.0398; IfTf = 9860; Li = 522; % Specification

D = 1500; Hs = 45; S = 144; B13 = 1.7816; dc = 85; Bc = 1.1;
Pts = 27.4; bt0 = 14.2326; % Specification

Vex = 220; ti = 0.01; Tcf = 1.9; cdfw = 3.3; efEx = 0.88;
% Assumption
```

```
B = [0.8 1.2 1.6 2 2.4]; WpKg = [7 15 24 34 50];
Vc = 0.8*Vex/P; Lmtf = 2*(Lp + Wp)/1000 + pi*(df + 2*ti);
af = 0.02*Lmtf*IfTf/Vc; Wcf = ceil(af/Tcf); af = Wcf*Tcf*0.98;
If = af*cdfw; Tf = ceil(IfTf/If); Rf = 0.02*Lmtf*Tf/af;
Pcuf = If^2*Rf*P; Wcuf = Lmtf*af*Tf*P*8.9e-3; Pbr = 2*1*If;
PcufT = (Pcuf + Pbr)/1e3; InpEx = PcufT/efEx;
Pex = InpEx - PcufT; Pm = 0.01*KVA*pf; D12 = D + 2*Hs/2;
sp12 = pi*D12/S; bt12 = sp12-Ws;

KgT = 7.8e - 6*S*bt12*Li*Hs;

PitpKg = interp1 (B, WpKg, B13, 'spline');Pit = PitpKg*KgT/1e3;

Dmc = D + 2*Hs + dc; KgC = 7.8e-6*(dc*Li)*(pi*Dmc);

PicpKg = interp1 (B, WpKg, Bc, 'spline'); Pic = PicpKg*KgC/1e3;
Pi = Pit + Pic; PT = Pts + Pi + PcufT + Pex + Pm; Po = KVA*pf;
EFF = Po/(Po + PT)*100; plot (B, WpKg); grid;
xlabel ('Flux Density(T)-->'); ylabel ('Loss(W/Kg)-->');
title ('Core Loss for 0.5mm Stampings');

% ------------------------------------------------------------
```

6.2.9 Calculation of Temp-Rises and Weights (Part-8)

Copper Losses in the slot portion (Pcust)

$$= \frac{Ls}{Lmt/2} \times Pcus = \frac{580}{2.3913/2} \times 19.8548 = 9.6314 \text{ KW}$$

Total Losses dissipated by stator core surface (Pscs)

$$= Pcust + Pi + Peddy = 9.6314 + 34.356 + 3.9675 = 47.955\text{KW}$$

(a) Area of outer cylindrical surface and End Surface (A1)

$$= (\pi \times D_0 \times L) + 2 \times \frac{\pi}{4}\left(D_0^{\,2} - D^2\right)$$

$$= ((\pi \times 1760 \times 670) + 2 \times \frac{\pi}{4}\left(1760^2 - 1500^2\right))/10^6 = 5.036 \text{ m}^2$$

Assuming Cooling Coeftt (cc1) = 0.032,

Dissipation Coefficient (WpC1) = A1/cc1 = 5.036/0.032 = 157.37 W/°C

(b) Area of inner cylindrical surface (A2)

$$= \pi \times D \times L \times 10^{-6} = \pi \times 1500 \times 670 \times 10^{-6} = 3.1573 \text{ m}^2$$

$$\text{Cooling Coeftt (cc2)} = \frac{0.031}{1 + (0.1 \times Vr)} = \frac{0.031}{1 + (0.1 \times 29.45)} = 0.0079$$

Dissipation Coefficient (WpC2) = A2/cc2 = 3.1573/0.0079 = 401.79 W/°C

(c) Duct surface (A3)

$$= \frac{\pi}{4}\left(D_0^{\ 2} - D^2\right) \times nv \times 10^{-6} = \frac{\pi}{4}\left(1760^2 - 1500^2\right) \times 9 \times 10^{-6} = 5.9913 m^2$$

$$\text{Cooling Coeftt (cc3)} = \frac{0.11}{0.1 \times Vr} = \frac{0.11}{0.1 \times 29.45} = 0.0374$$

Dissipation Coefficient (WpC3) = A3/cc3 = 5.9913/0.0374 = 160.4 W/°C

Total Dissipation coefft (WpCt)

= WpC1 + WpC2 + WpC3 = 157.37 + 401.79 + 160.4 = 719.57 W/°C

Stator Temp Rise (TrS) = Pscs × 1000/WpCt

= 47.955 × 1000/719.57 = **66.64** °C

Cooling Coeft of Field Winding (ccf)

$$= \frac{0.1}{1 + (0.1 \times Vr)} = \frac{0.1}{1 + (0.1 \times 29.45)} = 0.0253$$

Field Coil Surface (cs) = (2 × Lmtf × hf/1000) + (2 × Lmtf × df)

= (2 × 1.7307 × 209.547/1000) + (2 × 1.7307 × 0.0398) = 0.8631m^2

Dissipation Coefficient (WpCf) = cs/ccf = 0.8631/0.0253 = 34.05 W/°C

Field Wdg Temp Rise (Tr) = Pcuf/P/WpCf = 18136/16/34.05 = **33.29** °C

Total Weights of Acive Materials

Wt of Total Copper (Wcu) = Wcus + Wcuf = (546.62 + 741.1) = 1287.7 Kg

Wt of Active Iron (Wiron) = KgC + KgT + Wpoles + Wrot

= 1821.2 + 388 + 5611.6 + 3483.3 = 11304 Kg

Assuming 1 % weight for Insulation,

Weight of Active materials (Wtot)

= 1.01 × (Wcu + Wiron) = 1.01 × (1287.7 + 11304) = 12718 Kg

Weight/KVA (KgPkva) = Wtot/KVA = 12718/2000 = **6.36** Kg/KVA

6.2.9 *(a) Computer Program in "C" in MATLAB for Part-8*

```
% (7) Temp-Rise and Weights

KVA = 2000; Ls = 580; Lmt = 2.3913; Pcus = 19.8548;
Pi = 34.356; Peddy = 3.9675; % Specification

Do = 1760; L = 670; D = 1500; % Specification

P = 16; Lmtf = 1.7307; hf = 209. 547; df = 0.0398; cc1 = 0.032;
Vr = 29.45; nv = 9; Pcuf = 18136; % Specification
```

```
KgT = 388; KgC = 1821.2; Wcus = 546.62; Wpoles = 5611.6;
Wcuf = 741.1; Wrot = 3483.3; % Specification

Pcust = Ls/1000/(Lmt/2)*Pcus; Pscs = Pcust + Pi + Peddy;
A1 = (pi*Do*L + 2*pi/4*(Do^2 - D^2))/1e6; WpC1 = A1/cc1;
A2 = pi*D*L/1e6; cc2 = 0.031/(1 + 0.1*Vr); WpC2 = A2/cc2;

A3 = (pi/4*(Do^2 - D^2)*nv)/1e6; cc3 = 0.11/(0.1*Vr);
WpC3 = A3/cc3; WpCt = WpC1 + WpC2 + WpC3; TrS = Pscs*1e3/WpCt;
ccf = 0.1/(1 + 0.1*Vr); cs = 2*Lmtf*hf/1000 + 2*Lmtf*df;
WpCf = cs/ccf; TrF = Pcuf/P/WpCf;

%-----Total Weights------>

Wcu = Wcus + Wcuf; Wiron = KgC + KgT + Wpoles + Wrot; Wtot =
1.01*(Wcu + Wiron); KgPkva = Wtot/KVA;

%--------------------------------------------------------
```

6.2.10 Computer Program in “C” in MATLAB for Complete Design

```
% Salient Pole Alternator

KVA = 2000; V = 3300; N = 375; f = 50; pf = 0.85;
% Star connected Salient Pole Gen

KVAS = [ 100  200  500 1000 5000 10000 20000];

BavS = [0.52 0.54 0.56 0.58 0.61  0.63  0.65];

qS = [20e3 23e3 26e3 29e3 34e3 37e3 40e3];
% plot (KVAS, BavS); grid;

Bav = interp1(KVAS, BavS, KVA, 'spline');

q = interp1 (KVAS, qS, KVA, 'spline');

Kw = 0.955; LbyPP = 2.3; nv = 9; bv = 10; ki = 0.9; spp = 3;
PAbyPP = 0.7; % Assumptions Pole Arc = L;

cds = 4; PAtoPP = 0.7; Bt0 = 1.8; Bc = 1.1; Wins = 6;
Hins = 8; Hw = 4; HL = 1; Ks = 1; % Assumptions

%---------------------------------------------

KW = KVA*pf; P = 120*f/N; K = 11*Bav*q*Kw*1e-3; n = 2*f/P;
DsqL = KVA/(K*n); D1 = (DsqL*P/(LbyPP*pi))^(1/3)*1000;
D = ceil(D1/10)/100; L1 = DsqL/D^2; L = ceil(L1*100)*10;
D = D*1000; Ls = (L - nv*bv); Li = Ls*ki; PP =p i*D/P;

Vr = pi*D*n/1000; flux1 = Bav*pi*D*L/P/1e6;

Tphi=V/sqrt(3)/4.44*f*Kw*flux1); Zph = Tphi*2; S = spp*P*3;
sp= pi*D/S; Zsi = Zph*3/S; Zs = ceil(Zsi/2)*2; Tph = Zs*S/6;
```

```
Iph = KVA*1e3/(1.7321*V); SLoad = Iph*Zs;

FI = V/sqrt(3)/(4.44*f*Kw*Tph);

% ----------------------------------------------------------

% (2) Design of Stator Winding, Reactance, Core-------→

As = Iph/cds; TpP = ceil(PAtoPP*PP/sp);

bt0 = FI*1e6/(Bt0*Li*TpP); Ws1 = sp-bt0;

Wc = ceil((Ws1)-Wins); Hc = ceil(As/Wc*2)/2;

Ws = 1*Wc + Wins; As = Wc*Hc*0.98; cds = Iph/As;

Hs = Zs*Hc + Zs*2*0.5 + Hins + Hw + HL; rat1 = Hs/Ws;

Lmt = (2*L + 2.5*PP + 50*V/1000 + 150)/1000;

Rph = 0.021*Lmt*Tph/As; Pcus = 3*Iph^2*Rph/1e3;

Wcus    =    3*Lmt*As*Tph*8.9e-3;    alfa    =    sqrt(Wc/Ws);
Kdav = 1 + (alfa*Hc/10) ^4*Zs^2/9;

Peddy = (Kdav - 1)*Pcus; Pts = (Pcus + Peddy)*1.15;

Reff = Iph*Rph*Kdav/(V/sqrt(3));

%------- Leakage Reactance --------->

h1=Zs*Hc + 4 + 2 + 2; h2 = 2 + 2; h3 = Hw; h4 = HL; Ws0 = 9;

Lmdas = h1/(3*Ws) + h2/Ws + 2*h3/(Ws + Ws0) + h4/Ws0;

Fis = 2*sqrt(2)*4*pi*1e-7*Iph*Ls/1000*Lmdas*Zs;

CSbPP = [0.5 0.6 0.7 0.8 0.9 1];

ks = [0.5 0.63 0.76 0.84 0.92 1];

L0Lmd0 = Ks*PP^2/(pi*sp*1000); T0 = Zs;
Fio = 2*sqrt(2)*4*pi*1e-7*Iph*T0*L0Lmd0;

FIL=FIs+FIo;  XL=FIL/FI;  FIc=FI/2;  dc1  =  FIc*1e6/(Bc*Li);
Do1   =   D   +   2*(Hs   +   dc1);   Do   =   ceil   (Do1/10)*10;
dc = (Do - D)/2 - Hs; Ata = 1.35*Tph*Iph*Kw/(P/2);

% ------------------------------------------------------

f3 = fopen ('Final_Total_SPG_Output.m', 'w');

fprintf (f3, 'Design of Salient-Pole Generator:-\n');

fprintf (f3,'Parameter                      VALUES\n');
```

```
fprintf (f3,'---------                              ------');
fprintf (f3,'\nKVA                                  %5.0f',KVA);
fprintf (f3,'\nKW                                   %5.0f',KW);
fprintf (f3,'\nVolts                                %5.0f',V);
fprintf (f3,'\nPF                                   %5.2f',pf);
fprintf (f3,'\nHZ                                   %5.0f',f);
fprintf (f3,'\nR.P.M                                %5.0f',N);
fprintf (f3,'\nNo of Poles                          %5.0d',P);
fprintf (f3,'\nFlux/Pole              Wb            %6.5f',FI);
fprintf (f3,'\nGapFlux-den(Av)        Wb/m2         %6.4f',Bav) ;
fprintf (f3,'\nSp.Elec Loading        Ac/m          %6.0f',q) ;
fprintf (f3,'\n-----------------STATOR--------------->');
fprintf (f3,'\nStator Current         Amps          %5.1f',Iph);
fprintf (f3,'\nCORE LENGTH(tot)       mm            %5.0f',L);
fprintf (f3,'\nIron LENGTH            mm            %5.0f',Ls);
fprintf (f3,'\nVent Ducts No X        mm         %2d X%3d',nv,bv);
fprintf (f3,'\nCORE OD                mm            %5.0f',Do);
fprintf (f3,'\nCORE ID                mm            %5.0f',D) ;
fprintf (f3,'\nPOLE PITCH             mm            %5.1f',PP);
fprintf (f3,'\nTURNS/Ph                             %5.0d',Tph);
fprintf (f3,'\nSTATOR SLOTS                         %5.0d',S);
fprintf (f3,'\nSLOT PITCH             mm            %5.2f',sp);
fprintf (f3,'\nCore depth             mm            %5.1f',dc);
fprintf (f3,'\nCoreFlux-den           Wb/m2         %5.1f',Bc) ;
fprintf (f3,'\nSLOT WIDTH             mm            %5.1f',Ws);
fprintf (f3,'\nSLOT Height(tot)       mm            %5.1f',Hs);
fprintf (f3,'\nHt (Lip + Wedge)       mm     %2.0f+%2.0f',HL,Hw);
fprintf (f3,'\nFlux-den(tooth-tip)    Wb/m2         %5.2f',Bt0) ;
fprintf (f3,'\nConductors/Slot                      %5.0f',Zs);
```

```
fprintf (f3,'\nAmp-Conductors/Slot                    %5.0f',SLoad);
fprintf (f3,'\nBare cond Wdth        mm        %5.1f',Wc);
fprintf (f3,'\nBare Cond Ht          mm        %5.1f',Hc);
fprintf (f3,'\nCurrent density       A/mm2     %5.3f',cds) ;
fprintf (f3,'\nAv.Loss Ratio         --        %5.4f',Kdav);
fprintf (f3,'\nMean-Turn Length      m         %5.4f',Lmt);
fprintf (f3,'\nSt Wdg Res/Ph/20d     ohm       %5.4f',Rph);
fprintf (f3,'\nPephoral Speed        m/s       %5.1f',Vr) ;
fprintf (f3,'\nArmature Reaction     AT/P      %5.0f',ATa);
fprintf (f3,'\nLeakage Reactance     Perc      %5.2f',xL*100);
% (3) Rotor and AT Calc of Salient Pole TG
scr=0.9; Kg= 1.17; Bp = 1.5; Sf = 0.75; Pf = 725; Bcr = 1.1;
ATf0 = scr*ATa; ATg = 0.75*ATf0; Bg = Bav/PAtoPP;
Lg1 = ATg/ (0.796*Bg*Kg*1e3);Lg = floor (Lg1*10)/10;
rat2=Lg/PP; Dr=D-2*Lg; Lp = L-15; Lpi = Lp*ki; FIp = 1.1*FI;
Ap = FIp/Bp; Wp = Ap/Lpi*1e6; Wpoles = Ap*Lp*P*7.8;
fprintf (f3,'\nAir-gap LENGTH        mm        %5.1f',Lg);
% <--Table for Depth of Field Winding------------>
PPS = [0.1 0.2 0.3 0.4 0.5];
DF = [0.03 0.036 0.04 0.045 0.05];
% ------------------------------------------------------
df = interp1(PPS, DF, PP/1e3, 'spline'); IfTf = 1.8*ATa;
hf = IfTf/(1e4*sqrt(df*Sf*Pf))*1000; Hp = hf + 40 + 0.12*PP;
dcr = FIp*1e6/2/Bcr/L;   Wrot = pi*(Dr - 2*Hp)^2/4*L*7.8e-6;
%----------------Ampere-Turns at 100% V-------------------->
%<---B-H Curve for St Tooth for ks = 1----->
BtS = [1.8  1.9   2   2.1  2.15   2.2  2.25   2.3];
attS=[9e3 14e3 24e3 43e3  60e3  80e3 110e3 152e3];% ks = 1.0
% ----Carters Coefft for Semi-closed slots--->
```

```
W0byLgS= [0  1   2   3   4   5  6   7   8   9   10  11  12];
% Slot opening/Gap

CC = [0 .18 .33 .45 .53 .6 .66 .71 .75 .79 .82 .86 .89];

%<---B-H Curve common for St Core, Rt core and Poles---->

BS = [0 0.25 0.5 0.75  1.0  1.25  1.5 1.75 2.0 2.25];

atS = [0  80  100  150  300   580  2e3  8e3 3e4 16e4];

WsbyLg=Ws/Lg; CCg = interp1 (W0byLgS, CC, WsbyLg, 'spline');
kgs = sp/(sp - Ws*CCg);DbyG = bv/Lg;

CCv = interp1 (W0byLgS, CC, DbyG, 'spline');

kgv = L/(L -nv*bv*CCv); kg = kgs*kgv;

%------------------------------------------------------

ATg = 0.796*Bg*kg*Lg*1e3; D13 = D + 2*Hs/3; sp13 = pi*D13/S;
bt13 = sp13-Ws; B13 = FI/(Li*bt13*TpP)*1e6;

att = interp1 (BtS, attS, B13, 'spline'); ATt = att*Hs/1000;

Lc = pi* (D + 2*Hs + dc)/P/2;

atc = interp1 (BS, atS, Bc, 'spline'); ATc = atc*Lc/1000;
ATL = ATg + ATt + ATc; FILs = 19e-8*ATL; FILp = 72e-8*ATL;
Fipmin = FI + FILs; Fipmax = Fipmin + FILp;

Bpmin= FIpmin/Ap; atp1 = interp1 (BS, atS, Bpmin, 'spline');

Bpmax= FIpmax/Ap; atp2 = interp1 (BS, atS, Bpmax, 'spline');

ATp=(2/3*atp1+1/3*atp2)*Hp/1e3;    Lcr=pi*(Dr-2*Hp-dcr)/P/2;

atrc= interp1(BS, atS, Bcr, 'spline'); ATrc = atrc*Lcr/1000;

ATf0 = ATg + ATt + ATc + ATp + ATrc;

%------------------------------------------------------------

E = [100 110 120 130];

for i = 1:4;

Bgx(i) = Bg*E(i)/100; ATgx(i) = 0.796*Bgx(i)*kg*Lg*1e3;

Bstx(i) = B13*E(i)/100;

atST(i) = interp1(BtS, attS, Bstx(i), 'spline');
ATstx(i) = atST(i)*Hs/1000;

Bscx(i) = Bc*E(i)/100;
```

```
atSC(i) = interp1(BS, atS, Bscx(i), 'spline');
ATscx(i) = atSC(i)*Lc/1000;
Bpminx(i) = Bpmin*E(i)/100;
atP1(i) = interp1(BS, atS, Bpminx(i), 'spline');
Bpmaxx(i) = Bpmax*E(i)/100;
atP2(i) = interp1 (BS, atS, Bpmaxx(i), 'spline');
ATpx(i) = (2/3*atP1(i) + 1/3*atP2(i))*Hp/1000;
Bcrx(i) = Bcr*E(i)/100;
atRC(i) = interp1(BS, atS, Bcrx(i), 'spline');
ATrcx(i) = atRC(i)*Lcr/1000;
ATTx(i)=ATgx(i)+ATstx(i)+ATscx(i) + ATpx(i) + ATrcx(i); end;
%------------------------->
% (5) LMC->Load Magnetization Curve of Salient Pole TG
E = [0 100 110 120 130]; ATT = [0 5551 7052 10165 16902];
V = 3300; pf = 0.85; % Specification
%<---Kr ---->
PAbyPPS = [0.5 0.6 0.7 0.8 0.9 1.0]; % Pole Arc by PP Ratio
KrS = [0.2 0.26 0.35 0.5 0.68 0.98];
Kr = interp1(PAbyPPS, KrS, PAbyPP, 'spline');
A = acos(pf); Eph = V/sqrt(3); Zpu = Eph/Iph; X = Zpu*XL;
Eg= sqrt((Eph*cos(A) + Iph*Rph)^2 + (Eph*sin(A) + Iph*X)^2);
Egpu = Eg/Eph; ATx = interp1(E, ATT, Egpu*100, 'spline');
AT1 = sqrt((ATx*cos(A))^2 + (ATx*sin(A) + Kr*ATa)^2);
AA = asin(ATx*sin (pi/2 + A)/AT1);
AT2 = (1 - Kr)*ATa*cos(AA); ATFL = AT1 + AT2;
E0 = interp1 (ATT, E, ATFL, 'spline'); reg = E0 - E(2);
%----------------------------->
% (6) Field Winding Design + Losses + Efficiency
Vex1 = 220; ti = 0.01; Tcf = 1.9; cdfw = 3.3; efEx = 0.88;
% Assumption
```

```
B=[0.8   1.2   1.6   2   2.4];   WpKg=[7   15   24   34   50];
Vc = 0.8*Vex1/P;

Lmtf = 2*(Lp + Wp)/1000 + pi*(df + 2*ti);
af = 0.02*Lmtf*ATFL/Vc;

Wcf  =  ceil(af/Tcf);  af  =  Wcf*Tcf*0.98;  If  =  af*cdfw;
Tf=ceil(ATFL/If); If0 = ATTx(1)/Tf;  Rf = 0.02*Lmtf*Tf/af;
Vfn = If*Rf*(235. + 120.)/255; Vex2 = 1.1*Vfn;

Vex=ceil(Vex2/5)*5; Iex1 = 1.1*If; Iex = ceil (Iex1/10)*10;
KWex=Vex*Iex/1e3; Pcuf=If^2*Rf*P; Wcuf=Lmtf*af*Tf*P*8.9e-3;
Pbr = 2*1*If; PcufT=(Pcuf + Pbr)/1e3; InpEx = PcufT/efEx;
Pex=InpEx-PcufT;        Pm=0.01*KVA*pf;        D12=D+2*Hs/2;
sp12 = pi*D12/S; bt12 = sp12-Ws; KgT = 7.8e-6*S*bt12*Li*Hs;
PitpKg = interp1 (B, WpKg, B13, 'spline');

Pit = PitpKg*KgT/1e3; Dmc = D + 2*Hs + dc;

KgC = 7.8e-6*(dc*Li)*(pi*Dmc);

PicpKg    =    interp1    (B,    WpKg,    Bc,    'spline');
Pic = PicpKg*KgC/1e3; Pi = Pit + Pic;

PT = Pts + Pi + PcufT + Pex + Pm; EFF = KW/(KW + PT)*100;

% (7)Temp-Rise and Weights-------->

cc1 = 0.032; Pcust = Ls/1000/(Lmt/2)*Pcus;

Pscs = Pcust + Pi + Peddy;

A1 = (pi*Do*L + 2*pi/4*(Do^2 - D^2))/1e6; WpC1 = A1/cc1;

A2 = pi*D*L/1e6; cc2 = 0.031/(1 + 0.1*Vr); WpC2 = A2/cc2;

A3  =  (pi/4*(Do^2  -  D^2)*nv)/1e6;  cc3  =  0.11/(0.1*Vr);
WpC3   =   A3/cc3;WpCt   =   WpC1   +   WpC2   +   WpC3;
TrS = Pscs*1e3/WpCt; ccf = 0.1/(1 + 0.1*Vr);

cs   =   2*Lmtf*hf/1000   +   2*Lmtf*df;   WpCf   =   cs/ccf;
TrF = Pcuf/P/WpCf; Wcu = Wcus + Wcuf;

Wiron= KgC + KgT + Wpoles + Wrot; Wtot = 1.01*(Wcu + Wiron);
KgPkva = Wtot/KVA;

fprintf (f3,'\n----Fied Winding and Other data------>');

fprintf (f3,'\nPole Dim (L × B × H) mm × mm × mm  % 5.1fx%
5.1fx% 5.1f', Lp, Wp, Hp);
```

```
fprintf (f3,'\nNo of Turns/Pole          --      %5.0f',Tf);
fprintf (f3,'\nDepth of core             mm      %5.1f',dcr);
fprintf (f3,'\nSCR                       pu      %5.3f',scr);
fprintf (f3,'\nNo Ld Rot Current         Amps    %5.1f',If0);
fprintf (f3,'\nRot Current at FL         Amps    %5.1f',If);
fprintf (f3,'\nVoltage Regulation        (perc)  %5.1f',reg);
fprintf (f3,'\nRot Volts at FL           Volts   %5.1f',Vfn);
fprintf (f3,'\nExciter Rating:   %4.1fKW,% 4.1fV, % 4.1fA',
KWex, Vex, Iex);
%-----------------------------------------------------
fprintf (f3,'\n----Weights, Losses and Efficiency:------>');
fprintf(f3,'\n   VARIENT                          VALUES');
fprintf (f3,'\n---------------------');
fprintf (f3,'\nWt of St.Core             Kg      %6.0f',KgC);
fprintf (f3,'\nWt of St.Teeth            Kg      %6.0f',KgT);
fprintf (f3,'\nWt of Poles               Kg    %6.0f',Wpoles);
fprintf (f3,'\nWt of St.Copper           Kg      %6.0f',Wcus);
fprintf (f3,'\nWt of Fld-Copper          Kg      %6.0f',Wcuf);
fprintf (f3,'\nTotal Wt                  Kg      %6.0f',Wtot);
fprintf (f3,'\nKg/KVA                          %6.2f',KgPkva);
fprintf (f3,'\n---------------------');
fprintf (f3,'\nMech Losses               KW      %6.1f',Pm);
fprintf (f3,'\nSt- Cu Losses             KW      %6.1f',Pts);
fprintf (f3,'\nIron Losses               KW      %6.1f',Pi);
fprintf (f3,'\nFld Copper Losses         KW     %6.1f',PcufT);
fprintf (f3,'\nTotal Losses              KW      %6.1f',PT);
fprintf (f3,'\nEfficiency                Perc    %6.2f',EFF);
fprintf (f3,'\nTemp-Rise of St           Wdg(deg)%6.2f',TrS);
fprintf (f3,'\nTemp-Rise of Fld          Wdg(deg)%6.2f',TrF);
fprintf (f3,'\n<------End of Output:------>'); fclose (f3);
```

6.2.10 *(a) Computer Output Results for Complete Design*

Design of Salient-Pole Generator

Parameter		VALUES
---------		------
KVA		2000
KW		1700
Volts		3300
PF		0.85
HZ		50
R.P.M		375
No. of Poles		16
Flux/Pole	Wb	0.09361
GapFlux-den(Av)	Wb/m2	0.6089
Sp.Elec Loading	Ac/m	33400
----------------STATOR-------------->		
Stator Current	Amps	349.9
CORE LENGTH (tot)	mm	670
Iron LENGTH	mm	580
Vent Ducts	No. X mm	9 X 10
CORE OD	mm	1760
CORE ID	mm	1500
POLE PITCH	mm	294.5
TURNS/Ph		96
STATOR SLOTS		144
SLOT PITCH	mm	32.72
Core depth	mm	85.0
CoreFlux-den	Wb/m2	1.1
SLOT WIDTH	mm	19.0
SLOT Height(tot)	mm	45.0
Ht (Lip+Wedge)	mm	1+ 4
Flux-den(tooth-tip)	Wb/m2	1.80
Conductors/Slot		4

Amp-Conductors/Slot		1400
Bare cond Wdth	mm	13.0
Bare Cond Ht	mm	7.0
Current density	A/mm2	3.924
Av.Loss Ratio	--	1.1998
Mean-Turn Length	m	2.3913
St Wdg Res/Ph/20d	ohm	0.0541
Pephoral Speed	m/s	29.5
Armature Reaction	AT/P	5413
Leakage Reactance	Perc	8.30
Air-gap LENGTH	mm	4.5
----Fied Winding and Other data------>		
Pole Dim(LxBxH)	mmXmmXmm	655.0x116.5x284.9
No. of Turns/Pole	--	95
Depth of core	mm	69.9
SCR	pu	0.900
No. Ld Rot Current	Amps	58.4
Rot Current at FL	Amps	104.5
Voltage Regulation	(perc)	19.5
Rot Volts at FL	Volts	15.1
Exciter Rating: 2.4KW, 20.0V, 120.0A		
----Weights, Losses and Efficiency:------>		
VARIANT	VALUES	
-------------	--	------
Wt. of St.Core	Kg	1821
Wt. of St.Teeth	Kg	388
Wt. of Poles	Kg	5612
Wt. of St.Copper	Kg	547
Wt. of Fld-Copper	Kg	741
Total Wt.	Kg	12718
Kg/KVA		6.36
-------------	--	------

Mech Losses	KW	17.0
St- Cu Losses	KW	27.4
Iron Losses	KW	34.4
Fld Copper Losses	KW	18.3
Total Losses	KW	99.6
Efficiency	Perc	94.47
Temp-Rise of St	Wdg(deg)	66.64
Temp-Rise of Fld	Wdg(deg)	33.29

<-------End of Output:--------->

6.2.11 Modifications to be done in the above Program to get Optimal Design

1. Insert "for" Loops for the following parameters to iterate the total program between min and max permissible limits for selecting the feasible design variants:
   (a) Ratio of Length to Pole pitch
   (b) Short Circuit Ratio
   (c) Core flux density
   (d) Arme Wdg Current densities.
2. Insert also minimum or maximum range of required objective functional values as constraint values, for example (a) Efficiency (b) Kg/KVA, (c) Temp-rise of Stator Winding
   (a) Temp-rise of Rotor Winding
3. Run the program to get various possible design variants.

Note: From the feasible design variants printed in the output, select that particular design fulfilling the objective of optimal parameter for the design.

6.2.12 Computer Program in "C" in MATLAB for Optimal Design

```
% Salient Pole Alternator

KVA=2000; V = 3300; N = 375; f = 50; pf = 0.85; KW = KVA*pf;
% Star connected Salient Pole Gen

KVAS = [ 100  200  500 1000 5000 10000 20000];

BavS = [0.52 0.54 0.56 0.58 0.61  0.63  0.65];

qS = [20e3 23e3 26e3 29e3 34e3  37e3  40e3];

% plot (KVAS, BavS); grid;

Bav = interp1 (KVAS, BavS, KVA, 'spline');
```

```
q = interp1 (KVAS, qS, KVA, 'spline');

Kw = 0.955; LbyPP = 2.3; nv = 9; bv = 10; ki = 0.9; spp = 3;
PAbyPP = 0.7; % Assumptions Pole Arc = L;

Cds = 4; PAtoPP = 0.7; Bt0 = 1.8; Bc = 1.1; Wins = 6;
Hins = 8; Hw = 4; HL = 1;   Ks = 1; % Assumptions

% ---------------------------------------------------

f2 = fopen ('Final_Optimal_SPG_Output.m','w');

fprintf (f2,'Design of Salient-Pole Alternator of % 4d KVA,
% 5d V, % 4d RPM, % 3d hz, % 5.2f PF:-\n', KVA, V, N, f,
pf);

fprintf (f2,' Sn    D      L    L/PP   m/s    S    cds Zs Ac/S
St-Cond  St-Slot  Bc  SCR  EFF Kg/KVA  TrS   TrF\n');

fprintf (f2,' -- ----- ----- ----  ---- --- ---- -- ---- ---
----- --------- --- --- ----- ----  ----- -----\n');

sn = 0; M1 = 0; M2 = 0; M3 = 0; M4 = 0; EFFmax = 90;
minKgPkva = 9; minTrS = 75; min TrF = 50; % Assumptions

for LbyPP=2.1:0.2:2.5;%(Range:0.8to3.0for Rectangular Poles)

for scr = 0.9: 0.1: 1.1; % (Range: 0.9 to 1.3

for Bc = 1.1: 0.1: 1.3; % (Range: 1.1 to 1.3)

for cds = 3:0. 5:4; % Range: 3 to 5)

P=120*f/N; K = 11*Bav*q*Kw*1e-3; n = 2*f/P; DsqL = KVA/K*n);

D1 = (DsqL*P/(LbyPP*pi))^(1/3)*1000; D = ceil(D1/10)/100;

L1=DsqL/D^2; L=ceil(L1*100)*10; D= D*1000; Ls = (L - nv*bv);
Li = Ls*ki; PP = pi*D/P;

Vr = pi*D*n/1000; if Vr > = 30.0 continue; end;

flux1= Bav*pi*D*L/P/1e6; Tphi = V/sqrt(3)/(4.44*f*Kw*flux1);

Zph = Tphi*2; S = spp*P*3; sp = pi*D/S; Zsi = Zph*3/S;
Zs = ceil(Zsi/2)*2; Tph = Zs*S/6;  Iph=KVA*1e3/(1.7321*V);
SLoad = Iph*Zs; FI = V/sqrt(3)/(4.44*f*Kw*Tph);

% ------------------------------------------------------------

% (2) Design of Stator Winding, Reactance, Core
```

```
As=Iph/cds;   TpP=ceil(PAtoPP*PP/sp);bt0=FI*1e6/(Bt0*Li*TpP);
Ws1 = sp-bt0; Wc = ceil((Ws1) - Wins); Hc = ceil(As/Wc*2)/2;
Ws  =  1*Wc  +  Wins;  As  =  Wc*Hc*0.98;  cds  =  Iph/As;
Hs = Zs*Hc + Zs*2*0.5 + Hins + Hw + HL;      rat1 = Hs/Ws;

Lmt=(2*L+2.5*PP+50*V/1000+150)/1000;    Rph=0.021*Lmt*Tph/As;
Pcus=3*Iph^2*Rph/1e3;           Wcus=3*Lmt*As*Tph*8.9e-3;

alfa= sqrt(Wc/Ws); Kdav = 1 + (alfa*Hc/10) ^4*Zs^2/9;

Peddy = (Kdav - 1)*Pcus; Pts = (Pcus + Peddy)*1.15;

Reff = Iph*Rph*Kdav/(V/sqrt(3)); % Reff,

%-------Leakage Reactance--------->

h1=Zs*Hc + 4 + 2 + 2; h2 = 2 + 2; h3 = Hw; h4 = HL; Ws0 = 9;

Lmdas = h1/(3*Ws) + h2/Ws + 2*h3/(Ws + Ws0) + h4/Ws0;

Fis       =       2*sqrt(2)*4*pi*1e-7*Iph*Ls/1000*Lmdas*Zs;
% FIs, % Ls = 0.5835; Fis = 0.0036

CSbPP = [0.5 0.6 0.7 0.8 0.9 1];

ks = [0.5 0.63 0.76 0.84 0.92 1];

L0Lmd0 = Ks*PP^2/(pi*sp*1000); T0 = Zs;

Fio = 2*sqrt(2)*4*pi*1e - 7*Iph*T0 *L0Lmd0;

FIL=FIs+FIo; XL  = FIL/FI; FIc= FI/2; dc1=FIc*1e6/(Bc*Li);
Do1=D+2*(Hs + dc1);  Do= ceil(Do1/10)*10;  dc=(Do-D)/2-Hs;

%-------------------------------------------------------

% (3) Rotor and AT Calc of Salient Pole TG

Scr=0.9; Kg=1.17; Bp = 1.5; Sf = 0.75; Pf = 725; Bcr = 1.1;
Ata=1.35*Tph*Iph*Kw/(P/2);  ATf0=scr*ATa;   ATg = 0.75*ATf0;
Bg=Bav/PAtoPP;             Lg1=ATg/(0.796*Bg *Kg*1e3);

Lg  =  floor  (Lg1*10)/10;  rat2  =  Lg/PP;  Dr  =  D  -  2*Lg;
Lp = L - 15; Lpi = Lp*ki; FIp = 1.1*FI;

Ap = FIp/Bp; Wp = Ap/Lpi*1e6; Wpoles = Ap*Lp*P*7.8;

%<--Table for Depth of Field Winding------------>

PPS=[0.1 0.2 0.3 0.4 0.5]; DF=[0.03 0.036 0.04 0.045 0.05];

% -------------------------------------------------------

df = interp1(PPS, DF, PP/1e3, 'spline'); IfTf = 1.8*ATa;
```

```
hf = IfTf/(1e4*sqrt(df*Sf*Pf))*1000; Hp = hf + 40 + 0.12*PP;
dcr = FIp*1e6/ 2/Bcr/L;Wrot = pi*(Dr - 2*Hp)^2/4*L*7.8e-6;

%-------------------Ampere-Turns at 100% V---------------->

%<---B-H Curve for St Tooth for ks=1----->

BtS = [1.8  1.9   2   2.1  2.15   2.2  2.25   2.3];

attS=[9e3 14e3 24e3 43e3  60e3  80e3 110e3 152e3]; %ks = 1.0

% ----Carters Coefft for Semi-closed slots--->

W0byLgS = [0 1   2   3   4   5  6   7   8   9   10  11  12];
% Slot opening/Gap

CC =  [0 .18 .33 .45 .53 .6 .66 .71 .75 .79 .82 .86 .89];

% <--- B-H Curve common for St Core, Rt core and Poles---->

BS = [0 0.25 0.5 0.75  1.0  1.25  1.5 1.75 2.0 2.25];

atS = [0  80  100  150  300   580  2e3  8e3 3e4 16e4];

WsbyLg=Ws/Lg; CCg = interp1 (W0byLgS, CC, WsbyLg, 'spline');
kgs = sp/(sp -Ws*CCg);       byG = bv/Lg;

CCv = interp1(W0byLgS, CC, DbyG,'spline');

kgv = L/(L -nv*bv *CCv); kg = kgs*kgv;

% ----------------------------------------------------

ATg = 0.796*Bg*kg*Lg*1e3; D13 = D + 2*Hs/3; sp13 = pi*D13/S;
bt13    =    sp13    -Ws;    B13    =    FI/(Li*bt13*TpP)*1e6;
att = interp1(BtS, attS, B13, 'spline'); ATt = att*Hs/1000;

Lc=pi*(D+2*Hs+dc)/P/2; atc = interp1(BS, atS, Bc, 'spline');
ATc = atc*Lc/1000; ATL = ATg + ATt + ATc;

FILs = 19e - 8*ATL; FILp = 72e - 8*ATL; Fipmin = FI + FILs;
Fipmax = Fipmin + FILp;

Bpmin= FIpmin/Ap; atp1 = interp1 (BS, atS, Bpmin, 'spline');

Bpmax= FIpmax/Ap; atp2 = interp1 (BS, atS, Bpmax, 'spline');

Atp=(2/3*atp1+1/3*atp2)*Hp/1e3;   Lcr=pi*(Dr-2*Hp- dcr)/P/2;

Atrc=interp1 (BS, atS, Bcr, 'spline'); ATrc = atrc*Lcr/1000;

ATf0 = ATg + ATt + ATc + ATp + ATrc;

% ----------------------------------------------------------
```

```
E = [100 110 120 130];

for i = 1:4;

Bgx(i) = Bg*E(i)/100; ATgx(i) = 0.796*Bgx(i)*kg*Lg*1e3;

Bstx(i) = B13*E(i)/100;

atST(i) = interp1 (BtS, attS, Bstx(i), 'spline');

ATstx(i) = atST(i)*Hs/1000;

Bscx(i) = Bc*E(i)/100;

atSC(i) = interp1 (BS, atS, Bscx(i), 'spline');

ATscx(i) = atSC(i)*Lc/1000;

Bpminx(i) = Bpmin*E(i)/100;

atP1(i) = interp1 (BS, atS, Bpminx(i), 'spline');

Bpmaxx(i) = Bpmax*E(i)/100;

atP2(i) = interp1 (BS, atS, Bpmaxx(i), 'spline');

ATpx(i) = (2/3*atP1(i) + 1/3*atP2(i))*Hp/1000;

Bcrx(i) = Bcr*E(i)/100;

atRC(i) = interp1 (BS, atS, Bcrx(i), 'spline');

ATrcx(i) = atRC(i)*Lcr/1000;

ATTx(i)=ATgx(i)+ATstx(i)+ATscx(i)+ATpx(i) + ATrcx(i); end;

% -------------------------------->

% (5) LMC->Load Magnetization Curve of Salient Pole TG

E = [0 100 110 120 130]; ATT = [0 5551 7052 10165 16902];
V = 3300; pf = 0.85; % Specification

% <---Kr---->

PAbyPPS=[0.5 0.6 0.7  0.8  0.9  1.0]; % Pole Arc by PP Ratio

KrS = [0.2 0.26 0.35 0.5 0.68 0.98];

Kr = interp1 (PAbyPPS, KrS, PAbyPP, 'spline');

A = acos(pf); Eph = V/sqrt(3); Zpu = Eph/Iph; X = Zpu*XL;

Eg= sqrt((Eph*cos(A) + Iph*Rph)^2 + (Eph*sin(A) + Iph*X)^2);

Egpu = Eg/Eph; ATx = interp1 (E, ATT, Egpu*100, 'spline');
```

```
AT1 = sqrt((ATx*cos(A))^2 + (ATx*sin(A) + Kr*ATa)^2);
AA = asin(ATx*sin (pi/2 + A)/AT1);

AT2 = (1 - Kr)*ATa*cos(AA); ATFL = AT1 + AT2;
E0 = interp1(ATT, E, ATFL, 'spline'); reg = E0 - E(2);

% ----------------------------->

% (6) Field Winding Design + Losses + Efficiency

Vex1 = 220; ti = 0.01; Tcf = 1.9; cdfw = 3.3; efEx = 0.88;
% Assumption

B=[0.8 1.2 1.6 2 2.4]; WpKg=[7 15 24 34 50]; Vc=0.8*Vex1/P;

Lmtf=2*(Lp+Wp)/1000+ pi*(df + 2*ti); af = 0.02*Lmtf*ATFL/Vc;

Wcf = ceil(af/Tcf); af = Wcf*Tcf*0.98; If = af*cdfw;
Tf = ceil(ATFL/If); If0 = ATTx(1)/Tf; Rf=0.02*Lmtf*Tf/af;
Vfn = If*Rf*(235. + 120.)/255; Vex2 = 1.1*Vfn;

Vex=ceil(Vex2/5)*5; Iex1 = 1.1*If; Iex = ceil(Iex1/10)*10;
KWex=Vex*Iex/1e3; Pcuf=If^2*Rf*P; Wcuf=Lmtf*af*Tf*P*8.9e-3;
Pbr = 2*1*If; PcufT = (Pcuf + Pbr)/1e3; InpEx = PcufT/efEx;
Pex=InpEx-PcufT; Pm=0.01*KVA*pf; D12=D+2*Hs/2;
sp12=pi*D12/S; bt12=sp12-Ws; KgT = 7.8e-6*S*bt12*Li*Hs;
PitpKg=interp1(B,WpKg,B13,'spline'); Pit = PitpKg*KgT/1e3;
Dmc=D+ 2*Hs + dc; KgC = 7.8e-6*(dc*Li)*(pi*Dmc);

PicpKg=interp1(B,WpKg,Bc,'spline'); Pic = PicpKg*KgC/1e3;
Pi = Pit + Pic; PT = Pts + Pi + PcufT + Pex + Pm;
EFF = KW/(KW + PT)*100; if EFF < = 93.4 continue; end;

% (7) Temp-Rise and Weights-------->

cc1=0.032; Pcust=Ls/1000/(Lmt/2)*Pcus; Pscs=Pcust+Pi+Peddy;

A1 = (pi*Do*L + 2*pi/4*(Do^2 - D^2))/1e6; WpC1 = A1/cc1;

A2 = pi*D*L/1e6;cc2 = 0.031/(1 + 0.1*Vr); WpC2 = A2/cc2;

A3=(pi/4*(Do^2-D^2)*nv)/1e6; cc3=0.11/(0.1*Vr); WpC3=A3/cc3;

WpCt = WpC1 + WpC2 + WpC3; TrS = Pscs*1e3/WpCt;
if TrS > = 69 continue; end;

ccf=0.1/(1+0.1*Vr);cs=2*Lmtf/1000*hf+2*Lmtf*df; WpCf=cs/ccf;

TrF = Pcuf/P/WpCf; if TrF > = 40 continue; end;

Wcu = Wcus + Wcuf; Wiron = KgC + KgT + Wpoles + Wrot;
```

```
Wtot = 1.01*(Wcu + Wiron); KgPkva = Wtot/KVA;
if KgPkva > = 6.5 continue; end;
sn = sn + 1; if EFF > = EFFmax EFFmax = EFF; end;
if abs (EFF - EFFmax) < = 2e - 3 M2 = sn; end;
if KgPkva < = minKgPkva  minKgPkva = KgPkva; end;
if abs (KgPkva -minKgPkva) < = 0.001 M1 = sn; end;
if TrS < = minTrS minTrS = TrS; end;
if abs (TrS - minTrS) < = 0.0001 M3 = sn; end;
if TrF < = minTrF  minTrF = TrF; end;
if abs (TrF - minTrF) < = 0.0001 M4 = sn; end;
fprintf (f2,' % 3 d% 6.0 f% 6.0 f% 5.2 f% 6.1 f% 4 d% 5.2 f%
3 d% 5.0 f% 5.1 fX% 3.1 f% 5.1 fX% 3.1 f% 4.1 f% 4.1 f% 6.2
f% 5.2 f% 7.2 f% 6.2 f\n', sn, D, L, L by PP, Vr, S, cds,
Zs, SLoad, Wc, Hc, Ws, Hs, Bc, scr, EFF, KgPkva, TrS, TrF);
end; end; end; end;
fprintf (f2,' -- ----- ----- ----  ---- --- ---- -- ---- ---
----- --------- --- --- ----- ----  ----- ----\n');
fprintf  (f2,'Selection  of  Design  Variant  based  on
Optimization Criteria:');
fprintf  (f2,'\nIf  Maximum  Efficiency  is  Required,  Select
Variant (Sn) = %3d  (% 5.2f perc)', M2, EFFmax);
fprintf (f2,'\nIf Minimum Kg/KVA is Required, Select Variant
(Sn) = % 3d  (%5.2f)', M1, minKgPkva);
fprintf  (f2,'\nIf  Minimum  St-Wdg  Temp-Rise  is  Reqd,  Select
Variant (Sn) = %3d  (%4.2f)', M3, min TrS);
fprintf  (f2,'\nIf  Minimum  Rt-Wdg  Temp-Rise  is  Reqd,  Select
Variant (Sn) = % 3d  (%4.2f)', M4, min TrF);
fclose (f2);
```

6.2.12 (a) Computer Output Results for Optimal Design

Design of Salient-Pole Alternator of 2000 KVA, 3300 V, 375 RPM, 50 hz, 0.85 PF

Sn	D	L	L/PP	m/s	S	cds	Zs	Ac/S	St-Cond	St-Slot	Bc	SCR	EFF	Kg/KVA	TrS	TrF
1	1500	670	2.30	29.5	144	3.43	4	1400	13.0×8.0	19.0×49.0	1.1	0.9	94.43	6.43	68.73	33.29
2	1500	670	2.30	29.5	144	3.92	4	1400	13.0×7.0	19.0×45.0	1.1	0.9	94.47	6.36	66.64	33.29
3	1500	670	2.30	29.5	144	3.43	4	1400	13.0×8.0	19.0×49.0	1.1	0.9	94.43	6.43	68.73	33.29
4	1500	670	2.30	29.5	144	3.92	4	1400	13.0×7.0	19.0×45.0	1.1	0.9	94.47	6.36	66.64	33.29
5	1500	670	2.30	29.5	144	3.43	4	1400	13.0×8.0	19.0×49.0	1.1	0.9	94.43	6.43	68.73	33.29
6	1500	670	2.30	29.5	144	3.92	4	1400	13.0×7.0	19.0×45.0	1.1	0.9	94.47	6.36	66.64	33.29
7	1460	710	2.50	28.7	144	3.92	4	1400	13.0×7.0	19.0×45.0	1.1	0.9	94.41	6.48	67.89	33.90
8	1460	710	2.50	28.7	144	3.92	4	1400	13.0×7.0	19.0×45.0	1.1	0.9	94.41	6.48	67.89	33.90
9	1460	710	2.50	28.7	144	3.92	4	1400	13.0×7.0	19.0×45.0	1.1	0.9	94.41	6.48	67.89	33.90

Selection of Design Variant based on Optimization Criteria:

If Maximum Efficiency is Required, Select Variant (Sn) = 6 (94.47 perc)

If Minimum Kg/KVA is Required, Select Variant (Sn) = 6 (6.36)

If Minimum ratio of I0/I2 is Required, Select Variant (Sn) = 6 (66.64)

If Minimum Volume of Tank if Required, Select Variant (Sn) = 6 (33.29)

6.3 Non-Salient Pole (Cylindrical Solid Rotor) Type

Total design is split into eight parts in a proper sequence. Design calculations are given for a given Rating of Generator, followed by Computer Program written in "C" language using MATLAB software for each part. Finally all the Programs are added together to get the total Program by running which we get the total design. Computer output of total design is given.

This design may not be the optimum one. Now optimization objective and design constraints are inserted into this total Program. When this program is run we will get various alternative feasible designs from which the selected variant based on the Optimization Criteria can be picked up. Computer output showing the important design parameters for various feasible alternatives is given in the end of this chapter along with a logic diagram.

6.3.1 Sequential Steps for Design of Each Part and Programming Simultaneously

(a) Calculate Rotor dia, No. of Slots, Output Coefficient, Air gap length, Main dimensions of Stator Core (viz) D, L and Flux/Pole

(b) Calculate size of slot, conductor size, bar size, checking current density, slot balance. Calculate tooth flux density, Depth of core, Wt. of Copper, Copper losses and Leakage reactance, Core losses, weights of copper and core

(c) Calculate Rotor Slots, Cond size, Cond/slot, and check slot balance. Calculate Field current at rated load and pf, Weight of copper, Copper and mechanical losses

(d) Calculate Carter Coefft and Ampere-turns for Air gap, Stator tooth, Stator core, Rotor tooth, Rotor core and Total No-load AT at 100%En

(e) Calculation and plotting of Open Circuit Characteristic (OCC)

(f) Calculate of Field AT at rated Load and PF and check the rotor current density. Calculate Voltage regulation, Exciter rating, Total Weight and KG/KVA, Temp-rise of air.

Note: By adding programs established for each part sequentially we get the Program for complete design.

Problem:

Design a 30000 KVA, 11000 V, 3000 RPM, 50 Hz, 3 ph, 0.8 pf Star connected Generator

6.3.2 Calculation of Stator Main Dimensions (Part-1)

SPECIFIC MAGNETIC LOADING and SPECIFIC ELECTRIC LOADING

KVA	100	200	500	1000	5000	10000	20000	30000
Bav (T)	0.54	0.56	0.58	0.60	0.63	0.65	0.67	0.69
Amp -Cond/m (q)	30000	33000	36000	40000	45000	48000	53000	58000

Following Data is assumed:

Width of ventilating ducts (bv) = 0.01m; Winding factor (Kw) = 0.955;

Slots/pole/ph (spp) = 9;

Iron factor (k_i) = 0.92;

Values of Specific magnetic loading (B_{av})

= 0.69 T and Specific electric loading (q) = 58000 ac/m

(read from the Table given above).

Calculations

$$\text{Phase Current (Iph)} = \frac{\text{KVA} \times 1000}{\text{V} \times \sqrt{3}} = \frac{30000 \times 1000}{11000 \times \sqrt{3}} = 1574.5 \text{ amps}$$

Rated KW output = KVA × rpf = 30000 × 0.8 = 24000 KW

$$\text{No. of Poles} = \text{P} = \frac{120 \times \text{f}}{\text{N}} = \frac{120 \times 50}{3000} = 2$$

Speed (n) = N/60 = 3000/60 = 50 rps

Assuming Rotor Velocity (Vr) = 125 m/s, Rotor diameter (Dr)

$$= \frac{\text{Vr}}{\pi \times \text{n}} = \frac{125}{\pi \times 50} = 0.7958 \text{ m} \cong 800 \text{ mm}$$

Assuming Slots/Pole/Ph(spp) = 9; No. of Slots (S) = spp × P × 3 = 9 × 3 × 2 = 54

Assuming Bars/Slot(Zs) = 2, No. of Turns/Ph(Tph) = S × Zs/6 = 54 × 2/6 = 18

Armature Reaction AT (ATa) = 1.35 × Tph × Iph × Kw/(P/2)

= 1.35 × 18 × 1574.5 × 0.955/1 = 36540

Full pitch of Armature Wdg = S/P = 54/2 = 27 slots

Slot angle (alfac) = 2 × π/S = 0.1164 rad

Assuming Armature winding is short pitched by one slot,

Pitch Factor (Kc) = cos (alfac × 1) = cos(0.1164) = 0.9983

Output Coefft (K) = 11 × Bav × q × Kw × Kc × 1e-3

$= 11 \times 0.69 \times 58000 \times 0.955 \times 0.9983 \times 1 \times 10^{-3} = 419.7$

Assuming Short circuit ratio (scr) = 0.6,

AmpTurns on No-Load (ATf0) = scr × ATa = 0.6 × 36540 = 21924

Airgap AmpTurns (ATg) ≈ 0.75 × ATf0 ≈ 0.75 × 21294 = 16443

Max Flux density in Airgap (Bg) ≈ Bav × 1.5 ≈ 0.69 × 1.5 = 1.035 T

Air Gap length (Lg)

$$= \frac{\text{ATg}}{0.796 \times \text{Bg} \times \text{Kg} \times 1000} = \frac{\text{ATg}}{0.796 \times 1.035 \times 1.15 \times 1000} = 17.36 \approx 18 \text{ mm}$$

Stator core Inner Diameter (Di) = Dr + (2 × Lg) = 800 + (2 × 18) = 836 mm

$$Di^2L = \frac{KVA}{K \times n} = \frac{30000}{419.7 \times 50} = 1.4296 \quad m^3$$

$$\text{Total Core length(L)} = \frac{Di^2L}{Di^2} = \frac{1.4296}{(836/1000)^2} = 2.0455\,m \approx 2045.5 \text{ mm} \approx 2050 \text{ mm}$$

(by rounding off)

$$\text{Pole Pitch (PP)} = \frac{\pi \times Di}{P} = \frac{\pi \times 836}{2} = 1313.2\,mm$$

Assuming 82% of Total core length = Gross iron length (Ls)

= 0.82 × 2050 = 1681 mm

Assuming Width of Ventilating duct = 10mm,

No. of Vent ducts (nv) = (L – Ls1)/bv = (2050 – 1681)/10

= 36.9 ≈ 37 (should be integer)

L_s = (L – nv × bv) = [2050 – (37 × 10)] = 1680 mm

Assuming iron factor of 92%,

Net Iron length of core (L_i) = Ls × k_i = 1680 × 0.92 =1545.6 mm

$$\text{Flux/Pole } (\Phi) = \frac{B_{av} \times \pi \times Di \times L}{P} = \frac{0.69 \times \pi \times 836 \times 2050}{2 \times 10^6} = 1.8575 \text{ Wb}$$

6.3.2 *(a) Computer Program in "C" in MATLAB for Part-1*

```
% Design of KVA = 30000; V = 11000; N = 3000; f = 50; pf = 0.8;
%Non-Salient Pole Generator

KVA  =  30000;  V  =  11000;  N  =  3000;  f  =  50;  rpf  =  0.8;
% Star connected Non-Salient Pole Gen

Kw=0.955; bv = 10; ki = 0.92; Vr = 125; R1 = 0.82; % Input Data

spp=9; CP=26; Zs=2; Kg=1.15; spp=9; CP=26; scr=0.6;% Input Data

% <----Bav and q →Tables------->

KVAS = [ 100  200  500 1000 5000 10000 20000 30000];

BavS = [0.54 0.56 0.58 0.60 0.63  0.65  0.67  0.69];

qS = [30e3 33e3 36e3 40e3 45e3  48e3  53e3  58e3];

%-------------Main Dimensions----------------------->

Bav = interp1(KVAS,BavS,KVA,'spline');

q=interp1(KVAS,qS,KVA,'spline');KW=KVA*rpf;P=120*f/N; n= 2*f/P;
```

```
Iph=KVA*1e3/(1.7321*V);   Dr1=Vr/(pi*n);   Dr= ceil(Dr1*100)*10;
S = spp*P*3; Tph = S*Zs/6; Ata = 1.35*Tph*Iph*Kw/(P/2);
alfac = P*pi/S; Kc = cos(alfac/2);   K = 11*Bav*q*Kw*Kc*1e-3;
ATf0=scr*ATa;ATg=0.75*ATf0;Bg=Bav*1.5;
Lg1=ATg/(0.796*Bg*Kg*1e3); Lg = ceil(Lg1);   Di = Dr + 2*Lg;
DsqL = KVA/(K*n);   L1 = DsqL/(Di/1e3)^2; L = ceil(L1*100)*10;
PP = pi*Di/P; Ls1= R1*L; nv1=(L - Ls1)/bv; nv = ceil(nv1);
Ls = (L - nv*bv);  Li = Ls*ki; FI = Bav*pi*Di*L*1e-6/P;
% -----------------------------------------------------
```

6.3.3 Design of Armature Winding and Core (Part-2)

Slot pitch (sp) = $\pi \times$ Di/S = $\pi \times 836/54 = 48.6365$ mm

Tooth width at air gap (bt0) = $\dfrac{\Phi \times 10^6}{B_{t0} \times Li \times S/P} = \dfrac{1.8575}{1.6 \times 1545.6 \times 54/2} = 28.1365\ \text{mm}$

Slot width (Ws) = sp – bt0 = 48.6365 – 28.1365 = 20.5 mm

Assuming current density (cds1) = 3 A/mm^2;

Cond Area of CS (As) = $\dfrac{Iph}{cds} = \dfrac{1574.5}{3} = 524.849\ \text{mm}^2$

Assuming 2 conductors width-wise and thickness of HV insulation as 3 mm for 11 KV

Cond Width (Wc) = ((Ws – 2 × HVins – 2.5)/2) = (20.5– 2 × 3 – 2.5)/2

= 6.1585 ≈ 6 mm

Assuming Thickness of bare conductor as 1.8 mm,

No. of conductors /Bar height-wise (Ncv) = $\dfrac{As}{Hc \times Wc \times 2} = \dfrac{524.849}{1.8 \times 6 \times 2} = 24.3 \approx 25$

Assuming 2% reduction due to corners rouding of conductors,

Actual copper area in a bar (As) = 0.98 × Ncv × Hc × Wc × 2

= 0.98 × 25 × 1.8 × 6 × 2 = 529.3;

Corrected current density (cds1) = $\dfrac{Iph}{As} = \dfrac{1574.5}{529.3} = 2.9753\ \text{A}/\text{mm}^2$

Corrected width of slot (Ws) = 2 × Wc + 2 × HVins + 2.5

= 2 × 6 + 2 × 3 + 2.5 = 20.5 mm

Assuming Insulation thickness of conductor as 0.25 mm, of Height of each Bar (Hbar) = (Ncv + 1) × (Hc + Cins) + 2 × HVins = (25 + 1) × (1.8 + 0.25) + 2 × 3

= 59.3 mm

Assuming Ht of wedge (Hw) = 9 mm, Lip ht (HL) = 25mm and

Other Insulation in the slot (Hins) = 12 mm,

Height of Slot (Hs) = 2 × Hbar + Hins + Hw + HL = 2 × 59.3 + 12 + 9 + 25

= 164.6 mm ≈ 165mm

Conductors (ht-wise) in total slot(p) = 2 × 25 = 50

$$\alpha = \sqrt{Wc / Ws} = \sqrt{2 \times 6 / 20.5} = 0.7651$$

Loss Factor of Top conductor (KdTL)

Av Loss Factor (Kdav)

$$= 1 + (\alpha Hc)^4 \times \frac{(2 \times Ncv)^2}{9} = 1 + (0.7651 \times 1.8 / 10)^4 \times \frac{(2 \times 25)^2}{9} = 1.0999$$

Mean Length of Turn (Lmt1) = 2 × L/10 + 2.5 × PP/10 + 5 × KV + 15

= (2 × 2050/10) + (2.5 × 1313.2/10) + (5 × 11000/1000) + 15 = 808.3cm = 8.083 m

$$\text{Stator Wdg Res/ph (Rph)} = 0.021 \times Lmt \times \frac{Tph}{As} = 0.021 \times 8.083 \times \frac{18}{529.2} = 0.0058\,\Omega$$

DC Copper Loss (Pcus) = 3 × Iph^2 × Rph × 10^{-3}

= 3 × 1574.5^2 × 0.0058 × 10^{-3} = 42.9413 KW

Eddy Current Loss (Peddy) = (Kdav-1) × Pcus = 0.0999 × 42.9413 = 4.291 KW

Assuming Stray Losses = 15%,

Total Copper Losses in Stator Wdg (Pts) = (Pcus + Peddy) × 1.15

= (42.9413 + 4.291) × 1.15 = 54.317 KW

Effective PU Resistance (Reff) = Iph × Rph × Kdav × V/ √3

= 1574.5 × 0.0058 × 1.0999 /(11000/√3) = 0.0016 pu

Tooth width at air gap (bt0) = sp – Ws = 48.6365 – 20.5 = 28.1365 mm

Flux-density at Tooth tip at air gap (Bt0)

$$= \frac{\Phi \times 10^6}{b_{t0} \times Li \times S / P} = \frac{1.8575}{28.1365 \times 1545.6 \times 54 / 2} = 1.582\ \text{T}$$

Tooth width at 1/3 ht from tooth tip (bt1b3) = π × (Di + Hs × 2/3)/S – Ws

= π × (836 + 165 × 2/3)/54 – 0.5 = 34.536mm

Flux-density at 1/3 ht (Bt1b3)

$$= \frac{\Phi \times 10^6}{b_{t1b3} \times Li \times S / P} = \frac{1.8575}{34.356 \times 1545.6 \times 54 / 2} = 1.2888\ \text{T}$$

$$\text{Flux in core}\,(\phi_c) = \frac{\phi}{2} = \frac{1.8575}{2} = 0.9288\ \text{Wb}$$

Assuming Flux density ic core (Bc) = 1.1T, Depth of Core (dc)

$$= \frac{\phi_c}{B_c \times L_i} = \frac{0.9288 \times 10^6}{1.1 \times 1545.6} = 546.3\ \text{mm}$$

Outer diameter of Core (D_0) = Di + 2 × (Hs + dc) = 836 + 2 × (165 + 546.3)

= 2258.5 mm ≈ 2260 mm (Rounded off)

Revised (dc) = (2260 – 836)/2 – 165 = 547 mm

Actual Flux density in core (Bc) = $\dfrac{\phi_c}{dc \times L_i} = \dfrac{0.9288 \times 10^6}{547 \times 1545.6} = 1.0985$ T

Cross-Section of Stator Slot (dimensions in mm)

Height-Wise			
Lip			25.0
Wedge			9.0
Ins.Under Wedge			1.3
Cu.Coductor	bare + insulation	1.8 + 0.25 = 2.05	
Top- Bar	(25 + 1)*conductor	26*2.05 = 53.3	
	HV Insulation	2*3 = 6	59.3
Ins.between bars			7.0
Bottom- Bar			59.3
Bottom Insulation			1.3
Slack			2.8
Total			**165.0**

Width-Wise			
Cu.Conductor	bare + insulation	6 + 0.25 = 6.25	
Bar	2*conductor	2*6.25 = 12.5	
	HV Insulation	2*3 = 6	18.5
Insulation on sides	2*0.5		1.0
Slack			1.0
Total			**20.5**

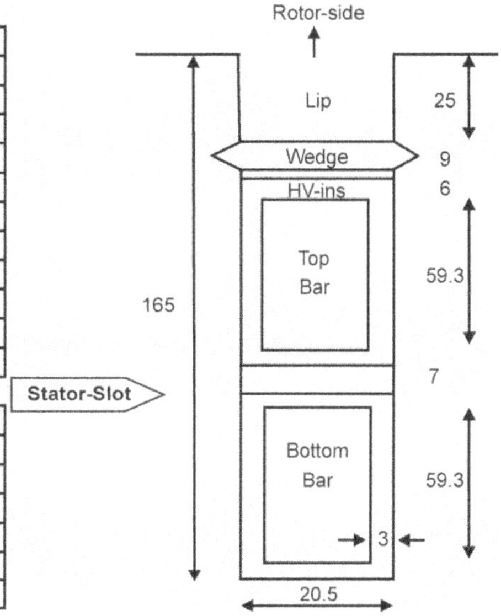

Slot – Balance Ht-wise (Stsbh) =

Hs – [Hw + HL + 7.5 + Zs × {(Ncv + 1) × (Hc + 0.25) + 7.06}]

= 165 – [9 + 25 + 7.5 + 2 × {(25 + 1) × (1.8 + 0.25) + 7.06}] = 2.78 mm → (OK)

Slot – Balance Width – wise (Stsbw) = Ws – [2 × (Wc + 0.25) + 2 × HVins + 1]

= 20.5 – [2 × (6 + 0.25) + 2 × 3 + 1] = 1 → (OK)

For Leakage Reactance Calculation

h1 = (Ncv + 1) × Zs × (Hc + Cins) + Zs × 2 × HVins + 6

= (25 × 1) × 2 × (1.8 + 0.25) + (2 × 2 × 3) + 6 = 124.6 mm

h2 = (Hs – h1 – HL – Hw)/2 = (165 – 124.6 – 25 – 9)/2 = 3.2 mm

h3 = Hw = 9 mm; h4 = HL = 25 mm; Width at opening of slot (Ws_0) = 20.5 mm;

Specific Permeance of Slot $(\lambda_s) = \dfrac{h_1}{3 \times W_s} + \dfrac{h_2}{W_s} + \dfrac{2h_3}{W_s + W_{s0}} + \dfrac{h_4}{W_{s0}}$

$= \dfrac{124.6}{3 \times 20.5} + \dfrac{3.2}{20.5} + \dfrac{2 \times 9}{20.5 + 20.5} + \dfrac{25}{20.5} = 3.8407$

Slot Leakage Flux (ϕ_s) = 2 × √2 × μ0 × Iph × Zs × Ls/1000 × λ_s

= 2 × √2 × 4 × π × 10^{-7} × 1574.5 × 4 × 1680/1000 × 3.8407 = 0.0722 Wb

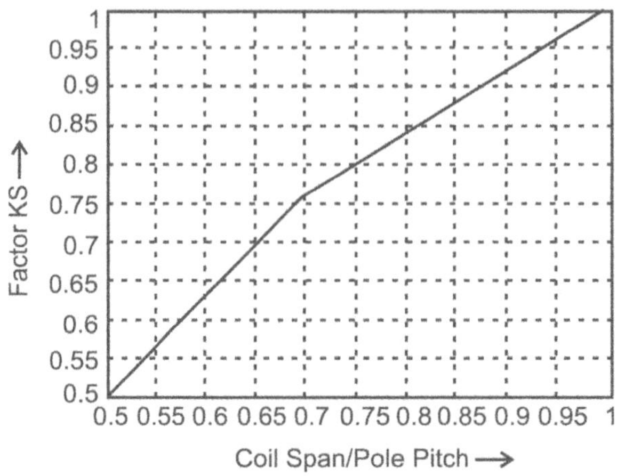

Fig. 6.9 Leakage factor.

Corresponding to Coil span/Pole pitch ratio = 26/27 = 0.963, $K_s = 0.9728$ from above graph (Fig. 6.2)

$$L_0\lambda_0 = \frac{K_s \times \tau_p^{\ 2}}{\pi\tau_s} = \frac{0.9728 \times 1313.2^2}{\pi \times 48.6365 \times 1000} = 10.979$$

where ($\tau_p \rightarrow$ Pole pitch and $\tau_s \rightarrow$ Slot Pitch)

Overhang Leakage Flux $(\phi_0) = 2 \times \sqrt{2} \times \mu_0 \times I_{Ph} \times Z_s \times L_0\lambda_0$

$$= 2 \times \sqrt{2} \times 4 \times \pi \times 10^{-7} \times 1574.5 \times 4 \times 10.979 = 0.1229$$

Total Leakage Flux $= \phi_L = \phi_S + \phi_0 = 0.0722 + 0.1229 = 0.1951$

Leakage Rectance $(X_L) = \dfrac{\phi_L}{\phi} = \dfrac{0.1951}{1.8575} = 0.105\text{pu} = 10.5\%$

Weights and Losses

Weight of Core (Kgcore) = $7.85 \times 10^{-6} \times Li \times \pi/4 \times (Do^2 - (Di + 2 \times Hs)^2)$

$= 7.85 \times 10^{-6} \times 1545.6 \times pi/4 \times (2260^2 - (836 + 2 \times 165)^2) = 35716$ Kg

Weight of Stator Steel (Kgstst) = $7.85 \times 10^{-6} \times Li \times (\pi \times (Do^2 - Di^2)/4 - S \times Ws \times Hs)$

$= 7.85 \times 10^{-6} \times 1545.6 \times (\pi \times (2260^2 - 836^2)/4 - 54 \times 20.5 \times 165) = 39795$ Kg

Weight of Teeth (Kgteth) = Kgstst – Kgcore = 39795 – 35716 = 4079 Kg;

Wt of Stator Copper (Kgstcu) = 3 × Lmt1 × As × Tph × 8.9/1000

$= 3 \times 8.083 \times 529.2 \times 18 \times 8.9/1000 = 2055.8$Kg

Stator DC copper Loss (Pcu1) = $3 \times Iph^2 \times Rph \times 1.22/1000$

$= 3 \times 1574.5^2 \times 0.005774 \times 1.22/1000 = 52.39$ KW

$$\text{Ampcond/m(ac)} = \frac{\text{Iph} \times \text{Tph} \times 2 \times 3}{\pi \times \text{Di}/1000} = \frac{1574.5 \times 18 \times 6}{\pi \times 836/1000} = 64746$$

$$\text{Addl Loss in End Wdg (Padd)} = 90 \times \text{Di}^2 \times \sqrt{\text{L}/1000} \times (\text{ac}/750)^2 \times 10^{-10}$$

$$= 90 \times 836^2 \times \sqrt{2050/1000} \times (64746/750)^2 \times 10^{-10} = 67.12 \text{ Kw.}$$

Assuming Stator Core is made with Silicon Steel Laminations of Specific Loss (1.8 W/Kg at 1 Tesla),

Iron Loss (Pi) = (1.8 × Kgteth × Bt1b3^2 + 1.8 × Kgcore × Bc^2)/1000

= (1.8 × 4079 × 1.2888^2 + 1.8 × 35716 × 1.0985^2)/1000 = 89.77 KW

6.3.3 *(a) Computer Program in "C" in MATLAB for Part-2*

```
% ----- (Part-2) Stator Winding and Core ---------------------
>

Di = 836; L = 2050; FI = 1.8575; Iph = 1574.5; Hc = 1.8;
Hins = 12; Hw = 9; HL = 25; % Input Data

Cins = 0.25; HVins = 3; P = 2; CP = 26; Bt0 = 1.6; cds1 = 3;
Bc = 1.1; S = 54; Li = 1545.6; % Input Data

PP=1313.2; V = 11000; Tph = 18; Zs = 2; Ls = 1680; % Input Data

sp=pi*Di/S; bt0=FI*1e6/(Bt0*Li*S/P); Ws=sp-bt0; As1 = Iph/cds1;

Wc1=((Ws-2*HVins-2.5)/2); Wc=floor(Wc1); Ncv1 = As1/(Hc*Wc*2);
Ncv = ceil(Ncv1); As = 0.98*Ncv*Hc*Wc*2; cds1 = Iph/As;
Ws = 2*Wc + 2*HVins + 2.5;

Hbar=(Ncv + 1)*(Hc + Cins)+ 2*HVins; Hs1=2*Hbar+Hins + Hw + HL;

Hs = ceil(Hs1); p = 2*Ncv; alfa = sqrt(2*Wc/Ws);
KdTL=1+(alfa*Hc/10)^4*p*(p-1)/3;
Kdav=1+(alfa*Hc/10)^4*(2*Ncv)^2/9;

Lmt1 = (2*L/10 + 2.5*PP/10 + 5*V/1e3 + 15)/100;

Rph=0.021*Lmt1*Tph/As;Pcus=3*Iph^2*Rph/1e3;Peddy=(Kdav-1)*Pcus;

Pts = (Pcus + Peddy)*1.15; Reff = Iph*Rph*Kdav/(V/sqrt(3));

bt0=sp-Ws; Bt0=FI*1e6/(bt0*Li*S/P); bt1b3=pi*(Di+Hs*2/3)/S- Ws;

Bt1b3=FI*1e6/(bt1b3*Li*S/P);FIc=FI/2;       dc1=FIc/(Bc*Li)*1e6;
Do1 = Di+ 2*(Hs + dc1); Do=ceil(Do1/10)*10; dc=(Do-Di- 2*Hs)/2;
Bc = FIc/(dc*Li)*1e6;

Stsbh=Hs - (Hw + HL + 7.5 + Zs*((Ncv + 1)*(Hc + 0.25) + 7.06));

Stsbw = Ws - (2*(Wc + 0.25) + 2*HVins + 1);
```

```
% ----------Reactances--------------------------------------->
h1=(Ncv+1)*Zs*(Hc+Cins)+Zs*2*HVins+6;        h2=(Hs-h1-HL-Hw)/2;
h3= Hw; h4 = HL; Ws0 = Ws;
lamdas = (h1/3/Ws) + (h2/Ws) + 2*h3/(Ws + Ws0) + h4/Ws0;
phis = 2.*sqrt(2.)*4*pi*1e-7*Iph*Zs*Ls/1000.*lamdas;
ks=(CP*P/S-0.66)*0.7354+0.75; %From Fig:6.9,Csp=0.66andks= 0.75
L0Ld0=ks*PP*PP/pi/sp/1000;
phi0=2*sqrt(2)*4*pi*1e- 7*Iph*Zs*L0Ld0;
phil=phis + phi0; Lflux = phil/FI; XLpu = Lflux; XL = XLpu*100;
%----------------Weights and Losses--------------------
Kgrtcu = 8.9e - 6*2.*Tp2*Lmt2*1e3*Wcu2*Hcu2;
Kgrtst = 7.85E -6*pi*Dr*Dr/4* (L + 80);
Pmech = 1.15*Dr^4*sqrt(L)*1E - 11; Pbrg = 3.8e - 6*Dbrg^2*Lbrg;
Pvent = Pmech - Pbrg;
Kgcore = 7.85E - 6*Li*pi/4*(Do^2 - (Di + 2*Hs)^2);
Kgstst = 7.85E - 6*Li*(pi*(Do^2 - Di^2)/4 - S*Ws*Hs);
Kgteth = Kgstst - Kgcore; Kgstcu = 3*Lmt1*As*Tph*8.9e - 3;
Pcu1=3*Iph^2*Rph*1.22/1e3;      ac = Iph*Tph*2*3/(pi*Di/1000);
Padd = 90*Di^2*sqrt(L/1000.)*(ac/750)^2*1e - 10;
Pi = (1.8*Kgteth*Bt1b3^2 + 1.8*Kgcore*Bc^2)/1000;
```

6.3.4 Design of Rotor-Winding (Part-3)

Assuming rotor slot pitch (Sp2) = 1/36 and 2/3 rd area is occupied by slots,

No. of Rotor Slots (S2) = $2/3 \times 36 = 24$

Assuming Exciter Rated Voltage (Vex1) = 220 V and taking 20% reserve,

Voltage across each Field Coil (Vc) = $0.8 \times \text{Vex1}/P = 0.8 \times 220/2 = 88$ V

AT on Full-Load (ATFL) $\approx 1.5 \times \text{ATa} = 1.5 \times 36540 = 54810$

Mean Length of Turn (Lmt2) = $(2 \times L/10 + 1.8 \times PP/10 + 25)/100$

$= (2 \times 2050/10 + 1.8 \times 1313.2/10 + 25)/100 = 6.7138$ m

Area of CS of conductor (Af) = $0.021 \times \text{Lmt2} \times \text{ATFL/Vc}$

$= 0.021 \times 6.7138 \times 54810/88 = 87.81 \text{ mm}^2$

Assuming Current density (cds2) = 4 A/ mm^2,

Conductors/Slot (Zs2) = $2 \times P \times \text{ATFL}/(\text{cds2} \times \text{Af} \times \text{S2})$

$$= \frac{2 \times P \times ATFL}{cds2 \times Af \times S2} = \frac{2 \times 2 \times 54810}{4 \times 87.81 \times 24} = 26.01 \cong 26 \text{ mm}$$

Turs/Pole (Tp2) = Zs2 × S2/(2 × P) $= \frac{Zs_2 \times S_2}{2 \times P} = \frac{26 \times 24}{2 \times 2} = 156$

Height of Copper cond (Hcu2) = $\frac{A_f}{W_{cu2}} = \frac{87.814}{28} = 3.1362 \text{ mm} \approx 3.2 \text{ mm}$

Revised Area of CS of conductor (Af) = Wcu2 × Hcu2 = 28 × 3.2 = 89.6 mm^2

Resistance of winding/Pole (Rfp) = 0.021 × Lmt2 × Tp2/Af

= 0.021 × 6.7138 × 156/89.6 = 0.2455 Ω

Field Current at Full-Load (IfFL) = $\frac{V_c}{R_{fp}} = \frac{88}{0.2455} = 358.5$ Amps

Wound Slot ht (Hs2) = Zs2 × Hcu2 + (Zs2 – 1) × 0.4 + 14

= 26 × 3.2 + (26 – 1) × 0.4 + 14 = 107.2 ≈ 108 mm

Ht-wise Slot balance (Rtsbh) = Hs2 – (Zs2 × Hcu2) – (Zs2 – 1) × 0.4 – 14

= 108 – (26 × 3.2) – (26 – 1) × 0.4 – 14 = 0.8 mm → (OK)

Width-wise Slot balance (Rtsbw) = Ws2 – Wcu2 – 2.9

= 31 – 28 – 2.9 = 0.1mm → (OK)

Cross-Section of Rotor Slot (dimensions in mm)

Height-Wise		
Wedge		25.0
Ins.under wedge		6.0
Cu.Coductors	26 × 3.2	83.2
Inter-turn Insulation	(26 – 1) × 0.4	10.0
Bottom - Insulation		8.0
Slack		0.8
Total		**133.0**

Width-Wise		
Cu.Conductor	1 × 28	28.0
Insulation on sides	2 × 1.45	2.9
Slack		0.1
Total		**31.0**

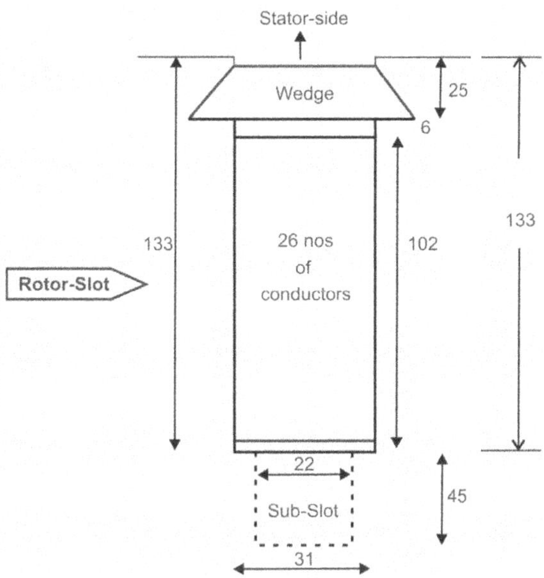

Wpoles = Ap × Lp × P × 7800 = 0.0686 × 0.6585 × 16 × 7800 = 5637.6 Kg

Diameter at bottom of Sub – Slot (dk) = Dr – 2 × (Hs2 + Hw2 + Hss)

= 800 –2 × (108 + 25 + 45) = 444 mm

Wide Tooth Pitch (wtp) = (S2/4 + 1)/SP2 = (24/4 + 1)/36 = 0.1944

Wt. of Rotor Copper (Kgrtcu) $= 8.9 \times 10^{-3} \times 2. \times Tp2 \times Lmt2 \times Wcu2 \times Hcu2$

$= 8.9 \times 10^{-3} \times 2. \times 156 \times 6.7138 \times 28 \times 3.2 = 1670.4$ Kg

Wt. of Rotor Steel (Kgrtst) $= 7.85 \times 10^{-6} \times \pi \times Dr^2/4 \times (L + 80)$

$= 7.85 \times 10^{-6} \times \pi \times 800^2/4 \times (2050 + 80) = 8404.6$ Kg

Mechanical Loss (Pmech) $= 1.15 \times Dr4 \times \sqrt{L} \times 10^{-11}$

$= 1.15 \times \sqrt{2050} \times 10^{-11} = 213.3$ KW

Bearing Loss (Pbrg) $= 3.8 \times 10^{-6} \times Dbrg^2 \times Lbrg$

$= 3.8 \times 10^{-6} \times 250^2 \times 250 = 59.4$ KW

Ventilation Loss (Pvent) = Pmech-Pbrg = 213.3 – 59.4 = 153.9 KW

6.3.4 *(a) Computer Program in "C" in MATLAB for Part-3*

```
% ---------------Part-3 (Rotor Winding)---------------->.
P = 2; ATa = 36540; L = 2050; PP = 1313.2;
Dri = 60; SP2 = 36; Ws2 = 31; Hs2 = 116; Hw2 = 25; Wss = 22;
Hss = 45; Wcu2 = 28; % Input Rotor Data
Vex1 = 220; cds2 = 4 % Input Rotor Data
S2 = 2/3*SP2; Vc = 0.8*Vex1/P; ATFL = 1.5*ATa;
Lmt2 = (2*L/10 + 1.8*PP/10 + 25)/100;
Af1 = 0.021*Lmt2*ATFL/Vc; Zs2a = 2*P*ATFL/(cds2*Af1*S2);
Zs2 = ceil(Zs2a);if (Zs2 - Zs2a)> = 0.5 Zs2 = floor(Zs2a); end;
Tp2 = Zs2*S2/(2*P); Hcu2a = Af1/Wcu2; Hcu2 = ceil(Hcu2a*10)/10;
Af=Wcu2*Hcu2; Rfp=0.021*Lmt2*Tp2/Af; Rf20=Rfp*P; IfFL = Vc/Rfp;
Hs2a = Zs2*Hcu2 + (Zs2 - 1)*0.4 + 14; Hs2 = ceil (Hs2a);
Rtsbh=Hs2-Zs2*Hcu2-(Zs2-1)*0.4 - 14; Rtsbw = Ws2 - Wcu2 - 2.9;
dk = Dr - 2*(Hs2 + Hw2 + Hss); wtp = (S2/4. + 1)/SP2;
------------------------------------------------------------
```

6.3.5 Ampere Turns for Various Parts of Magnetic Circuit at 100%V (Part-4)

Rotor Length $\approx$ L + 80 = 2050 + 80 =2130 mm

Wide tooth width in rotor (tzi) $= \pi \times Di \times wtp = \pi \times 836 \times 0.1944 = 510.57$ mm

$$\text{ratio1} = \frac{\text{St} - \text{SlotWidth(Ws)}}{\text{Airgap(Lg)}} = \frac{20.5}{18} = 1.1389$$

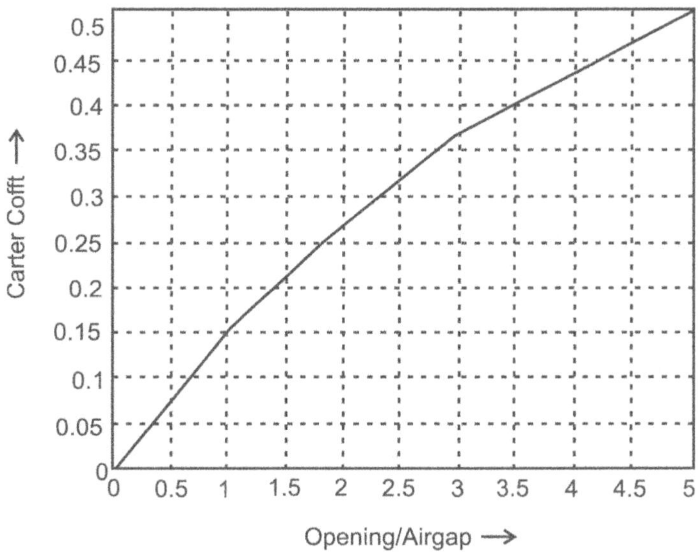

Fig 6.10 Carter Coefft. for open slots

Carters Coefficient (k01) read from graph (Fig. 6.10) corresponding to 1.1389 is 0.1681

$$\text{ratio2} = \frac{\text{Rt} - \text{SlotWidth(Ws2)}}{\text{Airgap(Lg)}} = \frac{31}{18} = 1.7222$$

Carters Coefficient (k02) read from graph (Fig. 6.10) corresponding to 1.7222 is 0.2387

$$\text{ratiod} = \frac{\text{St} - \text{Vent} - \text{ductWidth(bv)}}{\text{Airgap(Lg)}} = \frac{10}{18} = 0.5556$$

Carters Coefficient (k1) read from graph (Fig. 6.10) corresponding to 0.5556 is 0.0881

$$\text{St} - \text{SlotCoefft(ks1)} = \frac{\text{SlotPitch}}{\text{SlotPitch} - \text{Slotwidth} \times \text{CarterCoefft}} = \frac{48.6365}{48.6365 - 20.5 \times 0.1681} = 1.0762$$

$$\text{Rt} - \text{SlotCoefft(ks2)} = \frac{\text{SlotPitch}}{\text{SlotPitch} - \text{Slotwidth} \times \text{CarterCoefft}} = \frac{69.8132}{69.8132 - 31 \times 0.2397} = 1.1185$$

$$\text{Duct} = \text{Coefft(kdl)} = \frac{\text{bv}}{\text{bv} - \text{k1} \times \text{bv}} = \frac{10}{10 - 0.0881 \times 10} = 1.0966$$

kgs = ks1 × kd1 = 1.0762 × 1.0966 = 1.1802

kgs = (ks1 + ks2 – 1.) × kd1 = (1.0762 + 1.1185 – 1) × 1.0966 = 1.3102

Length of Flux path in stator (Lfps)

$$= \frac{2/3 \times \pi \times (D_0 - dc)}{1000 \times 2} = \frac{\pi \times (2260 - 547)}{1000 \times 2} = 1.7938\,\text{m}$$

Area for flux in rotor (Smax) = (Lr + 0.3333 × dk) × dk/10^6

= (2130 + 0.3333 × 444) × 444/10^6 = 1.0114 m^2

$$\text{Length of Flux path (Lsmax)} = \frac{dk - Dri}{2 \times 1000} = \frac{444 - 60}{2 \times 1000} = 0.192\ \text{m}$$

At Epu = 1 and ELL = Epu × V = 1 × 11000 = 11000 V

Flux = Epu × FI = 1 × 1.8575 = 1.8575 Wb

(a) Airgap AT (ATgap) = 8 × 10^5 × Bg × kgus × Lg/1000

= 0.8 × 10^5 × 1.035 × 1.1802 × 18/ 1000 = **17589**

(b) Stator Teeth

Flux-density at 1/3 ht from tooth tip (Bst1b3)

$$= \frac{\Phi \times 10^6}{b_{t1b3} \times Li \times S/P} = \frac{1.8575}{34.356 \times 1545.6 \times 54/2} = 1.2888\text{T}$$

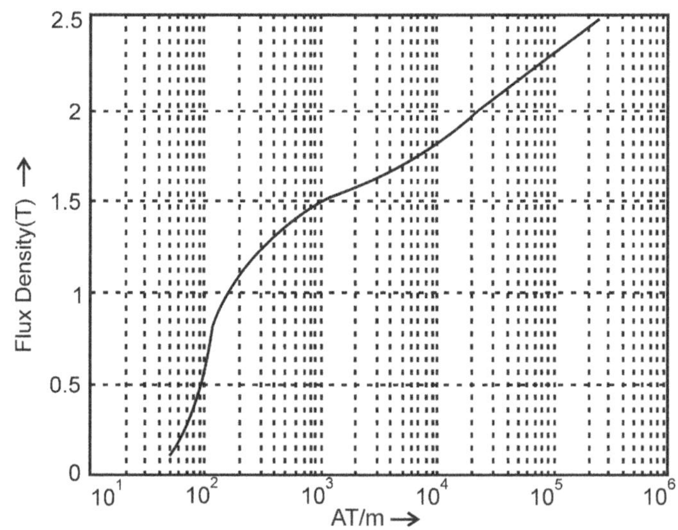

Fig. 6.11 Magnetization curve for Lowhys stamping steel.

AT/m for tooth corresponding to Bst1b3 = 1.2888 T as read from magnetization curve (Fig. 6.11) = 385.1

AT required for the **Stator tooth** (for total tooth height of 165mm)

= ATst = 385.1 × 165/1000 = **63.5**

(c) Stator Core

Actual Flux density in core (Bstc) $= \dfrac{\phi \times 0.5 \times 10^6}{dc \times L_i} = \dfrac{1.8575 \times 0.5 \times 10^6}{547 \times 1545.6} = 1.0985$ T

AT/m for Core corresponding to Bstc = 1.0985T as read from magnetization curve (Fig. 6.11) = 219.17

AT required for the **Stator Core** (ATsc) = 219.17 × 165/1000 = **393.2**

(d) Rotor Teeth

Flux-density at 1/3 ht from tooth tip (Brt1b3)

$$= \frac{Epu \times Bg \times tzi \times Li \times 1.36 \times 10^6}{Lr \times (\pi \times (Dr - 4/3 \times (Hs2 + Hw2 + Hss)) \times wtp)}$$

$$= \frac{1 \times 1.035 \times 510.57 \times 1545.6 \times 1.36 \times 10^6}{2130 \times (\pi \times (800 - 4/3 \times (108 + 25 + 45)) \times 0.1944)} = 1.5176 \text{ T}$$

AT/m for Rotor tooth corresponding to 1.5176T as read from magnetization curve (Fig. 6.11) = 1159.8

(Magnetization curve for Rotor is presumed same as that of Lowhys Steel.)

AT required for the **Rotor tooth (ATrt)** of (Hs2 + Hw2 + Hss) = 1159.8 × (108 + 25 + 45)/1000 = 1159.8 × 178/1000 = 206.45

(e) Rotor Core

Flux density in the Rotor Core (Brtc)

= Epu × 1.05 × FI/Smax = 1 × 1.05 × 1.8575/ 1.0114 = 1.9283 T

AT/m for Rotor Core (atrc) corresponding to 1.9283 T as read from magnetization curve (Fig. 6.11) = 17110

(Magnetization curve for Rotor is presumed same as that of Lowhys Steel.)

AT required for the **Rotor tooth (ATrc)** = atrc × Lsmax = 17110 × 0.1920 = 3285.1

Total No-Load AT for the Magnetic Circuit at 100% rated Voltage (ATtot)

= ATgap + ATst + ATsc + ATrt + ATrc = 17589 + 63.5 + 393.2 + 206.45 + 3285.1 = **21538**

No-Load Field Current at 100% V (Ifocc) = ATtot/Tp2 = 21538/156 = 138.06 Amp

Air-Gap Field Current at 100% V (Ifgap) = ATgap/Tp2 = 17589/156 = 112.75 Amp

6.3.5 *(a) Computer Program in "C" in MATLAB for Part-4*

```
%--------------- O.C.C Calculation:--------------------*/

P=2;L=2050;Di=836;wtp=0.1944;Ws=20.5;Lg=18;Ws2=31; %Input Data

sp=48.6365;Dr=800;SP2=36;Do=2260;dc=547;dk=444;Dri=60;
%Input Data

Li=1545.6;Hs=165;FI=1.8575;Bg=1.035;Hs2=108;Hw2=25; %Input Data

Hss=45;V=11000;S=54;bt1b3=34.536;Tp2=156;bv=10;%Input Data

%-------------------------------------------------------------

Lr=L+80;   tzi=pi*Di*wtp;

R=[0 1 2 3 4 5]; CC=[0 .15 0.27 .37 .44 .5];

ratio1=Ws/Lg;      k01=interp1(R,CC,ratio1,'spline');

ratio2=Ws2/Lg;     k02=interp1(R,CC,ratio2,'spline');

ratiod=bv/Lg;      k1=interp1(R,CC,ratiod,'spline');

ks1=sp/(sp-k01*Ws); ys2=pi*Dr/SP2;     ks2=ys2/(ys2-k02*Ws2);

kd1=bv/(bv-k1*bv);    kgus=ks1*kd1;        kgs=(ks1+ks2-1.)*kd1;
Lfps=2/3*pi*(Do-dc)/1e3/P;Smax=(Lr+0.3333*dk)*dk/1e6;
Lsmax=(dk-Dri)/2e3;

%------------------------------------------------------------>

Epu(1)=1.0;Epu(2)=1.1;Epu(3)=1.2;Epu(4)=1.3;
for ij=1:25;BB(ij)=0.1*ij;end;

%------B-H Curve From graph(Lohys)-------->

HH=[50 65 70 80 90 100 110 120 150 180 220 295 400 580 1e3 2400
5e3 8900 15e3 24e3 40e3 65e3 10e4 15e4 24e4];

for j=1:4;

ELL(j)=Epu(j)*V; flux(j)=Epu(j)*FI;

ATgap(j)=8e5*Epu(j)*Bg*Lg /1e3*kgus;

Bst1b3(j)=Epu(j)*FI*1e6/(bt1b3*Li*S/P);
atst(j)=interp1(BB,HH,Bst1b3(j),'spline');
ATst(j)=atst(j)*Hs/1e3;  Bstc(j)=flux(j)*0.5*1e6/(dc*Li);
```

```
atsc(j)=interp1(BB,HH,Bstc(j),'spline');  ATsc(j)=atsc(j)*Lfps;
Brt1b3(j)=Epu(j)*Bg*tzi*Li*1.36/Lr/(pi*(Dr-4./3.*(Hs2+Hw2+Hss))
*wtp);
atrt(j)=interp1(BB,HH,Brt1b3(j),'spline');
ATrt(j)=atrt(j)* (Hs2+Hw2+Hss)/1e3;Brtc(j)=Epu(j)*1.05*FI/Smax;
atrc(j)=interp1(BB,HH,Brtc(j),'spline'); ATrc(j)=atrc(j)*Lsmax;
ATtot(j)=ATgap(j)+ATst(j)+ATsc(j)+ATrt (j)+ATrc(j);
Ifocc(j)=ATtot(j)/Tp2;Ifgap(j)=ATgap(j)/Tp2;
IFOCC(j+1)=Ifocc(j);EPU(j+1)=Epu(j);end;IFOCC(1)=0;EPU(1)=0;
plot(IFOCC,EPU);grid;xlabel('If(amps)-->');
ylabel('Volts(PU)-->');
title('OCC of Non-Salient Pole Generator');
%--------------------------------------------------
```

6.3.6 Open Circuit Characteristic Curve (OCC) Part-5

A plot of Ampere Turns on X-Axis and Various Terminal voltages (0% to 130% of rated Votage)

Proceedure for Plotting Open Circuit Characteristic Curve (OCC)

1. Consider Terminal Voltages of 100%, 110%, 120% and 130% of rated voltage.
2. Fill up all the details pertaining to Air-Gap, Stator tooth, Stator Core, Rotor Tooth and Rotor Core calculated above under the 100% Voltage Column in the table.
3. Calculate the Flux densities in various parts at other 3 Voltages which are linearly proportional to Voltage (Ex: If Flux density = 1.1T at 100% Volts, then at 110%Volts it will be (110/100) × 1.1 = 1.21T).
4. At these values of Flux densities measure the "**AT/m**" from Magnetization curve (as was referred for calculation for 100%V).
5. Calculate Values of **"AT"** for each part by directly multiplying the **"AT/m"** value with corresponding **"length"** of each part.
6. Add values of "AT" of all the parts under each %Voltage Colum to get the **Total AT** required for each voltage.
7. Plot Values of AT on X-Axis and % Rated Voltages on Y-Axis to get **OCC**.

Open Circuit Characteristic Curve (OCC) of Non-Salient Pole Generator

Part	Length	100%(	3300	Volts)	110%(	3630	Volts)	120%(	3960	Volts)	130%(	4290	Volts)
	m	B(tesla)	AT/m	AT	B(tesla)	AT/m	AT	B(tesla)	AT/m	AT	B(tesla)	AT/m	AT
Air-Gap	0.018	1.035		17589	1.1385		19347.9	1.242		21106.8	1.3455		22865.7
Stator Teeth	0.165	1.2888	385.1	64	1.41768	619.6	102.234	1.54656	1502.6	248	1.67544	4260.8	703
Stator-Core	1.7938	1.0985	219.17	393	1.20835	302.25	542.1761	1.3182	426.93	766	1.42805	645.92	1159
Rotor Teeth	0.178	1.5176	1160	206	1.66936	4085	727.13	1.82112	9998	1780	1.97288	21038	3745
Rotor-Core	0.192	1.9283	17110	3285	2.12113	44500	8544	2.31396	105720	20298	2.50679	248380	47689
Total-AT=				**21537**			**29263.44**			**44198.8**			**76161.7**
Field-Amps	(OCC)			**138.1**			**187.6**			**283.3**			**488.2**

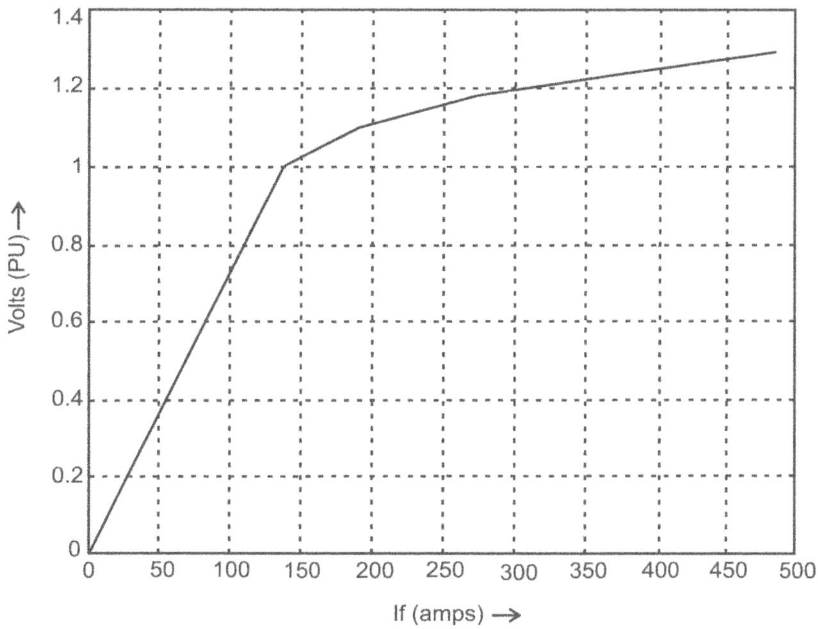

Fig. 6.12 OCC of non-salient pole generaotor.

6.3.6 *(a) Computer Program in "C" in MATLAB for Part-5*

```
%-------------------(OCC Calculation)------------------------->
P=2;L=2050;Di=836;wtp=0.1944;Ws=20.5;Lg=18;Ws2=31;sp=48.6365;
 Dr=800;SP2=36;Do=2260;dc=547;dk=444;Dri=60;%Input Data
Li=1545.6;Hs=165;FI=1.8575;Bg=1.035;Hs2=108;Hw2=25; %Input Data
Hss=45;V=11000;S=54;bt1b3=34.536;Tp2=156;bv=10; %Input Data
%-------------------------------------------------------
Lr=L+80;tzi=pi*Di*wtp;ratio1=Ws/Lg;ratio2=Ws2/Lg;ratiod=bv/Lg;
R=[0 1 2 3 4 5]; CC=[0 .15 0.27 .37 .44 .5];
k01=interp1(R,CC,ratio1,'spline');
k02=interp1(R,CC,ratio2,'spline');
k1=interp1(R,CC,ratiod,'spline');
ks1=sp/(sp-k01*Ws);ys2=pi*Dr/SP2;ks2=ys2/(ys2-k02*Ws2);
kd1=bv /(bv-k1*bv);   kgus=ks1*kd1;kgs=(ks1+ks2-1.)*kd1;
Lfps=2/3*pi*(Do-dc)/1e3/P;   Smax=(Lr+0.3333*dk)*dk/1e6;
Lsmax=(dk-Dri)/2e3;
```

```
Epu(1)=1.0;Epu(2)=1.1;Epu(3)=1.2;Epu(4)=1.3;

for ij=1:25;BB(ij)=0.1*ij;end;

%-------------B-H Curve From graph for Lohys steel

HH=[50 65 70 80 90 100 110 120 150 180 220 295 400 580 1e3 2400
5e3 8900 15e3 24e3 40e3 65e3 10e4 15e4 24e4];

%semilogx(HH,BB);grid;xlabel('AT/m-->');

ylabel('Flux density(T)-->');

%title('Magnetization Curve for Lowhys Stamping Steel');

for j=1:4;ELL(j)=Epu(j)*V;flux(j)=Epu(j)*FI;

ATgap(j)=8e5*Epu(j)*Bg*Lg/1e3*kgus;

Bst1b3(j)=Epu(j)*FI*1e6/(bt1b3*Li*S/P);

atst(j)=interp1(BB,HH,Bst1b3(j),'spline');

ATst(j)=atst(j)*Hs /1e3;

Bstc(j)=flux(j)*0.5*1e6/(dc*Li);

atsc(j)=interp1(BB,HH,Bstc(j),'spline');ATsc(j)=atsc(j)*Lfps;

Brt1b3(j)=Epu(j)*Bg*tzi*Li*1.36/Lr/

(pi*(Dr-4./3.*(Hs2+Hw2+Hss))*wtp);

atrt(j)=interp1(BB,HH,Brt1b3(j),'spline');

ATrt(j)=atrt(j)*(Hs2+Hw2+Hss)/1e3;Brtc(j)=Epu(j)*1.05*FI/Smax;

atrc(j)=interp1(BB,HH,Brtc(j),'spline');ATrc(j)=atrc(j)*Lsmax;

ATtot(j)=ATgap(j)+ATst(j)+ATsc(j)+ATrt(j)+ATrc(j);

Ifocc(j)=ATtot(j)/Tp2;Ifgap(j)=ATgap(j)/Tp2;

IFOCC(j+1)=Ifocc(j);EPU(j+1)=Epu(j);end;IFOCC(1)=0;EPU(1)=0;

%plot(IFOCC,EPU);grid;xlabel('If(amps)-->');ylabel('Volts(PU)--
>');

%title('OCC of Non-Salient Pole Generator');
```

6.3.7 Calculation of Field AT at Rated Load and PF and Other Calculations (Part-6)

Armature reaction Field Current (Ifar) = ATa/Tp2 = 36540/156 = 234.23 A

Field Current on Short Circuit at Rated Armature current (Ifsc)

$$= \frac{\text{ATa} + \text{XLpu} \times \text{ATgap(1)}}{\text{Tp2}} = \frac{36540 + 0.105 \times 17589}{156} = 246.07\text{Amps}$$

No load Field Current (If0) = Ifocc (1) = 138.06 A

Short Circuit Ratio (scr) = If0/Ifsc = 138.06/246.07 = 0.5611

$$\text{pfr} = \sqrt{1 - \text{pf}^2} = \sqrt{1 - 0.8^2} = 0.6$$

$$\text{E1} = \sqrt{1 \times \text{pf}^2 + (1 \times \text{pfr} + \text{XLpu})^2} = \sqrt{1 \times 0.8^2 + (1 \times 0.6 + 0.105)^2} = 1.0663 \text{ pu}$$

Field Current (If1) corresponding to E1 = 1.0663pu read from OCC = 168.9 amps

Field Current at rated Load and Pf (IfFL)

$$= \sqrt{\text{If}_1^{\,2} + \text{If}_{ar}^{\,2} - 2 \times \text{If}_1 \times \text{If}_{ar} \times \cos(\pi / 2 \times \cos^{-1}(\text{pf})}$$

$$= \sqrt{168.9^2 + 234.23^2 - 2 \times 168.9 \times 234.23 \times \cos(\pi / 2 \times \cos^{-1}(0.8)} = 361.7 \text{ A}$$

Open circuit voltage (E0) corresponding to IfFL

= 361.7A, read from OCC = 1.2327 pu

$$\text{Voltage Regulation} = \frac{E_0 - V}{V} = \frac{1.2327 - 1}{1} = 0.23267\text{pu} = 23.267\%$$

Rotor Current Density (cds2) = IfFL/Af = 361.7/89.6 = 4.04 A/mm^2

Voltage Across Rotor Wdg at Rated Ld and PF (Vfn)

= IfFL × Rf20 × (235. + 120.)/255

= 361.7 × 0.4909 × (355/255) = 247.21 Volts

Voltage rating of Exciter (Vex) = Vfn × 1.1 = 247.21 × 1.1 = 271.9 V(Say 275V)

Current Rating of Exciter (Iex) = Ifn × 1.1 = 361.7 × 1.1 = 397.9 (Say 400A)

KW rating of Exciter (KWex) = Vex × Iex = 275 × 400 = 110 KW

Rotor Copper Loss (Pcu2) = IfFL^2 × Rf20 × (235. + 75.)/255 =

361.7^2 × 0.4909 × (310/ 255)/1000 = 78.09 KW

Total Weight of Active Material (KgT) = Kgstst + Kgstcu + Kgrtcu + Kgrtst

= 39795 + 2055.8 + 1670.4 + 8404.6 = 51926 Kg

Considering Insulation Wt as1%, Total Weight of Active Material (KgT)

= 1.01 × (Kgstst + Kgstcu + Kgrtcu + Kgrtst)

= 1.01 × (39795 + 2055.8 + 1670.4 + 8404.6) = 52445 Kg

Kgpkva = KgT/KVA = 52445/30000 = 1.748

Total Losses (Pt) = Pcu1 + Padd + Pmech + Pcu2 + Pi

$$= 52.39 + 67.12 + 213.3 + 78.1 + 89.77 = 500.65 \text{ KW};$$

Efficiency at rated Ld and PF (Eff)

$$= \text{KW}/(\text{KW} + \text{Pt}) \times 100 = 24000/(24000 + 500.65) \times 100 = 97.96\ \%$$

Assuming Temp-Rise of cooling Air as 24° C, Quantity of Cooling Air (Qair)

$$= \frac{\text{Pcool} \times 1000}{1140 \times 24} = \frac{441.3 \times 1000}{1140 \times 24} = 16.13 \text{ (say 17 m}^3\text{/s)}$$

$$\text{Air Temp Rise} = \frac{\text{Pcool} \times 1000}{1140 \times \text{Qair}} = \frac{441.3 \times 1000}{1140 \times 17} = 22.77\ ^\circ\text{C}$$

6.3.7 *(a) Computer Program in "C" in MATLAB for Part-6*

```
%----------LMC and Other Calculations:------------------->
ATa=36540;XLpu=0.105;ATgap(1)=17589;Tp2=156;
Epu=[1 1.1 1.2 1.3]; %Input Data
Ifocc=[138.0616  187.5910  283.3260  488.2209];
Ifgap(1)=112.75; %Input Data
rpf=0.8;Af=89.6;Rf20=0.4909; %Input Data
%-----------------------------------------------
Ifar=ATa/Tp2;    Ifsc=(ATa+XLpu*ATgap(1))/Tp2;    If0=Ifocc(1);
scr=If0/Ifsc; pf=rpf;                         pfr=sqrt(1-pf*pf);
E1=sqrt(1.0*pf*pf+(1*pfr+XLpu)^2);
If1=interp1(Epu,Ifocc,E1,'spline');
IfFL=sqrt(If1^2+Ifar^2-2*If1*Ifar*cos(pi/2+acos (pf)));
E0=interp1(Ifocc,Epu,IfFL,'spline');reg=(E0-Epu(1))*100;
cds2=IfFL/Af;    Vfn=IfFL*Rf20*(235.+120.)/255;    Vex1=1.1*Vfn;
Vex=ceil(Vex1/5)*5;    Iex1=1.1*IfFL;         Iex=ceil(Iex/10)*10;
KWex=Vex*Iex/1e3; Pcu2=IfFL*IfFL*Rf20* (235.+75.)/255e3;
KgT=1.01*(Kgstst+Kgstcu+Kgrtcu+Kgrtst); Kgpkva= KgT/KVA;
Pt=Pcu1+Padd+Pmech+Pcu2+Pi; Eff=KW/(KW+Pt)*100.;
Qair=ceil(Pcool*1e3/1140./24.)*1.0; Tair=Pcool*1e3/1140./Qair;
%----------------------------------------------------------
```

6.3.8 Computer Program in "C" in MATLAB for Complete Design

```
%Design   of   KVA=30000;V=11000;N=3000;f=50;pf=0.8;Non-Salient
Pole Generator

f3=fopen('NSP_Gen_Output.m','w');sn=1;

KVA=30000;V=11000;N=3000;f=50;rpf=0.8;

%Star connected Non-Salient Pole Gen

Kw=0.955;bv=10;ki=0.92;Vr=125;R1=0.82;Hc=1.8;

Hins=12;Hw=9;HL=25;Cins=0.25;HVins=3;Kg=1.15;

Stsbh=2.88;Stsbw=1.14 ; Dbrg=250;Lbrg=250;Zs=2;

spp=9;CP=26;scr=0.6;Bt0=1.6;cds1=3;Bc=1.1;

Dri=60;SP2=36;Ws2=31;Hw2=25;Wss=22;Hss=45;Wcu2=28;Vex1=220;
cds2=4;%Rotor Data

%<----Bav and q------->

KVAS=[ 100  200  500 1000 5000 10000 20000 30000];

BavS=[0.54 0.56 0.58 0.60 0.63  0.65  0.67  0.69];

qS=[30e3 33e3 36e3 40e3 45e3  48e3  53e3  58e3];

%-------------Main Dimensions----------------------->

Bav=interp1(KVAS,BavS,KVA,'spline');

q=interp1(KVAS,qS,KVA,'spline');

KW=KVA*rpf;  P=120*f/N;n=2*f/P;  Iph=KVA*1e3/(1.7321*V);

 Dr=Vr/(pi*n);  Dr=ceil(Dr*100)*10;  Vr=pi*Dr/1000*n;

S=spp*P*3;   Tph=S*Zs/6;  ATa=1.35*Tph*Iph*Kw/(P/2);

Kc=sin(CP/(S/2)*pi/2);    K=11*Bav*q*Kw*Kc*1e-3;    ATf0=scr*ATa;
ATg=0.75*ATf0;        Bg=Bav*1.5;        Lg=ATg/(0.796*Bg*Kg*1e3);
Lg=ceil(Lg);  Di=Dr+2*Lg;    DsqL=KVA/(K*n);  L=DsqL/(Di/1e3)^2;
L=ceil(L*100)*10;     PP=pi*Di/P;          Ls=R1*L;nv=(L-Ls)/bv;
FI=Bav*pi*Di*L*1e-6/P;  nv=ceil(nv);Ls=(L-nv*bv);      Li=Ls*ki;
rat2=Lg/PP;

%-----(Part-2)Stator Winding and Core------------------->

Di=836;L=2050;FI=1.8575;Iph=1574.5;Hc=1.8;Hins=12;Hw=9;HL=25;
Cins=0.25; HVins=3; P=2; CP=26; Bt0=1.6; cds1=3; Bc=1.1; S=54;
Li=1545.6; PP=1313.2; V=11000; Tph=18;Zs=2; Ls=1680;
```

```
sp=pi*Di/S;  bt0=FI*1e6/(Bt0*Li*S/P);  Ws=sp-bt0;  As1=Iph/cds1;
Wc1=((Ws-2*HVins-2.5)/2); Wc=floor(Wc1); Ncv1=As1/(Hc*Wc*2);

Ws=2*Wc+2*HVins+2.5;            Hbar=(Ncv+1)*(Hc+Cins)+2*HVins;
Hs1=2*Hbar+Hins+Hw+HL;                  Hs=ceil(Hs1);      p=2*Ncv;
alfa=sqrt(2*Wc/Ws);  KdTL=1+(alfa*Hc/10)^4*p*(p-1)/3;

Kdav=1+(alfa*Hc/10)^4*(2*Ncv)^2/9;

Lmt1=(2*L/10+2.5*PP/10+5*V/1e3+15)/100;

Rph=0.021*Lmt1*Tph/As;Pcus=3*Iph^2*Rph/1e3;Peddy=(Kdav-1)*Pcus;

Pts=(Pcus+Peddy)*1.15; Reff=Iph*Rph*Kdav/(V/sqrt(3));

bt0=sp-Ws; Bt0=FI*1e6/(bt0*Li*S/P); bt1b3=pi*(Di+Hs*2/3)/S-Ws;

Bt1b3=FI*1e6/(bt1b3*Li*S/P);    FIc=FI/2;    dc1=FIc/(Bc*Li)*1e6;
Do1=Di+2*(Hs+dc1);       Do=ceil(Do1/10)*10;    dc=(Do-Di-2*Hs)/2;
Bc=FIc/(dc*Li)*1e6;

Stsbh=Hs-(Hw+HL+7.5+Zs*((Ncv+1)*(Hc+0.25)+7.06));

Stsbw=Ws-(2*(Wc+0.25)+2*HVins+1);

%----------Reactances---------------------------------------->

h1=(Ncv+1)*Zs*(Hc+Cins)+Zs*2*HVins+6;h2=(Hs-h1-HL-Hw)/2;  h3=Hw;
h4=HL;Ws0=Ws;     lamdas=(h1/3/Ws)+(h2/Ws)+2*h3/(Ws+Ws0)+h4/Ws0;

phis=2.*sqrt(2.)*4*pi*1e-7*Iph*Zs*Ls/1000.*lamdas;

ks=(CP*P/S-0.66)*0.7354+0.75;          L0Ld0=ks*PP*PP/pi/sp/1000;
phi0=2*sqrt(2)*4*pi*1e-7*Iph*Zs* L0Ld0;

phil=phis+phi0;  Lflux=phil/FI;  XLpu=Lflux;  XL=XLpu*100;

%---------WeightsandLosses---------------------------------------
--

Kgcore=7.85E-6*Li*pi/4*(Do^2-(Di+2*Hs)^2);

Kgstst=7.85E-6*Li*(pi*(Do^2-Di^2)/4-S*Ws*Hs);
Kgteth=Kgstst-Kgcore;      Kgstcu=3*Lmt1*As*Tph*8.9e-3;

Pcu1=3*Iph^2*Rph*1.22/1e3;      ac=Iph*Tph*2*3/(pi*Di/1000);

Padd=90*Di^2*sqrt(L /1000.)*(ac/750)^2*1e-10;

Pi=(1.8*Kgteth*Bt1b3^2+1.8*Kgcore*Bc^2)/1000;

%*****************************************************

fprintf(f3,'Design of Non-Salient-Pole Generator:-\n\n');

fprintf(f3,'Parameter                         VALUES\n');
```

```
fprintf(f3,'---------                                  ------');
fprintf(f3,'\nKVA                                      %5.0f',KVA);
fprintf(f3,'\nKW                                       %5.0f',KW);
fprintf(f3,'\nVolts                                    %5.0f',V);
fprintf(f3,'\nPF                                       %5.3f',rpf);
fprintf(f3,'\nHZ                                       %5.0f',f);
fprintf(f3,'\nR.P.M                                    %5.0f',N);
fprintf(f3,'\nNo of Poles                              %5.0d',P);
fprintf(f3,'\nFlux/Pole             Wb                 %5.4f',FI);
fprintf(f3,'\nGapFlux-den(Av)       Wb/m2              %6.4f',Bav) ;
fprintf(f3,'\nSp.Elec Loading       Ac/m               %6.0f',q) ;
fprintf(f3,'\n-----------------STATOR--------------->');
fprintf(f3,'\nStator Current        Amps               %5.1f',Iph);
fprintf(f3,'\nCORE LENGTH(tot)      mm                 %5.0f',L);
fprintf(f3,'\nIron LENGTH           mm                 %5.0f',Ls);
fprintf(f3,'\nVent Ducts No X       mm                %2d X%3d',nv,bv);
fprintf(f3,'\nCORE OD               mm                 %5.0f',Do);
fprintf(f3,'\nCORE ID               mm                 %5.0f',Di) ;
fprintf(f3,'\nPOLE PITCH            mm                 %6.1f',PP);
fprintf(f3,'\nTURNS/Ph                                 %5.0d',Tph);
fprintf(f3,'\nSTATOR SLOTS                             %5.0d',S);
fprintf(f3,'\nChording Pitch        --                %2d/%2d',CP,S/2);
fprintf(f3,'\nSLOT PITCH            mm                 %5.2f',sp);
fprintf(f3,'\nCore depth            mm                 %5.1f',dc);
fprintf(f3,'\nCoreFlux-den          Wb/m2              %6.4f',Bc) ;
fprintf(f3,'\nSLOT WIDTH            mm                 %5.1f',Ws);
fprintf(f3,'\nSLOT Height(tot)      mm                 %5.1f',Hs);
fprintf(f3,'\nHt(Lip+Wedge)         mm              %2.0f+%2.0f',HL,Hw);
fprintf(f3,'\nFlux-den(tooth-tip)Wb/m2                 %6.4f',Bt0) ;
fprintf(f3,'\nConductors/Slot                          %5.0f',Zs);
```

```
fprintf(f3,'\nBare cond Wdth      mm          %5.1f',Wc);
fprintf(f3,'\nBare Cond Ht        mm          %5.1f',Hc);
fprintf(f3,'\nStrip/bar(Ht-W)                 %5.0f',Ncv);
fprintf(f3,'\nCurrent density     A/mm2       %6.4f',cds1) ;
fprintf(f3,'\nAv.Loss Ratio       --          %5.4f',Kdav);
fprintf(f3,'\nMean-Turn Length    m           %5.4f',Lmt1);
fprintf(f3,'\nSt Wdg Res/Ph/20d   ohm         %6.4f',Rph);
fprintf(f3,'\nAir-gap LENGTH      mm          %5.1f',Lg);
fprintf(f3,'\nPephoral Speed      m/s         %5.1f',Vr) ;
fprintf(f3,'\nArmature Reaction   AT/P        %6.0f',ATa);
fprintf(f3,'\nLeakage Reactance   Perc        %6.1f',XL);
%=============Part-3 (Rotor Winding)======================>
P=2;Dr=800;ATa=36540;L=2050;PP=1313.2;Dbrg=250;Lbrg=250;
Dri=60;  SP2=36;  Ws2=31;  Hw2=25;  Wss=22;  Hss=45;  Wcu2=28;
Vex1=220; cds2=4;%Input Rotor Data
S2=2/3*SP2;Vc=0.8*Vex1/P;ATFL=1.5*ATa;
Lmt2=(2*L/10+1.8*PP/10+25)/100;         Af1=0.021*Lmt2*ATFL/Vc;
Zs2a=2*P*ATFL/(cds2*Af1*S2); Zs2=ceil(Zs2a);
if (Zs2-Zs2a)>=0.5 Zs2=floor(Zs2a); end;
Tp2=Zs2*S2/(2*P);Hcu2a=Af1/Wcu2;   Hcu2=ceil(Hcu2a*10)/10;
Af=Wcu2*Hcu2; Rfp=0.021*Lmt2*Tp2/Af; Rf20=Rfp*P; IfFL=Vc/Rfp;
Hs2a=Zs2*Hcu2+(Zs2-1)*0.4+14;Hs2=ceil(Hs2a);
dk=Dr-2*(Hs2+Hw2+Hss);  wtp=(S2/4.+1)/SP2;
Kgrtcu=8.9e-6*2.*Tp2*Lmt2*1e3*Wcu2*Hcu2;
Kgrtst=7.85E-*pi*Dr*Dr/4*(L+80);
Pmech=1.15*Dr^4*sqrt(L)*1E-11;
Pbrg=3.8e-6*Dbrg^2*Lbrg;  Pvent=Pmech-Pbrg;
%-----------Checking of Slot-Balances------>
Rtsbh=Hs2-Zs2*Hcu2-(Zs2-1)*0.4-14;Rtsbw=Ws2-Wcu2-2.9;
Stsbh=Hs-(Hw+HL+7.5+Zs*((Ncv+1)*(Hc+0.25)+7.06));
Stsbw=Ws-(2*(Wc+ 0.25)+2*HVins+1);
```

```
fprintf(f3,'\n  Parameter    Unit   Permissible     Designed');
fprintf(f3,'\n---------------    ----   -----------     -------
-');
fprintf(f3,'\nSt.Slot Bal(ht)     mm      > 2.0    %6.2f',Stsbh);
fprintf(f3,'\nSt.Slot Bal(wdth)   mm      > 1.0    %6.2f',Stsbw);
fprintf(f3,'\nRt.Slot Bal(ht)     mm      > 0.5    %6.2f',Rtsbh);
fprintf(f3,'\nRt.Slot Bal(wdth)   mm      > 0.1    %6.2f',Rtsbw);
%----------------------------------------------------------->
fprintf(f3,'\n------------------ROTOR--------------->');
fprintf(f3,'\nRotor Dia          mm          %5.0f',Dr);
fprintf(f3,'\nRotor bore Dia     mm          %5.0f',Dri);
fprintf(f3,'\nSlot pitches      ----         %5.0f',SP2);
fprintf(f3,'\nSlot Width         mm          %5.1f',Ws2);
fprintf(f3,'\nWound Slot height  mm          %5.1f',Hs2);
fprintf(f3,'\nWedge height       mm          %5.1f',Hw2);
fprintf(f3,'\nSub Slot Width     mm          %5.1f',Wss);
fprintf(f3,'\nSub Slot height    mm          %5.1f',Hss);
fprintf(f3,'\nCu strip Width     mm          %5.1f',Wcu2);
fprintf(f3,'\nCu strip height    mm          %5.1f',Hcu2);
fprintf(f3,'\nConductors/slot    nos         %5.1f',Zs2);
fprintf(f3,'\nTurns/Pole         nos         %5.0f',Tp2);
fprintf(f3,'\nLgth of Mean turn  m           %5.3f',Lmt2);
fprintf(f3,'\nRot Wdg Res@20d   ohm          %6.4f',Rf20);
fprintf(f3,'\n--------------O.C.C Calculation:------->');
P=2;L=2050;Di=836;wtp=0.1944;Ws=20.5;Lg=18;Ws2=31; %Input Data
sp=48.6365; Dr=800; SP2=36; Do=2260; dc=547; dk=444; Dri=60;
Li=1545.6; Hs=165; FI=1.8575; Bg=1.035; Hs2=108;  Hw2=25;
Hss=45; V=11000; S=54; bt1b3=34.536; Tp2=156; bv=10;%Input Data
%-----------------------------------------------------------
Lr=L+80;tzi=pi*Di*wtp;ratio1=Ws/Lg;ratio2=Ws2/Lg;ratiod=bv/Lg;
R=[0 1 2 3 4 5]; CC=[0 .15 0.27 .37 .44 .5];
k01=interp1(R,CC,ratio1,'spline');
```

```
k02=interp1(R,CC,ratio2,'spline');
k1=interp1(R,CC,ratiod,'spline');
ks1=sp/(sp-k01*Ws);    ys2=pi*Dr/SP2;    ks2=ys2/(ys2-k02*Ws2);
kd1=bv/(bv-k1*bv);     kgus=ks1*kd1;     kgs=(ks1+ks2-1.)*kd1;
Lfps=2/3*pi*(Do-dc)/1e3/P;
Smax=(Lr+0.3333*dk)*dk/1e6;Lsmax=(dk-Dri)/2e3;
%---------------------------------------------------------------->
Epu(1)=1.0;Epu(2)=1.1;Epu(3)=1.2;Epu(4)=1.3;
for ij=1:25;BB(ij)=0.1*ij;end;
%-------------B-H Curve From graph(for Lohys material→
HH=[50 65 70 80 90 100 110 120 150 180 220 295 400 580 1000
2400 5000 8900 15e3 24e3 40e3 65e3 10e4 15e4 24e4];
%semilogx(HH,BB);grid;xlabel('AT/m-->');
%ylabel('Flux density(T)-->');
%title('Magnetization Curve for Lowhys Stamping Steel');
for j=1:4;       ELL(j)=Epu(j)*V;flux(j)=Epu(j)*FI;
ATgap(j)=8e5*Epu(j)*Bg*Lg/1e3*kgus;
Bst1b3(j)=Epu(j)*FI*1e6/(bt1b3*Li*S/P);
atst(j)=interp1(BB,HH,Bst1b3(j),'spline');
ATst(j)=atst(j)*Hs/ 1e3;
Bstc(j)=flux(j)*0.5*1e6/(dc*Li);
atsc(j)=interp1(BB,HH,Bstc(j),'spline');ATsc(j)=atsc(j)*Lfps;
Brt1b3(j)=
Epu(j)*Bg*tzi*Li*1.36/Lr/(pi*(Dr-4./3.*(Hs2+Hw2+Hss))*wtp);
atrt(j)=interp1(BB,HH,Brt1b3(j),'spline');
ATrt(j)=atrt(j)*(Hs2+Hw2+Hss)/1e3; Brtc(j)=Epu(j)*1.05*FI/Smax;
atrc(j)=interp1(BB,HH,Brtc(j),'spline');
ATrc(j)=atrc(j)*Lsmax;
ATtot(j)=ATgap(j)+ATst(j)+ATsc(j)+ATrt(j)+ATrc(j);
Ifocc(j)=ATtot(j)/Tp2;Ifgap(j)=ATgap(j)/Tp2;
IFOCC(j+1)=Ifocc(j);EPU(j+1)=Epu(j);end;IFOCC(1)=0;EPU(1)=0;
%plot(IFOCC,EPU);grid;xlabel('If(amps)-->');
%ylabel('Volts(PU)-->');
```

```
%title('OCC of Non-Salient Pole Generator');

%--------------------------------------------------

%--------------- OCC--------------------*/

fprintf(f3,'\nE(L-L)pu          %6.2f          %6.2f          %6.2f
%6.2f',Epu(1),Epu(2),Epu(3),Epu(4));

fprintf(f3,'\n---------        -----  -----  -----  -----');

fprintf(f3,'\nE(L-L)      V      %6.0f       %6.0f        %6.0f
%6.0f',ELL(1),ELL(2),ELL(3), ELL(4));

fprintf(f3,'\nFlux(Wb)          %6.2f          %6.2f          %6.2f
%6.2f',flux(1),flux(2),flux(3),flux(4));

fprintf(f3,'\nSt.Core(Wb/m2)%6.3f%6.3f%6.3f%6.3f',Bstc(1),Bstc(
2),Bstc(3), Bstc(4));

fprintf(f3,'\nSt.Core-AT/Mt        %6.0f       %6.0f        %6.0f
%6.0f',atsc(1),atsc(2),atsc(3),atsc(4));

fprintf(f3,'-> for Lohys steel');

fprintf(f3,'\nSt.Core-AT%6.0f            6.0f                %6.0f
%6.0f',ATsc(1),ATsc(2),ATsc(3),ATsc(4));

intf(f3,'\nSt.Tooth(Wb/m2)%6.3f%6.3f%6.3f%6.3f',Bst1b3(1),Bst1b
3(2),st1b3(3),Bst1b3(4));

printf(f3,'\nSt.Tooth-AT/Mt            %6.0f     %6.0f     %6.0f
%6.0f',atst(1),atst(2),atst(3),atst(4));

fprintf(f3,'-> for Lohys steels');

fprintf(f3,'\nSt.Tooth-AT%6.0f%6.0f%6.0f
%6.0f',ATst(1),ATst(2),ATst(3),ATst(4));

fprintf(f3,'\nAir.Gap-
AT%6.0f%6.0f%6.0f%6.0f',ATgap(1),ATgap(2),ATgap(3), ATgap(4));

fprintf(f3,'\nRt.WTth(Wb/m2)%6.3f%6.3f%6.3f%6.3f',Brt1b3(1),Brt
1b3(2),Brt1b3 (3),Brt1b3(4));

fprintf(f3,'\nRt.Tooth-AT/Mt             %6.0f     %6.0f     %6.0f
%6.0f',atrt(1),atrt(2),atrt(3),atrt(4));

fprintf(f3,'-> for Lohys steels');

fprintf(f3,'\nRt.Tooth-AT%6.0f%6.0f%6.0f
%6.0f',ATrt(1),ATrt(2),ATrt(3),ATrt(4));

fprintf(f3,'\nRt.Core(Wb/m2)%6.3f%6.3f%6.3f%6.3f',Brtc(1),Brtc(
2),Brtc(3), Brtc(4));
```

```
fprintf(f3,'\nRt.Core-AT/Mt%6.0f          %6.0f          %6.0f
%6.0f',atrc(1),atrc(2),atrc(3),atrc(4));

fprintf(f3,'-> for Lohys steels');

fprintf(f3,'\nRt.Core-AT%6.0f%6.0f%6.0f%6.0f',ATrc(1),ATrc(2),
ATrc(3),ATrc(4));

fprintf(f3,'\nTotal-OCC-AT%6.0f%6.0f%6.0f%6.0f',ATtot(1),
ATtot(2),ATtot(3),ATtot(4));

fprintf(f3,'\nIf-OCC-Amps%6.1f%6.1f%6.1f%6.1f',
Ifocc(1),Ifocc(2),Ifocc(3), Ifocc(4));

fprintf(f3,'\nIf-Airgap-Amps%6.1f%6.1f%6.1f%6.1f',Ifgap(1),
Ifgap(2),Ifgap(3), Ifgap(4));

fprintf(f3,'\n--------------Other Calculations:----------------
--->');

Ifar=ATa/Tp2; Ifsc=(ATa+XLpu*ATgap(1))/Tp2; If0=Ifocc(1);

scr=If0/Ifsc;      pf=rpf;         pfr=sqrt(1-pf*pf);

E1=sqrt(1.0*pf*pf+(1*pfr+XLpu)^2);

If1=interp1(Epu,Ifocc,E1,'spline');

IfFL=sqrt(If1^2+Ifar^2-2*If1*Ifar*cos(pi/2+acos(pf)));

E0=interp1(Ifocc,Epu,IfFL,'spline');        reg=(E0-Epu(1))*100;
cds2= IfFL/Af;     Vfn=IfFL*Rf20*(235.+120.)/255;  Vex1=1.1*Vfn;
Vex=ceil(Vex1/5)*5;     Iex1=1.1*IfFL;     Iex=ceil(Iex1/10)*10;
KWex=Vex*Iex/1e3;     Pcu2=IfFL*IfFL*Rf20*(235.+75.)/255e3;

KgT=1.01*(Kgstst+Kgstcu+Kgrtcu+Kgrtst);     Kgpkva=KgT/KVA;

Pt=Pcu1+Padd+Pmech+Pcu2+Pi;Eff=KW/(KW+Pt)*100.; Pcool=Pt-Pbrg;

Qair=ceil(Pcool*1e3/1140./24.)*1.0; Tair=Pcool*1e3/1140./Qair;

%--------------------------------------------------------------

fprintf(f3,'\nIf-Arm reaction    Amps       %5.1f',Ifar);

fprintf(f3,'\nIfscc              Amps       %5.1f',Ifsc);

fprintf(f3,'\nSCR                 pu        %5.3f',scr);

fprintf(f3,'\nNo Ld Rot Current  Amps       %5.1f',If0);

fprintf(f3,'\nRot Current at FL  Amps       %5.1f',IfFL);

fprintf(f3,'\nVoltage Regulation(perc)      %5.1f',reg);

fprintf(f3,'\nFL-Rot-Cur-dens   A/mm2       %5.2f',cds2) ;
```

```
fprintf(f3,'\nRot Volts at FL    Volts      %5.1f',Vfn);
fprintf(f3,'\nExciter         Rating          %5.1fKW,%5.1fV,%5.
1fA',KWex,Vex,Iex);
%------------------------------------------------------------
fprintf(f3,'\n----Weights,LossesandEfficiency Calculations:----
----->');
fprintf(f3,'\n   VARIANT                          VALUES');
fprintf(f3,'\n-------------     --            ------');
fprintf(f3,'\nWt of St.Core     Kg            %6.0f',Kgcore);
fprintf(f3,'\nSt.Teeth          Kg            %6.0f',Kgteth);
fprintf(f3,'\nSt.Stampings      Kg            %6.0f',Kgstst);
fprintf(f3,'\nRotor Forging     Kg            %6.0f',Kgrtst);
fprintf(f3,'\nWt of St.Copper   Kg            %6.0f',Kgstcu);
fprintf(f3,'\nWt of Rt.Copper   Kg            %6.0f',Kgrtcu);
fprintf(f3,'\nTotal Wt          Kg            %6.0f',KgT);
fprintf(f3,'\nKg/KVA                          %6.2f',Kgpkva);
fprintf(f3,'\n-------------     --            ------');
fprintf(f3,'\nMech Losses       KW            %6.1f',Pmech);
fprintf(f3,'\nBearing Losses    KW            %6.1f',Pbrg);
fprintf(f3,'\nVentilation Loss  KW            %6.1f',Pvent);
fprintf(f3,'\nSt-DC Cu Losses   KW            %6.1f',Pcu1);
fprintf(f3,'\nAdd Losses        KW            %6.1f',Padd);
fprintf(f3,'\nIron Losses       KW            %6.1f',Pi);
fprintf(f3,'\nRot Copper Losses KW            %6.1f',Pcu2);
fprintf(f3,'\nTotal Losses      KW            %6.1f',Pt);
fprintf(f3,'\nEfficiency       Perc           %6.2f',Eff);
fprintf(f3,'\n<-------End of Output:--------->');
fclose(f3);
```

6.3.8 *(a) Computer Output Results for Complete Design*

Design of Non-Salient-Pole Generator

```
Parameter                     VALUES
---------                     ------
```

KVA		30000
KW		24000
Volts		11000
PF		0.800
HZ		50
R.P.M		3000
No. of Poles		2
Flux/Pole	Wb	1.8575
GapFlux-den(Av)	Wb/m2	0.6900
Sp.Elec Loading Ac/m		58000
----------------STATOR-------------->		
Stator Current Amps		1574.5
CORE LENGTH(tot)	mm	2050
Iron LENGTH	mm	1680
Vent Ducts No X	mm	37 X 10
CORE OD	mm	2260
CORE ID	mm	836
POLE PITCH	mm	1313.2
TURNS/Ph		18
STATOR SLOTS		54
Chording Pitch	--	26/27
SLOT PITCH	mm	48.64
Core depth	mm	547.0
CoreFlux-den	Wb/m2	1.0985
SLOT WIDTH	mm	20.5
SLOT Height(tot)	mm	165.0
Ht (Lip+Wedge)	mm	25+ 9
Flux-den(tooth-tip)	Wb/m2	1.5820
Conductors/Slot		2
Bare cond Wdth	mm	6.0
Bare Cond Ht	mm	1.8
Strip/bar(Ht-W)		25
Current density	A/mm2	2.9752

Av.Loss Ratio	--	1.0999
Mean-Turn Length	m	8.0830
St Wdg Res/Ph/20d	ohm	0.0058
Air-gap LENGTH	mm	18.0
Pephoral Speed	m/s	125.7
Armature Reaction	AT/P	36540
Leakage Reactance	Perc	10.5

Parameter	Unit	Permissible	Designed
St.Slot Bal(ht)	mm	> 2.0	2.78
St.Slot Bal(wdth)	mm	> 1.0	1.00
Rt.Slot Bal(ht)	mm	> 0.5	0.80
Rt.Slot Bal(wdth)	mm	> 0.1	0.10

------------------ROTOR--------------->

Rotor Dia	mm	800
Rotor bore Dia	mm	60
Slot pitches	----	36
Slot Width	mm	31.0
Wound Slot height	mm	108.0
Wedge height	mm	25.0
Sub Slot Width	mm	22.0
Sub Slot height	mm	45.0
Cu strip Width	mm	28.0
Cu strip height	mm	3.2
Conductors/slot	nos	26.0
Turns/Pole	nos	156
Lgth of Mean turn	m	6.714
Rot Wdg Res@20d	ohm	0.4909

--------------O.C.C Calculation:------------------->

E(L-L) pu	1.00	1.10	1.20	1.30
E(L-L) V	11000	12100	13200	14300
Flux(Wb)	1.86	2.04	2.23	2.41
St.Core (Wb/m2)	1.099	1.208	1.318	1.428

St.Core-AT/Mt	219	302	427	646
St.Core-AT	393	542	766	1159
St.Tooth(Wb/m2)	1.289	1.418	1.547	1.675
St.Tooth-AT/Mt	385	620	1503	4261
St.Tooth-AT	64	102	248	703
Air.Gap-AT	17589	19348	21107	22866
Rt.WTth(Wb/m2)	1.518	1.669	1.821	1.973
Rt.Tooth-AT/Mt	1160	4085	9998	21038
Rt.Tooth-AT	206	727	1780	3745
Rt.Core (Wb/m2)	1.928	2.121	2.314	2.507
Rt.Core-AT/Mt	17110	44501	105719	248382
Rt.Core-AT	3285	8544	20298	47689
Total-OCC-AT	21538	29264	44199	76162
If-OCC-Amps	138.1	187.6	283.3	488.2
If-Airgap-Amps	112.8	124.0	135.3	146.6

-------------Other Calculations:------------------>

If-Arm reaction	Amps	234.2
Ifscc	Amps	246.1
SCR	pu	0.561
No Ld Rot Current	Amps	138.1
Rot Current at FL	Amps	361.7
Voltage Regulation(perc)		23.3
FL-Rot-Cur-dens	A/mm2	4.04
Rot Volts at FL	Volts	247.2

Exciter Rating 110.0KW,275.0V,400.0A

----Weights, Losses and Efficiency Calculations:--------->

VARIANT		VALUES
-------------	--	------
Wt. of St.Core	Kg	35716
St.Teeth	Kg	4079
St.Stampings	Kg	39795
Rotor Forging	Kg	8405

Wt. of St.Copper	Kg	2056
Wt. of Rt.Copper	Kg	1670
Total Wt	Kg	52445
Kg/KVA		1.75
-------------	--	------
Mech Losses	KW	213.3
Bearing Losses	KW	59.4
Ventilation Loss	KW	153.9
St-DC Cu Losses	KW	52.4
Add Losses	KW	67.1
Iron Losses	KW	89.8
Rot Copper Losses	KW	78.1
Total Losses	KW	500.7
Efficiency	Perc	97.96

<-------End of Output:--------->aa

6.3.9 Modifications to be done in the above Program to get Optimal Design

1. Insert "for" Loops for the following parameters to iterate the total program between min and max permissible limits for selecting the feasible design variants:
   (a) Chording pitch
   (b) Short Circuit Ratio
   (c) Tooth flux density
   (d) Stator Wdg Current density
   (e) Core flux density.
2. Insert also minimum or maximum range of required objective functional values as constraint values, for example, (a) Kg/KVA, (b) Efficiency, (c) Short circuit ratio (d) Leakage reactance.
3. Run the program to get various possible design variants.

Note: From the feasible design variants printed in the output, select that particular design fulfilling the objective of optimal parameter for the design.

6.3.10 Computer Program in "C" in MATLAB for Optimal Design

```
%Design  og  KVA=30000;V=11000;N=3000;f=50;pf=0.8;Non-Salient
Pole Generator
```

```
f2=fopen('NSP_Gen_Optimal_Opt.m','w');

KVA=30000;V=11000;N=3000;f=50;rpf=0.8;  %Star connected Non-Salient Pole Gen

Kw=0.955;bv=10;ki=0.92;Vr=125;R1=0.82;Hc=1.8;Zs=2;

Hins=12; Hw=9; HL=25; Cins=0.25; HVins=3; Kg=1.15; Dbrg=250; Lbrg =250; Dri=60; SP2=36; Ws2=31; Hw2=25; Wss=22; Hss=45; Wcu2=28; Vex1=220; cds2=4; %Rotor Data

sn=0; M1=0; M2=0; M3=0; M4=0; EFFmax=90; minKgpkva=2; SCRmax=0.4; XLmax=9; %Assumptions

fprintf(f2,'Design of Non-Salient-Pole Alternator of %5d KVA,%5d V,%4d RPM,%3d hz,%5.2f PF:-\n',KVA,V,N,f,rpf);

fprintf(f2,'****************************************');

fprintf(f2,'\nsn   Di   L   nv bv  S CP cds1 Ncv Wc   Hc  Ws Hs  Bt0  Dr Zs2 Hcu2 Wcu2 cds2   scr  EFF  Kg/KVA KWex reg XL');

fprintf(f2,'\n--  --- ---- -- -- -- -- ---- --- --  --- ---- --- ----  --- -- ---- ---- ----  ---- ----- ------ ---- -------');

%<----Bav and q------->

KVAS=[ 100  200  500 1000 5000 10000 20000 30000];

BavS=[0.54 0.56 0.58 0.60 0.63  0.65  0.67  0.69];

qS=[30e3 33e3 36e3 40e3 45e3  48e3  53e3  58e3];

%------------Main Dimensions---------------------->

Bav=interp1(KVAS,BavS,KVA,'spline');

q=interp1(KVAS,qS,KVA,'spline');
KW=KVA*rpf;P=120*f/N;n=2*f/P;

Iph=KVA*1e3/(1.7321*V);   Dr=Vr/(pi*n);   Dr=ceil(Dr*100)*10; Vr=pi*Dr/1000*n;     for spp=9:11;%(Range:6 to 11)

S=spp*P*3;      Tph=S*Zs/6;      ATa=1.35*Tph*Iph*Kw/(P/2); CPmin=ceil(0.8*S/P);

for CP=CPmin:S/2;

Kc=sin(CP/(S/2)*pi/2); K=11*Bav*q*Kw*Kc*1e-3;

for scr=0.5:0.1:0.6;

ATf0=scr*ATa;           ATg=0.75*ATf0;          Bg=Bav*1.5; Lg=ATg/(0.796*Bg*Kg*1e3);     Lg=ceil(Lg);     Di=Dr+2*Lg; DsqL=KVA/(K*n); L=DsqL/(Di/1e3)^2; L=ceil(L*100)*10;
```

```
PP=pi*Di/P; Ls=R1*L; nv=(L-Ls)/bv; FI=Bav*pi*Di*L*1e-6/P;

nv=ceil(nv); Ls=(L-nv*bv); Li=Ls*ki;rat2=Lg/PP;

for Bt0=1.6:0.1:1.9;

%-----(Part-2)Stator Winding and Core---------------->

sp=pi*Di/S;bt0=FI*1e6/(Bt0*Li*S/P);Ws=sp-bt0;

for cds1=3:0.5:4%(Permissible Range:3 to 5 A/sq.mm)

As1=Iph/cds1;     Wc1=((Ws-2*HVins-2.5)/2);     Wc=floor(Wc1);
Ncv1=As1/(Hc*Wc*2);    Ncv=ceil(Ncv1);    As=0.98*Ncv*Hc*Wc*2;
cds1=Iph/As;                                 Ws=2*Wc+2*HVins+2.5;
Hbar=(Ncv+1)*(Hc+Cins)+2*HVins;          Hs1=2*Hbar+Hins+Hw+HL;
Hs=ceil(Hs1);            p=2*Ncv;           alfa=sqrt(2*Wc/Ws);
KdTL=1+(alfa*Hc/10)^4*p*(p-1)/3;

Kdav=1+(alfa*Hc/10)^4*(2*Ncv)^2/9;
Lmt1=(2*L/10+2.5*PP/10+5*V/1e3+15)/ 100;

Rph=0.021*Lmt1*Tph/As;                        Pcus=3*Iph^2*Rph/1e3;
Peddy=(Kdav-1)*Pcus;                        Pts=(Pcus+Peddy)*1.15;
Reff=Iph*Rph*Kdav/(V/sqrt(3));                           bt0=sp-Ws;
Bt0=FI*1e6/(bt0*Li*S/P);   bt1b3=pi*(Di+Hs*2/3)/S-Ws;

Bt1b3=FI*1e6/(bt1b3*Li*S/P); FIc=FI/2;for Bc=1.1:0.1:1.2;

dc1=FIc/(Bc*Li)*1e6; Do1=Di+2*(Hs+dc1);    Do=ceil(Do1/10)*10;

dc=(Do-Di-2*Hs)/2;Bc=FIc/(dc*Li)*1e6;

Stsbh=Hs(Hw+HL+7.5+Zs*((Ncv+1)*(Hc+0.25)+7.06));

Stsbw=Ws(2*(Wc+0.25)+2*HVins+1);

%----------Reactances-------------------------------------->

h1=(Ncv+1)*Zs*(Hc+Cins)+Zs*2*HVins+6;      h2=(Hs-h1-HL-Hw)/2;
h3=Hw;h4= HL;Ws0=Ws;

lamdas=(h1/3/Ws)+  (h2/Ws)+2*h3/(Ws+Ws0)+h4/Ws0;

phis=2.*sqrt(2.)*4*pi*1e-7*Iph*Zs*Ls/1000.*lamdas;

ks=(CP*P/S-0.66)*0.7354+0.75;       L0Ld0=ks*PP*PP/pi/sp/1000;
phi0=2*sqrt(2)*4*pi*1e-7*Iph*Zs*L0Ld0;        phil=phis+phi0;
Lflux=phil/FI;   XLpu=Lflux;  XL=XLpu*100;

if XL <=10.4 continue;end;

%---------WeightsandLosses--------------------------------

Kgcore=7.85E-6*Li*pi/4*(Do^2-(Di+2*Hs)^2);

Kgstst=7.85E-6*Li*(pi*(Do^2-Di^2)/4-S*Ws*Hs);

Kgteth=Kgstst-Kgcore;       Kgstcu=3*Lmt1*As*Tph*8.9e-3;
```

```
Pcu1=3*Iph^2*Rph* 1.22/1e3;    ac=Iph*Tph*2*3/(pi*Di/1000);

Padd=90*Di^2*sqrt (L/1000.)*(ac/750)^2*1e-10;

Pi=(1.8*Kgteth*Bt1b3^2+1.8*Kgcore*Bc^2)/1000;

%=============Part-3 (Rotor Winding)=======================>

S2=2/3*SP2;  Vc=0.8*Vex1/P;    ATFL=1.5*ATa;

Lmt2=(2*L/10+1.8*PP/10+25)/100;       Af1=0.021*Lmt2*ATFL/Vc;
Zs2a=2*P*ATFL/(cds2*Af1*S2);         Zs2=ceil(Zs2a);

if (Zs2-Zs2a)>=0.5 Zs2=floor(Zs2a); end;

Tp2=Zs2*S2/(2*P);  Hcu2a=Af1/Wcu2;  Hcu2=ceil(Hcu2a*10)/10;

Af= Wcu2*Hcu2;        Rfp=0.021*Lmt2*Tp2/Af;       Rf20=Rfp*P;
IfFL=Vc/Rfp;  Hs2a=Zs2*Hcu2+(Zs2-1)*0.4+14; Hs2=ceil(Hs2a);

dk=Dr-2*(Hs2+Hw2+Hss); wtp=(S2/4.+1)/SP2;

Kgrtcu=8.9e-6*2.*Tp2*Lmt2*1e3*Wcu2*Hcu2;

Kgrtst=7.85E-6*pi*Dr*Dr/4*(L+80);
Pmech=1.15*Dr^4*sqrt(L)*1E-11;

Pbrg=3.8e-6*Dbrg^2*Lbrg;         Pvent=Pmech-Pbrg;

Rtsbh=Hs2-Zs2*Hcu2-(Zs2-1)*0.4-14;   Rtsbw=Ws2-Wcu2-2.9;

%--------------- O.C.C Calculation:--------------------*/

Lr=L+80;    tzi=pi*Di*wtp;    ratio1=Ws/Lg;    ratio2=Ws2/Lg;
ratiod= bv/Lg;

R=[0 1 2 3 4 5]; CC=[0 .15 0.27 .37 .44 .5];

k01=interp1(R,CC,ratio1,'spline');

k02=interp1(R,CC,ratio2, 'spline');

k1=interp1(R,CC,ratiod,'spline');

ks1=sp/(sp-k01*Ws);   ys2=pi*Dr/SP2;   ks2=ys2/(ys2-k02*Ws2);
kd1=bv/(bv-k1*bv);

kgus=ks1*kd1;kgs=(ks1+ks2-1.)*kd1;Lfps=2/3*pi*(Do-dc)/1e3/P;

Smax=(Lr+0.3333*dk)*dk/1e6;   Lsmax=(dk-Dri)/2e3;

%--------------------(OCC Calculation)---------------------->

Epu(1)=1.0; Epu(2)=1.1; Epu(3)=1.2 ; Epu(4)=1.3;

for ij=1:25;BB(ij)=0.1*ij;end;

%-------------B-H Curve From graph for Lohys material---→

HH=[50 65 70 80 90 100 110 120 150 180 220 295 400 580 1000
2400 5000 8900 15e3 24e3 40e3 65e3 10e4 15e4 24e4];
```

```
%semilogx(HH,BB);grid;xlabel('AT/m-->');
%ylabel('Flux density(T)-->');
%title('Magnetization Curve for Lowhys Stamping Steel');
for j=1:4;  ELL(j)=Epu(j)*V;flux(j)=Epu(j)*FI;
ATgap(j)=8e5*Epu(j)*Bg*Lg/1e3*kgus;
Bst1b3(j)=Epu(j)*FI*1e6/(bt1b3*Li*S/P);
atst(j)=interp1(BB,HH,Bst1b3(j),'spline');
ATst(j)=atst(j)* Hs/1e3;     Bstc(j)=flux(j)*0.5*1e6/(dc*Li);
atsc(j)=interp1(BB,HH,Bstc(j),'spline');
ATsc(j)=atsc(j) *Lfps;
Brt1b3(j)=Epu(j)*Bg*tzi*Li*1.36/Lr/(pi*(Dr4./3.*(Hs2+Hw2+Hss
))*wtp);
atrt(j)=interp1(BB,HH,Brt1b3(j),'spline');
ATrt(j)=atrt(j)*(Hs2+Hw2+Hss)/1e3;
Brtc(j)=Epu(j)*1.05*FI/Smax;
atrc(j)=interp1(BB,HH,Brtc(j),'spline');
ATrc(j)=atrc(j)*Lsmax;
ATtot(j)=ATgap(j)+ATst(j)+ATsc(j)+ATrt(j)+ATrc(j);
Ifocc(j)=ATtot(j)/Tp2;  Ifgap(j)=ATgap(j)/Tp2;
IFOCC(j+1)=Ifocc(j);EPU(j+1)=Epu(j);end;
 IFOCC(1)=0;  EPU(1)=0;
%-------------LMC andOther Calculations:------------------->
Ifar=ATa/Tp2;  Ifsc=(ATa+XLpu*ATgap(1))/Tp2; If0=Ifocc(1);
scr2=If0 /Ifsc;    if scr2 <=0.50 continue;end;
pf=rpf;pfr=sqrt(1-pf*pf);E1=sqrt(1.0*pf*pf+(1*pfr+XLpu)^2);
If1=interp1(Epu,Ifocc,E1,'spline');
IfFL=sqrt(If1^2+Ifar^2-2*If1*Ifar*cos(pi/2+acos(pf)));
E0=interp1(Ifocc,Epu,IfFL,'spline');reg=(E0-Epu(1))*100;
cds2= IfFL/Af;       if cds2 >=4.2 continue;end;
Vfn=IfFL*Rf20*(235.+120.)/255;  Vex2=1.1*Vfn;
Vex=ceil (Vex2/5)*5; Iex1=1.1*IfFL;     Iex=ceil(Iex1/10)*10;
KWex=Vex*Iex/1e3;       Pcu2=IfFL*IfFL*Rf20*(235.+75.)/255e3;
KgT=1.01*(Kgstst+Kgstcu+Kgrtcu+Kgrtst);  Kgpkva=KgT/KVA;
if Kgpkva >=1.9 continue;end;
```

```
Pt=Pcu1+Padd+Pmech+Pcu2+Pi;  Pcool=Pt-Pbrg;
Eff=KW/(KW+Pt)*100.;  if Eff <=97.5 continue;end;
Qair=ceil(Pcool*1e3/1140./24.)*1.0;
Tair=Pcool*1e3/1140./Qair;                sn=sn+1;
if Kgpkva <=minKgpkva minKgpkva=Kgpkva;end;
if abs(Kgpkva-minKgpkva)<= 0.001 M1=sn; end;
if Eff >=EFFmax  EFFmax=Eff;end;
if abs(Eff-EFFmax)<=0.01 M2=sn; end;
if scr2 >=SCRmax  SCRmax=scr2;end;
if abs(scr2-SCRmax)<=0.001 M3=sn; end;
if XL>=XLmax XLmax=XL;end;
if abs(XL-XLmax)<=0.01 M4=sn; end;                    sn,
fprintf(f2,'\n%2d%5.0f%5.0f%3d%3d%3.0f%3d%5.2f%4.0f%3.0f%5.1
f%5.1f%4.0f%5.2f',sn,Di,L,nv,bv,S,CP,cds1,Ncv,Wc,Hc,Ws,Hs,Bt
0);
fprintf(f2,'%5d%3d%5.1f%5.1f%5.2f%6.2f%6.2f%6.2f%5.0f%6.1f%5
.1f',Dr,Zs2,Hcu2,Wcu2,cds2,scr2,Eff,Kgpkva,KWex,reg,XL);
end;end;end;end;end;end;
printf(f2,'\n--  --- ---- -- -- -- -- ---- --- --  --- ----
--- ----  --- -- ---- ---- ----  ---- ----- ------ ---- ---
--');
fprintf(f2,'\nSelection of Design Variant based on
Optimization Criteria:');
fprintf(f2,'\nIf Minimum Kg/KVA is Required, Select
Variant(Sn)=%3d  (%5.2f)',M1,minKgpkva);
fprintf(f2,'\nIf Maximum Efficiency is Required, Select
Variant(Sn)=%3d (%5.2f perc)',M2,EFFmax);
fprintf(f2,'\nIf Maximum SCR is Required, Select
Variant(Sn)=%3d (%5.2f pu)',M3,SCRmax);
fprintf(f2,'\nIf Maximum XL is Required, Select
Variant(Sn)=%3d (%5.1f perc)',M4,XLmax);fclose(f2);
```

6.3.10 (a) Computer Output Results for Optimized Design

Design of Non-Salient-Pole Alternator of 30000 KVA, 11000 V,3000 RPM, 50 hz, 0.80 PF

Sn	Di	L	nv	bv	S	CP	cds1	Ncv	Wc	Hc	Ws	Hs	Bt0	Dr	Zs2	Hcu2	Wcu2	cds2	scr2	EFF	Kg/KVA	KWex	reg	XL
1	836	2050	37	10	54	26	2.98	25	6	1.8	20.5	165	1.58	800	26	3.2	28.0	4.04	0.56	97.96	1.75	110	23.3	10.5
2	836	2050	37	10	54	26	2.98	25	6	1.8	20.5	165	1.58	800	26	3.2	28.0	4.05	0.56	97.95	1.62	110	23.1	10.5
3	836	2050	37	10	54	26	2.98	25	6	1.8	20.5	165	1.58	800	26	3.2	28.0	4.04	0.56	97.96	1.75	110	23.3	10.5
4	836	2050	37	10	54	26	2.98	25	6	1.8	20.5	165	1.58	800	26	3.2	28.0	4.05	0.56	97.95	1.62	110	23.1	10.5
5	836	2050	37	10	54	27	2.98	25	6	1.8	20.5	165	1.58	800	26	3.2	28.0	4.04	0.56	97.96	1.75	110	23.3	10.7
6	836	2050	37	10	54	27	2.98	25	6	1.8	20.5	165	1.58	800	26	3.2	28.0	4.06	0.56	97.95	1.62	110	23.1	10.7
7	836	2050	37	10	54	27	3.54	21	6	1.8	20.5	149	1.58	800	26	3.2	28.0	4.03	0.56	97.96	1.70	110	23.2	10.4
8	836	2050	37	10	54	27	3.54	21	6	1.8	20.5	149	1.58	800	26	3.2	28.0	4.05	0.56	97.96	1.57	110	23.1	10.4
9	836	2050	37	10	54	27	2.98	25	6	1.8	20.5	165	1.58	800	26	3.2	28.0	4.04	0.56	97.96	1.75	110	23.3	10.7
10	836	2050	37	10	54	27	2.98	25	6	1.8	20.5	165	1.58	800	26	3.2	28.0	4.06	0.56	97.95	1.62	110	23.1	10.7
11	836	2050	37	10	54	27	3.54	21	6	1.8	20.5	149	1.58	800	26	3.2	28.0	4.03	0.56	97.96	1.70	110	23.2	10.4
12	836	2050	37	10	54	27	3.54	21	6	1.8	20.5	149	1.58	800	26	3.2	28.0	4.05	0.56	97.96	1.57	110	23.1	10.4
13	834	2160	39	10	60	24	2.98	30	5	1.8	18.5	186	1.59	800	25	3.6	28.0	4.11	0.51	97.89	1.77	127	22.6	10.8
14	834	2160	39	10	60	24	3.57	25	5	1.8	18.5	165	1.59	800	25	3.6	28.0	4.08	0.51	97.89	1.84	127	22.6	10.4
15	834	2160	39	10	60	24	3.57	25	5	1.8	18.5	165	1.59	800	25	3.6	28.0	4.09	0.51	97.89	1.70	127	22.5	10.4
16	834	2160	39	10	60	24	2.98	30	5	1.8	18.5	186	1.59	800	25	3.6	28.0	4.11	0.51	97.89	1.77	127	22.6	10.8
17	834	2160	39	10	60	24	3.57	25	5	1.8	18.5	165	1.59	800	25	3.6	28.0	4.08	0.51	97.89	1.84	127	22.6	10.4
18	834	2160	39	10	60	24	3.57	25	5	1.8	18.5	165	1.59	800	25	3.6	28.0	4.09	0.51	97.89	1.70	127	22.5	10.4

Selection of Design Variant based on Optimization Criteria :

If Minimum Kg/KVA is Required, Select Variant (Sn) = 12 (1.57)

If Maximum Efficiency is Required, Select Variant (Sn) = 12 (97.96 perc)

If Maximum SCR is Required, Select Variant (Sn) = 12 (0.56 pu)

If Maximum XL is Required, Select Variant (Sn) = 16 (10.8 perc).

CHAPTER 7

Three-Phase Induction Motors

7.1 Introduction

Three-phase induction Motors are of two types (viz.) Squirrel Cage Rotor type and Slip-ring Rotor type. Theory portion of design is not given in this book, but necessary formulae, curves and tables given in standard books are made use of.

7.1.1 Squirrel Cage Motor

Total design is split into six parts in a proper sequence. Design Calculations are given for a given Rating of an Ind motor, followed by Computer Program written in "C" language using MATLAB software for each part. Finally all the Programs are added together to get the total Program by running which we get the total design. Computer output of total design is given.

This design may not be the optimum one. Now optimization objective and design constraints are inserted into this total Program. When this program is run we will get various alternative feasible designs from which the selected variant based on the optimization criteria can be picked up. Computer output showing the important design parameters for various feasible alternatives is given in the end of this chapter along with a logic diagram.

7.1.2 Sequential Steps for Design of Each Part and Programming Simultaneously

(a) Calculate Output Coefficient, Main dimensions of Stator Core (viz) D, L and Flux/Pole.

(b) Calculate Number and size of Stator slots, conductor size, by checking current density and slot balance. Calculate tooth flux density, Copper losses and Wt. of Copper, Core flux density, height of core, iron losses

(c) Calculate Air gap length, Rot dia, no. of Rotor Slots, Cond size, Copper Losses, Flux densities in tooth and Core, Weight of rotor copper.

(d) Calculate Carter Coefft and Ampere-turns for Air gap, Stator tooth, Stator core, Rotor tooth, Rotor core and Total No-load AT, Magnetizing current, No-Load PF.

(e) Calculation of Reactance, Short-circuit current, and Short-ckt PF.

(f) Calculation of total losses, Efficiency, Slip, Starting torque, Temp-rise, Total Weight and Kg/KW.

Note: By adding programs established for each part sequentially we get the Program for complete design.

Problem: Design a 30 KW, 440V, 50 Hz, delta connected, 3 ph Squirrel Cage Ind.Motor

7.1.3 Calculation of Stator Main Dimensions and Flux (Part-1)

Table 7.1 Approx. Values of Bav and q.

KW	1	2	5	10	20	50	100	500
Bav(T)	0.35	0.38	0.42	0.46	0.48	0.50	0.51	0.53
q(ac/m)	16000	19000	23000	25000	26000	29000	31000	33000

Table 7.2 Approx. Values of Power-Factor.

KW	5	10	20	50	100	200	500
1000rpm	0.82	0.83	0.85	0.87	0.89	0.90	0.92
1500rpm	0.85	0.86	0.88	0.90	0.91	0.92	0.93

Table 7.3 Approx. Values of Efficiency.

KW	5	10	20	50	100	200	500
1000rpm	0.83	0.85	0.87	0.89	0.91	0.92	0.93
1500rpm	0.85	0.87	0.88	0.90	0.91	0.93	0.94

From Table.7.1,Values of Bav= 0.4838 T and q = 26802 ac/m corresponding to 30 KW

Assuming No. of Poles (P) = 6,

Sync Speed (Ns) = 120 × f/P = 120 × 50/6 = 1000 RPM and n = 1000/60 = 16.667 rps

From Tables.7.2 and 7.3, Values of pf = 0.8621 and eff = 0.88 corresponding to 30 KW and 1000 RPM

KW input to motor = KW/eff = 30/0.88 = 34.0908

For Δ connection, Vph = V = 440 Volts

$$\text{Phase Current} = \frac{\text{KWinp} \times 1000}{3 \times \text{Vph} \times \text{pf}} = \frac{34.098 \times 1000}{3 \times 440 \times 0.8621} = 29.958 \text{ A}$$

Assuming Winding factor (Kw) = 0.955,

Output Coefft(C0) = 11 × Kw × Bav × q × eff × pf × 10^{-3}

= 11 × 0.955 × 0.4838 × 26802 × 0.88 × 0.8621 × 10^{-3} = 103.3417

$$D^2L = \frac{\text{KW}}{\text{C0} \times \text{ns}} = \frac{30}{103.3417 \times 16.667} = 0.0174$$

$$\text{Total Core Length (L)} = \sqrt{\frac{D^2L}{(0.135 \times P)^2}} = \sqrt{\frac{0.0174}{(0.135 \times 6)^2}}$$

$$= 0.1629\ \text{m} = 162.9\ \text{mm} \approx\ 160\ \text{mm (Rounded off)}$$

Assuming Ventilating ducts (nvd) = 2, each of length (bvd) = 10 mm,

Gross iron Length (Ls) = L – nvd × bvd = 160 – 2 × 10 = 140 mm

Assuming Iron factor (ki) = 0.92, Net Iron length (Li)

$$= \text{Ls} \times \text{ki} = 140 \times 0.92 = 128.8\ \text{mm}$$

$$\text{Core inner diameter (D)} = \sqrt{\frac{D^2L}{L}} = \sqrt{\frac{0.0174}{140/1000}} = 0.3299\ \text{m}$$

$$= 329.9\ \text{mm} \approx 330\ \text{mm (Rounded off)}$$

$$\text{Polepitch (PP)} = \frac{\pi \times D}{P} = \frac{\pi \times 330}{6} = 172.79\ \text{mm}$$

$$\frac{\text{Length}}{\text{Polepitch}} = \frac{140}{172.79} = 0.926\ \ (\approx 1 \text{ and hence OK})$$

$$\text{Periphoral Velocity(v)} = \frac{\pi \times D}{ns} = \frac{\pi \times 330}{16.67} = 17.28\ \ \text{m/s}\ \ (\le 30 \text{ and hence OK})$$

$$\text{Flux (FI)} = \frac{\pi \times D \times L \times Bav}{P \times 10^6} = \frac{\pi \times 330 \times 140 \times 0.4838}{6 \times 10^6} = 0.0134\ \text{Wb}$$

7.1.3 *(a) Computer Program in "C" in MATLAB for Part-1*

```
% 3ph, KW = 30;V = 440; P = 6; f = 50Sq.Cage IM
%------------Standard Curves/Tables for%Data-------------->
SKW=[1 2 5 10 20 50 100 500];
SBav=[0.35 0.38 0.42 0.46 0.48 0.50 0.51 0.53];
Sq=[16e3 19e3 23e3 25e3 26e3 29e3 31e3 33e3];
SKWa=[5 10 20 50 100 200 500];
SPF6P=[0.82 0.83 0.85 0.87 0.89 0.9 0.92]; %for 1000RPM
SEFF6P=[0.83 0.85 0.87 0.89 0.91 0.92 0.93];% for 1000RPM
SPF4P=[0.85 0.86 0.88 0.9 0.91 .92 .93];% for 1500RPM
SEFF4P=[.85 .87 .88 .9 .91 .93 .94];% for 1500RPM
%(1)<-----------Main Dimensions----------------------->
KW=30;V=440;f=50; P=6;  %Input Data------>
insW=3.4;Hw=4;HL=1;insH=6;nvd=2;bvd=0.01;ki=0.92;Bc=1.35;Vph=V;
%Assumptions
```

```
Zr=1;kwr=1;cdb=6;Tb=6;cde=6;dd=0.05;Brc=1.35;    %Assumptions
spp=3;Tstrip=1.9;Zsw=3; % Assumption
Kw=0.955;nvd=2;bvd=10;ki=0.92;Tstrip=1.9;insS=0.5;%Assumptions
insW=3.4;Hw=4;HL=1;insH=6;nvd=2;ki=0.92;Bc=1.35;%Assumptions
Bav=interp1(SKW,SBav,KW,'spline');
q=interp1(SKW,Sq,KW,'spline');
pf=interp1(SKWa,SPF6P,KW,'spline');
eff=interp1(SKWa,SEFF6P,KW,'spline');
if P=4 pf=interp1(SKWa,SPF4P,KW,'spline');end;
if P=4 eff=interp1(SKWa,SEFF4P,KW,'spline');end;
KWinp=KW/eff; Iph=KWinp*1e3/(3*Vph*pf); Ns=120*f/P;
ns=Ns/60; C0=11*Kw*Bav*q*eff*pf*1e-3; DsqL=1/C0*(KW/ns);
L1=sqrt(DsqL/(0.135*P)^2); L=floor(L1*100)*10;  Ls=(L-nvd*bvd);
Li=ki*Ls;       D1=sqrt(DsqL/(L/1000));       D=ceil(D1*100)*10;
PP=pi*D/P;  LbyPP=L/PP;  if LbyPP <0.8||LbyPP >2 continue; end;
v=pi*D*ns/1000;    if v >30 continue;end;
FI=pi*D/P*L*Bav/1e6;
%---------------------------------------------------------------
```

7.1.4 Design of Stator Winding (Part-2)

Assuming Slots/pole/ph(spp) = 3, No of slots(S) = spp × 3 × P = 3 × 3 × 6 = 54

$$\text{Slot pitch (sp)} = \frac{\pi \times D}{S} = \frac{\pi \times 330}{54} = 19.1986\text{mm}$$

(OK since it is between 18 and 25mm)

$$\text{Turns/ph (Tph)} = \frac{Vph}{4.44 \times f \times FI \times Kw} = \frac{440}{4.44 \times 50 \times 0.0134 \times 0.955} = 154.8$$

Conductors/ph (Zph) = Tph × 2 = 154.8 × 2 = 309.7

Slots/ph (sph) = 54/3 = 18

Conductors/slot (Zs) = Zph/sph = 309.7/18 = 17.2 ≈ 18(Rounded off to even integer)

Corrected Turns/ph (Tph) = Zs × sph/2 = 18 × 18/2 = 162

$$\text{Corrected flux (FI)} = \frac{Vph}{4.44 \times f \times Tph \times Kw} = \frac{440}{4.44 \times 50 \times 162 \times 0.955} = 0.01281\ \text{Wb}$$

Table 7.4 Approx. Values of Stator Current Density.

D (mm)	100	150	200	300	400	500	750	1000
A/mm²	4	3.8	3.6	3.5	3.5	3.5	3.5	3.5

From Table.7.4,Current Density (CDSW) = 3.5 A/mm^2 corresponding to D = 330 mm

$$\text{Area of CS of conductor (As)} = \frac{\text{Iph}}{\text{CDSW}} = \frac{29.958}{3.5} = 8.56 \text{ mm}^2$$

Assuming Thickness of strip/conductor (Tstrip) = 1.9 mm,

$$\text{Height of the strip (Hstrip)} = \frac{\text{As}}{\text{Tstrip}} = \frac{8.56}{1.9} = 4.5 \text{ mm} \approx 5\text{mm (Rounded off)}$$

Height/Width ratio of strip = 5/1.9 = 2.632 (Lies between 2.5 to 3.5 and hence OK)

Corrected Area of strip/conductor (As) = 0.967 × Hstrip × Tstrip

$$= 0.967 \times 5 \times 1.9 = 9.1865 \text{ mm}^2$$

Assuming width-wise Insulation (insW) = 3.4 mm,

Strip Insulation Thickness (insS) = 0.5 mm and width-wise No. of conductors (Zsw) = 3,

Slot width (Ws) = [Zsw × (Tstrip + insS) + insW] = (3 × (1.9 + 0.5) + 3.4) = 10.6 mm

No of strips/conductors height-wise in a slot (Zsh) = Zs/Zsw = 18/3 = 6

Assuming height of Lip (HL) = 1mm, height of Wedge (Hw) = 4 mm,

Slot Height (Hs) = [Zsh × (Hstrip + insS) + Hw + HL + insH]

$$= [6 \times (5 + 0.5) + 4 + 1 + 6 + 2] = 46 \text{ mm}$$

Cross-Section of Stator Slot (dimensions in mm)

Height-Wise			
Lip			1.0
Wedge			4.0
Ins.Under Wedge			1.0
Cu.Coductor	bare + insulation	5 + 0.5 = 5.5	
Top-Layer	for 3 conductors	3 × 5.5 = 16.5	16.5
Ins.between Layers			4.0
Bottom- Layer	for 3 conductors	3 × 5.5 = 16.5	16.5
Bottom Insulation			1.0
Slack			2.0
Total height			**46.0**

Width-Wise			
Cu.Conductor	bare + insulation	1.9 + 0.5 = 2.4	
	for 3 conductors	3 × 2.4 = 7.2	7.2
Insulation on sides	2 × 1.3 = 2.6		2.6
Slack			0.8
Total width			**10.6**

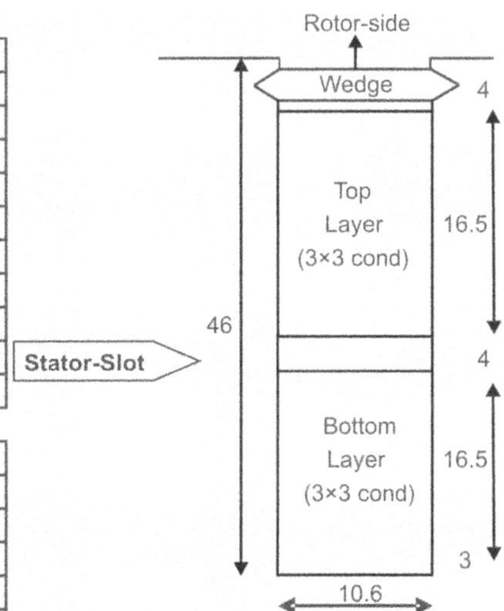

Dia at ⅓ ht from tooth tip (D13) = D + ⅔ × Hs = 330 + ⅔ × 46 = 360.67 mm

Slot pitch at dia D13 (sp13) = $\frac{\pi \times D13}{S} = \frac{\pi \times 360.67}{54}$ = 20.9827 mm

Tooth width at dia D13 (Wt13) = sp13 – Ws = 20.983 – 10.6 = 10.2837 mm

Flux density at ⅓ ht from tooth tip(B13) = $\frac{FI \times P \times 10^6}{Li \times Wt13 \times S}$

$$= \frac{0.01281 \times 6 \times 10^6}{128.8 \times 10.3827 \times 54} = 1.0644 \text{ T}$$

Max Flux density of the tooth (Btmax) = 1.5 × B13 = 1.5 × 1.0644 = 1.5966 T

Mean Length of turn (Lmt) = [2 × L + (2.3 × PP) + 240]/1000

= [2 × 160 + (2.3 × 172.79) + 240]/1000 = 0.9574m

Resistance/ph at 20ºC (Rph) = 0.021 × Lmt × Tph/As

= 0.021 × 0.9574 × 162/9.1865 = 0.3546Ω

Copper Loss (Pcus) = 3 × Iph^2 × Rph = 3 × 29.958^2 × 0.3546 = 954.6 W

Weight of Copper (Wcus) = Lmt × Tph × 3 × As × 8.9/1000

= 0.9574 × 162 × 3 × 9.1865 × 8.9 /1000 = 38.04Kg

Flux in core (FIc) = FI/2 = 0.01281/2 = 0.006405 T;

Assuming Flux density in the core (Bc) = 1.35 T,

Area of core (Ac) = FIc × 10^6/Bc = 0.006405 × 10^6/1.35 = 4744.8 mm^2

Height of the Core (Hc) = Ac/Li = 4744.8/128.8 = 36.84 mm

Core Outer Dia (D0) = D + 2 × (Hs + Hc) = 330 + 2 × (46 + 36.84)

= 495.7 ≈ 500 mm (Rounded off)

Corrected ht of core (Hc) = (D0 – D)/2 – Hs = (500 – 330)/2 – 46 = 39 mm

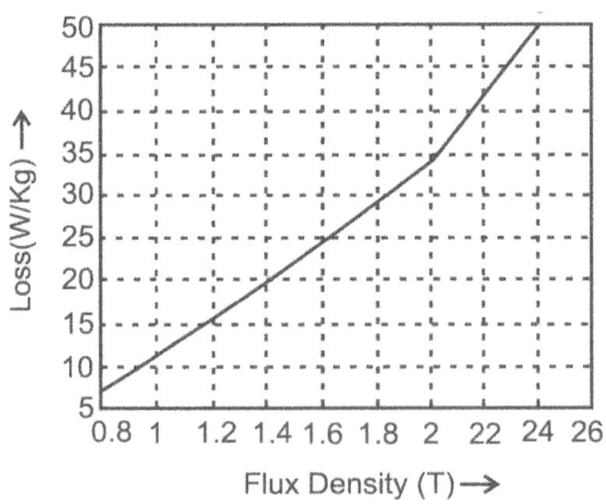

Fig. 7.1 Core loss for 0.5 mm stampings.

Iron loss in tooth (PitpKg) = 23.92 corresponding to Btmax = 1.5966T from Fig. 7.1

Iron loss in Core (PicpKg) = 18.325 corresponding to Bcmax = 1.35 T from Fig.7.1

Wt of tooth (Wt) = Li × Wt13 × S × Hs × 7.8 × 10^{-6}

= 128.8 × 10.383 × 54 × 46 × 7.8 × 10^{-6} = 25.91 Kg

Mean dia of the core (Dmcs) = D + (2 × Hs) + Hc = 330 + (2 × 46) + 39 = 461mm

Weight of core (Wc) = Ac × π × Dmcs × 7.8 × 10^{-6}

= 4744.8 × π × 461 × 7.8 × 10^{-6} = 53.56 Kg

Iron Loss in Tooth (Pit) = PitpKg × Wt = 23.92 × 25.91 = 619.9 W

Iron Loss in Core (Pic) = PicpKg × Wc = 18.325 × 53.56 = 982.2 W

7.1.4 *(a) Computer Program in "C" in MATLAB for Part-2*

```
%(2)<--------------Stator Slots and Winding----->
V=440;f=50;FI=0.0134; Kw=0.955;P=6;D=330;KW=30;eff=0.88;
pf=0.8621;Li=128.8; L=160; PP=172.79; Iph=29.958; %Input Data
spp=3; Tstrip=1.9; Zsw=3;insS=0.5;  %Assumptions
insW=3.4;Hw=4;HL=1;insH=6;nvd=2;ki=0.92;Bc=1.35; %Assumptions
SD=[0.1 0.15 0.2 0.3 0.4 0.5 0.75 1];
SCDSW=[4 3.8 3.6 3.5 3.5 3.5 3.5 3.5];
B=[0.8 1.2 1.6 2 2.4]; WpKg=[7 15 24 34 50];
Tphi=Vph/(4.44*f*FI*Kw); DSW=interp1(SD,SCDSW,D/1000,'spline');
As1=Iph/CDSW;S=spp*P*3;  SPitch=pi*D/S;
if SPitch<18||SPitch>=25 continue;end;
Zphi=2*Tphi; sph=S/3; Zs1=Zphi/sph; Zs=ceil(Zs1); Tph=Zs*sph/2;
FI=Vph/(4.44*f*Tph*Kw);                          Hstrip1=As1/Tstrip;
Hstrip=ceil(Hstrip1*2)/2;      WbyT=Hstrip/Tstrip;
if WbyT <2.5||WbyT >=3.5 continue;end;
As=0.967*Hstrip*Tstrip;             Ws=(Zsw*(Tstrip+insS)+insW);
Zsh=Zs/Zsw; Hs=(Zsh*(Hstrip+insS)+Hw+HL+insH+2);  D13=D+2/3*Hs;
sp13=pi*D13/S;     Wt13=sp13-Ws;       B13=FI*P*1e6/(Li*Wt13*S);
Btmax=1.5*B13; Lmt=(2*L+2.3*PP+240)/1000; Rph=0.021*Lmt*Tph/As;
Pcus=3*Iph^2*Rph;    Wcus=Lmt*Tph*3*As*8.9e-3;         FIc=FI/2;
Ac=FIc*1e6/Bc ;  Hc=Ac/Li;  D01=D+2*(Hs+Hc);
D0=ceil(D01/10)*10;   Hc=(D0-D)/2-Hs;
PitpKg=interp1(B,WpKg,Btmax,'spline');
PicpKg=interp1(B,WpKg,Bc,'spline');
Wt=Li*Wt13*S*Hs*7.8e-6; Dmcs=D+2*Hs+Hc;   Wc=Ac*pi*Dmcs*7.8e-6;
Pit=PitpKg*Wt;   Pic=PicpKg*Wc;
%----------------------------------------------------------
```

7.1.5 Design of Squirrel Cage Rotor (Part-3)

Air-gap length (Lg) = $0.2 + 2 \times \sqrt{D \times L / 10^6}$

$$= 0.2 + 2 \times \sqrt{330 \times 160 / 10^6} = 0.66 \text{ mm}$$

Rotor dia (Dr) = D – 2 × Lg = 330 – 2 × 0.66 = 328.68 mm

No. of Rotor Slots (Sr) should **not** be equal to the following:

(Ss-3 × P) = (54 – 3 × 6) = 36, (Ss – P) = (54 – 6) = 48,

(Ss – 2 × P) = (54 – 2 × 6) = 42, (Ss – 5 × P) = (54 – 5 × 6) = 24,

(Ss – 1) = (54 – 1) = 53, (Ss – 2) = (54 – 2) = 52, (Ss – 7) = (54 – 7) = 47, (Ss – 8) = (54 – 8) = 46,

Hence Sr = 45 Selected

(Note: A new symbol Ss = S = Stator slots and kws = Kw = Stator winding factor is used.)

Slot pitch (sp2) = $\dfrac{\pi \times Dr}{Sr} = \dfrac{\pi \times 328.68}{45} = 22.946$ mm

Equivalent Rotor current (Ir) = 0.85 × Iph = 0.85 × 29.958 = 25.464 A

Assuming Winding factor for Rotor (kwr) = 1 and Conductors/Slot (Zr) = 1,

Bar Current (Ib) = $\dfrac{Ir \times kws \times Ss \times Zs}{kwr \times Sr \times Zr} = \dfrac{25.464 \times 0.955 \times 54 \times 18}{1 \times 45 \times 1} = 525.28$ A

Assuming Current density in bar (cdb) = 6 A/mm^2 and Thickness of bar (Tb) = 6 mm

Area of cs of bar (Ab) = $\dfrac{Ib}{cdb} = \dfrac{525.28}{6} = 87.55$ m^2,

Width of bar (Wb) = $\dfrac{Ab}{Tb} = \dfrac{87.55}{6} = 14.59 \approx 15$ mm (Rounded off)

Corrected Area of cs of bar (Ab) = Wb × Tb × 0.98

= 15 × 6 × 0.98 = 88.2 mm^2

(Factor 0.98 is taken for edges rounding off)

Width of slot (Wsr) = Tb + 0.5 = 6 + 0.5 = 6.5 mm and

Ht of slot (Hsr) = Wb + 0.5 = 15 + 0.5 = 15.5 mm

Length of bar (Lb) = L + 50 = 160 + 50 = 210 mm

Resistance of bar (Rb) = 0.021 × Lb/1e3/Ab = $\dfrac{0.021 \times Lb}{Ab \times 1000} = \dfrac{0.021 \times 210}{87.55 \times 1000} = 5 \times 10^{-5}\,\Omega$

Copper loss in the bars (Pcub) = Ib^2 × Rb × Sr = 525.28^2 × 5 × 10^{-5} × 45 = 620.8 W

End Ring Current (Ie) = $\dfrac{Ib \times Sr}{\pi \times P} = \dfrac{525.28 \times 45}{\pi \times 6} = 1254$ A

Assuming Current density in end ring (cde) = 6 A/mm^2

Area of cs of end ring (Ae) = $\dfrac{Ie}{cde} = \dfrac{1254}{6} = 209$ mm^2,

Mean dia of end-ring (Dme) = Dr – 50 = 328.68 – 50 = 278.68 m

Resistance of end ring (Re) = $\dfrac{0.021 \times \text{Lme}}{\text{Ae}} = \dfrac{0.021 \times \pi \times \text{Dme}}{1000 \times \text{Ae}} = 8.8 \times 10^{-5}\Omega$

Copper loss in the 2 End-rings (Pcue) = $2 \times \text{Ie}^2 \times \text{Re} = 2 \times 1254^2 \times 8.8 \times 10^{-5} = 276.7$ W

Total Rotor copper loss (Pcur) = Pcub + Pcue = 620.8 + 276.7 = 897.5 W

Equivalent Rot res (Rr) = $\dfrac{\text{Pcur}}{3 \times \text{Ir}^2} = \dfrac{897.5}{3 \times 25.4645^2} = 0.4614\ \Omega$

Dia of rotor at ⅓tooth ht from tip (Dr13) = Dr – 2 × ⅔ × Hsr

= 328.68-2 × ⅔ × 15.5 = 308

Rotor slot pitch at Dr13 (spr13) = $\dfrac{\pi \times \text{Dr13}}{\text{Sr}} = \dfrac{\pi \times 308}{45} = 21.50\text{mm}$

Width of tooth at Dr13 (Wtr13) = spr13 – Wsr = 21.5 – 6.5 = 15 mm

Area of tooth at Dr13 (Atr) = Wtr13 × Li × Sr/P = 15 × 128.8 × 45/6 = 14493 mm^2

Flux density in tooth (Brt) = FI × 10^6/Atr = 0.01281 × 10^6/14493 = 0.8839 T

Max Flux density in tooth (Brtmax) = Brt × 1.5

= 0.8839 × 1.5 = 1.3259 T (lies bet 1.2 to 1.4 and hence OK)

Assuming Flux density in Rotor Core (Brc) = 1.35 T,

Area of Core (Ac) = $\dfrac{\text{FI} \times 10^6}{2 \times \text{Brc}} = \dfrac{0.01281 \times 10^6}{2 \times 1.35} = 4744.8\ \text{m}^2$

Depth of core (dcr) = $\dfrac{\text{Ac}}{\text{Li}} = \dfrac{4744.8}{128.8} = 36.84$ mm

Assuming Friction and Windage Loss (Pfw) 1%

= 0.01 × KW × 1000 = 0.01 × 30 × 1000 = 300 W

Noload Loss (PnL) = Pit + Pic + Pfw = 619.9 + 982.2 + 300 = 1902.1 W

Active/Wattful Component of No-load Current (Iw) = $\dfrac{\text{PnL}}{3 \times \text{V}} = \dfrac{1902.1}{3 \times 440} = 1.441$ A

Wt of Rotor Copper (Wcur) = Lb × Sr × Ab × 8.9 × 10^{-6}

= 210 × 45 × 88.2 × 8.9 × 10^{-6} = 7.42 Kg

Wt of Rotor End-Rings (Wcue) = Lme × 2 × Ae × 8.9 × 10^{-3}

= 0.8755 × 2 × 209 × 8.9 × 10^{-6} = 3.29 Kg

7.1.5 (a) Computer Program in "C" in MATLAB for Part-3

```
%(3)<--------Squirrel Cage Rotor------------------------------->
Kw=0.955; D=330; L=160; S=54; \P=6; Iph=29.96; Zs=18; Li=128.8;
```

```
FI=0.01281;KW=30;Pit=619.9;Pic=982.2;V=440; %Input data
Zr=1;kwr=1;cdb=6;Tb=6;cde=6;dd=50;Brc=1.35;    %Assumptions
kws=Kw;Ss=S;Lg1=0.2  +  2*sqrt(D*L/1e6);  Lg=ceil(Lg1*100)/100;
Dr=D-2*Lg;  d1=Ss-3*P;  d2=Ss-P;      d3=Ss-2*P;     d4=Ss-5*P;
d5=Ss-1;   d6=Ss-2;   d7=Ss-7; d8=Ss-8; Sr=Ss-9;  sp2=pi*Dr/Sr;
Ir=0.85*Iph;          Ib=Ir*kws*Ss*Zs/(kwr*Sr*Zr);   Abi=Ib/cdb;
Wb=ceil(Abi/Tb);    Ab=Tb*Wb*0.98;     Wsr=Tb+0.5;    Hsr=Wb+0.5;
Lb=L+50;      Rb=0.021*Lb/1e3/Ab;               Pcub=Ib^2*Rb*Sr;
Ie=Ib*Sr/P/pi;              Ae=Ie/cde;Dme=Dr-dd;Lme=pi*Dme/1000;
Re=0.021*Lme/Ae;    Pcue=2*Ie^2*Re;    Pcur=Pcub+Pcue;
Rr=Pcur/(3*Ir^2);     Dr13=Dr-2*2/3*Hsr;        spr13=pi*Dr13/Sr;
Wtr13=spr13-Wsr;       Atr=Wtr13*Li*Sr/P;       Brt=FI*1e6/Atr;
Brtmax=Brt*1.5;  Ac=FI*1e6/2/Brc;
dcr=Ac/Li;  Pfw=0.01*KW*1e3;  PnL=Pit+Pic+Pfw;  Iw=PnL/3/V;
Wcur=Lb*Sr*Ab*8.9e-6;  Wcue=Lme*2*Ae*8.9e-3;
%----------------------------------------------------------------
```

7.1.6 Total AT and Magnetizing Current (Part-4)

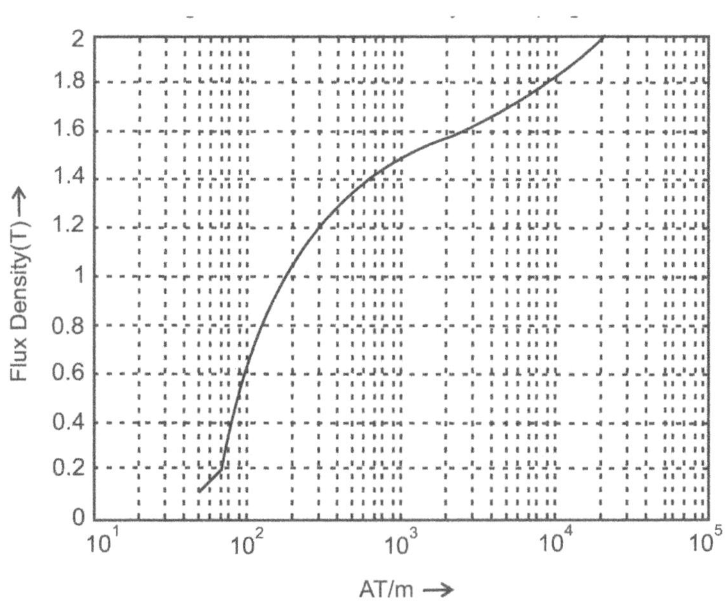

Fig.7.2 Magnetization curve for Lohys stamping steel.

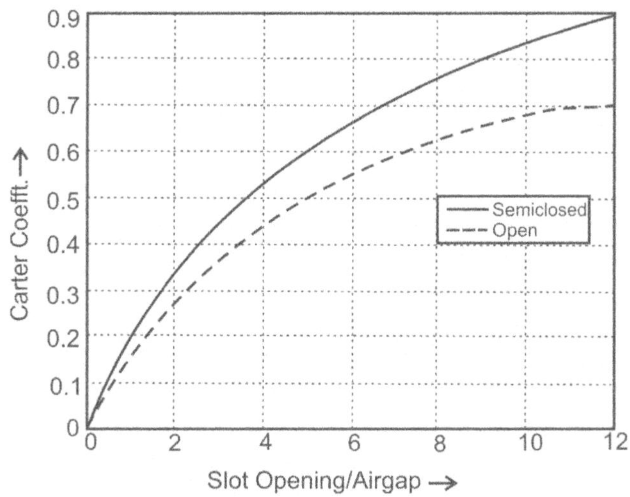

Fig. 7.3 Carter coefft. for slots.

From Fig. 7.2, Amp-Turns/m for Stator core (at Flux density Bc = 1.35 T),

$$\text{atsc} = 480.8$$

St-Core mean dia (Dcav) = D + (2 × Hs) + Hc = 330 + (2 × 46) + 39 = 461 mm

$$\text{AmpTurns for St-Core (ATSC)} = \frac{\pi \times \text{Dcav} \times \text{atsc}}{\text{P} \times 3 \times 1000} = \frac{\pi \times 461 \times 480.8}{6 \times 3 \times 1000} = 38.68$$

Flux density at 30º from the centre of the pole (Bt30)

$$= \text{B13} \times 1.36 = 1.0644 \times 1.36 = 1.4476 \text{ T}$$

From Fig. 7.2, Amp-Turns/m for Stator teeth (at Flux density Bt30)

$$= 1.4476\text{T}), \text{ atst} = 706.5$$

AmpTurns for St-tooth (ATST) = atst × Hs/1000 = 706.5 × 46/1000 = 32.5

Total AT for Stator (ATS) = ATSC + ATST = 38.68 + 32.5 = 71.18

$$\text{Assuming St-Slot opening (Wss0) = 4mm, } \frac{\text{Wss0}}{\text{Lg}} = \frac{4}{0.66} = 6.06$$

From Fig. 7.3, Cartr coeft corresponding to 6.06 (k01) = 0.6633

$$\text{Gap Coefft for St-Slots (kgs)} = \frac{\text{Spitch}}{\text{Spitch} - \text{Wss0} \times \text{k01}} = \frac{19.2}{19.2 - 4 \times 0.6633} = 1.1604$$

$$\text{Assuming Rt-Slot opening (Wss0) = 2mm, } \frac{\text{Wss0}}{\text{Lg}} = \frac{2}{0.66} = 3.03$$

From Fig. 7.3, Carter coeft corresponding to 3.03(k02) = 0.4529

Rotor Slotpitch near airgap (spr0) $= \dfrac{\pi \times Dr}{Sr} = \dfrac{\pi \times 328.68}{45} = 22.946$ mm

Gap Coefft for Rt-Slots (kgr) $= \dfrac{Spr0}{Spr0 - Wsr0 \times k02} = \dfrac{22.946}{22.946 - 2 \times 0.4529} = 1.1041$

Air gap Coefft (kg) = kgs × kgr = 1.1604 × 1.1041 = 1.2081

Effective air gap (Lgd) = Lg × kg = 0.66 × 1.2081 = 0.7973 mm

For Ventilating duct, $\dfrac{bvd}{Lg} = \dfrac{10}{0.66} = 15.15$

From Fig. 7.3 for open slots, Carter coeft corresponding to 15.15 (kv) = 0.7

Effective axial length (Ld) = L – kv × nvd × bvd = 160 – 0.7 × 2 × 10 = 146 mm

Air gap area/Pole (Aag) $= \dfrac{\pi \times D \times Ld}{P} = \dfrac{\pi \times 330 \times 146}{6} = 25227 \text{mm}^2$

Flux density in the air gap (Bg) $= \dfrac{FI \times 10^6}{Aag} = \dfrac{0.01281 \times 10^6}{25227} = 0.5078$ T

Gap flux density at 30° from the centre of the pole (B30d)

= 1.36 × Bg = 1.36 × 0.5078 = 0.6906 T

Air gap AT (ATg) = 0.796 × B30d × Lgd × 1000 = 0.796 × 0.6906 × 0.7973 = 438.3

Flux density In rotor tooth at 30°from the centre of the pole (Btr30)

= Brt × 1.36 = 0.8839 × 1.36 = 1.2021 T

From Fig. 7.2, Amp-Turns/m for Rotor teeth (at Flux density Btr30 = 1.2021T),

atrt = 296.8

AmpTurns for Rt-tooth (ATRT) = atrt × Hsr/1000 = 296.8 × 15.5/1000 = 4.6

From Fig. 7.2, Amp-Turns/m for Rotor core (at Flux density Bc = 1.35 T),

atrc = 480.8

Rt-Core mean dia (Dcrav) = Dr – (2 × Hsr) – dcr

= 328.68 – (2 × 15.5) – 36.84 = 260.84 mm

AmpTurns for Rt-Core (ATRC) $= \dfrac{\pi \times Dcrav \times atrc}{P \times 3 \times 1000} = \dfrac{\pi \times 260.84 \times 480.8}{6 \times 3 \times 1000} = 21.89$

Total AT for Rotor (ATR) = ATRC + ATRT = 21.89 + 4.6 = 26.49

Total AT for the motor (ATT) = ATS + ATR + ATg = 71.18 + 26.49 + 438.3 = 536

Magnetization AT of Sq-Cage 3ph,Ind.Motor

	Length	Flux=	0.01281	Wb
Part	M	B(tesla)	AT/m	AT
Stator-Core	0.08046	1.35	480.8	39
Stator Teeth	0.046	1.4466	706.5	32
Air-Gap	7.96E-04	0.6906		438
Rotor Teeth	1.55E-02	1.2021	296.8	5
Rotor-Core	0.04553	1.9283	480.8	22
Total-AT=				536

$$\text{Magnetizing current (Im)} = \frac{\text{P} \times \text{ATT}}{2 \times 1.17 \times \text{Kw} \times \text{Tph}} = \frac{6 \times 536}{2 \times 1.17 \times 0.955 \times 162} = 8.8835 \text{ A}$$

$$\text{No load Phase current (I0)} = \sqrt{\text{Iw}^2 + \text{Im}^2} = \sqrt{1.441^2 + 8.8835^2} = 9.0$$

$$\text{No load Power Factor (pf0)} = \frac{\text{Iw}}{\text{I0}} = \frac{1.441}{9} = 0.1601.$$

7.1.6 (a) Computer Program in "C" in MATLAB for Part-4

```
%(4)<------------AmpTurns and Magnetizing-Current---------->

Bc=1.35;D=330;Hs=46;Hc=39;P=6;B13=1.0644;Lg=0.66;SPitch=19.2;

Dr=328.68;Sr=45;bvd=10;nvd=2;L=160;FI=0.01281;Brt=0.8839;

Hsr=15.5; dcr=36.8385;Brc=1.35;Kw=0.955;Tph=162;Iw=1.441;

%Input Data

Wss0=4;Wsr0=2; %Assumptions

%-------------BH Curve for 0.5mm,LOHYS Quality--------->

BB=[.1 .2 .3 .4 .5 .6  .7  .8  .9   1  1.1 1.2 1.3 1.4 1.5  1.6
1.7  1.8   1.9   2];

H= [50 65 70 80 90 100 110 120 150 180 220 295 400 580 1000
2400 5000 8900 15000 24000];

semilogx(H,BB); grid;  xlabel('AT/m-->');

ylabel('Flux density(T)-->');

title('Magnetization Curve for Lohys Stamping Steel');

%-------Carters Coefft for Air Gap)--->

Ratio=[0  1   2   3   4   5  6   7   8   9   10  11  12 ];

CC= [0 .18 .33 .45 .53 .6 .66 .71 .75 .79 .82 .86 .89];
%Semiclosed Slots
```

```
CC1=[0 .14 .27 .37 .44 .5 .54 .58 .62 .65 .68 .69 .7];
%Open Slots
plot(Ratio,CC,Ratio,CC1);grid;xlabel('Slot Opening/Airgap-->');
ylabel('Carter Coefft-->');  title('Carter Coefft for slots');
legend('Semiclosed','Open');
semilogx(HH,BB);  grid;  xlabel('AT/m-->');
ylabel('Flux density(T)-->');
title('Magnetization Curve for Lowhys Stamping Steel');
atsc=interp1(BB,H,Bc,'spline');                     Dcav=D+2*Hs+Hc;
ATSC=pi*Dcav/P/3*atsc/1e3;                          Bt30=B13*1.36;
atst=interp1(BB,H,Bt30,'spline');   ATST=atst*Hs/1000 ;
ATS=ATSC+ATST;                                       rat1=Wss0/Lg;
k01=interp1(Ratio,CC,rat1,'spline');
kgs=SPitch/(SPitch-Wss0*k01);                        rat2=Wsr0/Lg;
k02=interp1(Ratio,CC,rat2,'spline');
spr0=pi*Dr/Sr;    kgr=spr0/(spr0-Wsr0*k02);          kg=kgs*kgr;
Lgd=Lg*kg;  rat3=bvd/Lg;   kv=interp1(Ratio,CC1,rat3,'spline');
if rat3 >=12 kv=0.7;end;
Ld=L-kv*nvd*bvd;  Aag=pi*D/P*Ld;  Bg=FI*1e6/Aag;  B30d=1.36*Bg;
ATg=0.796*B30d*Lgd*1e3;                              Btr30=Brt*1.36;
atrt=interp1(BB,H,Btr30,'spline');
ATRT=atrt*Hsr/1e3; atrc=interp1(BB,H,Brc,'spline');
Dcrav=Dr-2*Hsr-dcr;  ATRC=pi*Dcrav/1e3/P/3*atrc;  ATR=ATRC+ATRT;
ATT=ATS+ATR+ATg;  Im=P/2*ATT/(1.17*Kw*Tph);   I0=sqrt(Iw^2+Im^2);
pf0=Iw/I0;
%-----------------------------------------------------
```

7.1.7 Short-Circuit Current Calculation (Part-5)

h1 = Zsh × (Hstrip + insS) = 6 × (5 + 0.5) = 33mm, h3 = Hw = 4, h4 = HL = 1,

bs = Ws = 10.6, b0 = Wss0 = 4,

Assuming h2 = 1.6mm, Specific Permeance of Stator Slot (Lmdss)

$$= \frac{h_1}{3 \times b_s} + \frac{h_2}{b_s} + \frac{2h_3}{b_s + b0} + \frac{h_4}{b0} = \frac{33}{3 \times 10.6} + \frac{1.6}{10.6} + \frac{2 \times 4}{10.6 + 4} + \frac{1}{4} = 1.9866$$

h1r = Wb = 15, br = Wsr = 6.5, br0 = Wsr0 = 2 and Assuming h2r = h3r = 0,

h4r = 0.5, and Ks = 1,

Specific Permeance of Rotor Slot (Lmdsr)

$$= \frac{h_1 r}{3 \times br} + \frac{h_{2r}}{br} + \frac{2h_{3r}}{br + br0} + \frac{h_{4r}}{br0} = \frac{15}{3 \times 6.5} + \frac{0}{6.5} + \frac{2 \times 0}{6.5 + 2} + \frac{0.5}{2} = 1.0192$$

and same referred to stator (Lmddsr)

$$= Kw^2 \times S/Sr \times Lmdsr = 0.955^2 \times 54/45 \times 1.0192 = 1.1155$$

Specific Slot Permeance (ssp) = Lmdss + Lmddsr = 1.9866 + 1.1155 = 3.1021

gd = spp = 3, p = P/2 = 6/2 = 3,

Slot Reactance (Xs) = $15.8 \times f \times L \times ssp \times Tph^2/(p \times gd) \times 10^{-9}$

$$= 15.8 \times 50 \times 160 \times 3.1021 \times 162^2/(3 \times 3) \times 10^{-9} = 1.1434\ \Omega$$

For Over hang (L0Lmd0) $= \frac{Ks \times PP^2}{\pi \times Spitch \times 1000} = \frac{1 \times 172.79^2}{\pi \times 19.1986 \times 1000} = 0.495$

Overhang Reactance (X0) = $15.8 \times f \times L0Lmd0 \times Tph^2/(p \times gd) \times 10^{-6}$

$$= 15.8 \times 50 \times 0.495 \times 162^2/(3 \times 3) \times 10^{-6} = 1.1403\ \Omega$$

St.Slots/Pole(gs) = S/P = 54/6 = 9; Rt Slots/pole(gr) = 45/6 = 7.5

Magnetizing reactance (Xm) = Vph/Im = 440/8.8835 = 49.53

Zig-Zag Reactance(Xz) $= \frac{1}{gs^2} + \frac{1}{gr^2} = \frac{1}{9^2} + \frac{1}{7.5^2} = 1.2433\ \Omega$

Total Reactance/ph(X) = Xs + X0 + Xz = 1.1434 + 1.1403 + 1.2433 = 3.527 Ω

Resistance(R) = Rph + Rr = 0.3546 + 0.4614 = 0.8159 Ω

Impedance/ph(Z) $= \sqrt{R^2 + X^2} = \sqrt{0.8159^2 + 3.527^2} = 3.6202\ \Omega$

Short Circuit Current (Isc) $= \frac{Vph}{Z} = \frac{440}{3.6202} = 121.54 Amps = \frac{Isc}{Iph} = \frac{121.54}{29.958} = 4.057 pu$

Short Circuit PF (pfsc) = $\frac{R}{Z} = \frac{0.8159}{3.6202} = 0.2254$

7.1.7 (a) Computer Program in "C" in MATLAB for Part-5

```
%(5)<---------Short-Circuit-Current-------------------->');

Zsh=6; Hstrip=5; insS=0.5; Hw=4; Ws=10.6; Wb=15; Wsr=6.5; S=54;
Iph=29.958;  f=50;PP=172.79;HL=1;Vph=440;  Rph=0.3546;Rr=0.4614;
Wss0=4Tph=162;SPitch=19.1986;Im=8.8835;L=160; Wsr0=2; Kw=0.955;
Sr=45;P=6;%Input Data

h2=1.6;h2r=0;h3r=0;h4r=0.5;ks=1; %Assumptions
```

```
h1=Zsh*(Hstrip+insS); h3=Hw;  h4=HL;  bs=Ws;  b0=Wss0;
Lmdss=h1/3/bs +h2/bs +2*h3/(bs+b0) +h4/b0;   h1r=Wb;   br=Wsr;
br0=Wsr0;  Lmdsr=h1r/3/br  +h2r/br  +2*h3r/(br+br0)  +h4r/br0;
Lmddsr=Kw^2*S/Sr*Lmdsr;  ssp=Lmdss+Lmddsr;  gd=S/P/3;    p=P/2;
Xs=15.8*f*L*ssp*Tph^2/(p*gd)*1e-9;
L0Lmd0=ks*PP^2/pi/SPitch/1000;
X0=15.8*f*L0Lmd0*Tph^2/(p*gd)*1e-6; gs=S/P; gr=Sr/P; Xm=Vph/Im;
Xz=5/6*Xm*(1/gs^2+1/gr^2); X=Xs+X0+Xz;  R=Rph+Rr;
Z=sqrt(R^2+Xt^2);   Isc=Vph/Z;pfsc=R/Z;RAT=Isc/Iph;
%-----------------------------------------------------------
```

7.1.8 Performance Calculation (Part-6)

Total Losses (Pt) = PnL + Pcus + Pcur = 1902.1 + 954.63 + 897.49

= 3754.2 W = 3.7542 KW

$$\text{Efficiency (EFF)} = \frac{\text{KW}}{\text{KW} + \text{Pt}} \times 100 = \frac{30}{30 + 3.7542} \times 100 = 88.88\%$$

Rotor Input (Rinp) = KW × 1000 + Pfw + Pcur = 30 × 1000 + 300 897.49 = 31197 W

Slip at Full Load (SFL) = Pcur/Rinp × 100

= 897.49/31197 × 100 = 2.88% = 0.0288 pu

$$\text{Starting Tq (Tst)} = (\frac{\text{Isc}}{\text{Ir}})^2 \times \text{SFL} = (\frac{121.564}{25.4645})^2 \times 0.0288 = 0.6554 \text{ pu}$$

$$\text{Pmax} = \frac{3 \times \text{Vph} \times (\text{Isc} - \text{I0})}{2(1 + \text{pfsc}) \times 1000} = \frac{3 \times 440 \times (121.564 - 9)}{2(1 + 0.2254) \times 1000} = 60.616 \text{ KW}$$

Inner cooling area (Acool1) = $[\pi \times D \times (L \times 2.5) + 2 \times \pi \times (D + 50) \times 0.04]/10^6$

$= [\pi \times 330 \times (160 \times 2.5) + 2 \times \pi \times (330 + 50) \times 0.04]/10^6 = 0.4148 \text{ m}^2$

Acool2 = Acool1 × (1 + 0.1 × v) = Acool1 × (1 + 0.1 × 17.28) = 1.1315 m^2

Outer cooling area (Acool3) = $\pi \times D0 \times L/10^6 = \pi \times 500 \times 160/10^6 = 0.2513 \text{ m}^2$

Total cooling area (AcoolT) = Acool2 + Acool3

$1.1315 + 0.2513 = 1.3828 \text{ m}^2$

Total Stator Loss (Pst) = Pcus + Pit + Pic = 954.63 + 619.86 + 982.2 = 2557 W

Temp rise (Tr) = 0.03 × Pst/AcoolT = 0.03 × 2557/1.3828 = 55.47 °C

Area of Rotor slots (Ars) = Wsr × Hsr × Sr = 6.5 × 15.5 × 45 = 4534 m^2

Rotor inner dia (Dri) = Dcrav – dcr = 260.84 – 36.84 = 224 m

Weight of Rotor (Wri) = $[\pi \times (Dr^2 - Dri^2)/4 - Ars] \times L \times 7.8/10^6$

$= [\pi \times (328.68^2 - 224^2)/4 - 4534] \times 160 \times 7.8/10^6 = 51.05 \text{ Kg}$

Assuming Insulation Wt of 1%,

Total wt (Wtot) = 1.01 × (Wcus + Wt + Wc + Wri + Wcur + Wcue)

= 1.01 × (38.04 + 25.91 + 53.6 + 51.05 + 7.42 + 3.26) = 181.07 Kg

KgPKw = Wtot/KW = 181.07/30 = 6.04

7.1.8 (a) Computer Program in "C" in MATLAB for Part-6

```
%(6)<--------------Performance-------------------->');
PnL=1902.1;Pcus=954.63;Pcur=897.49;KW=30;Pfw=300;Isc=121.564;
Ir=25.4645;Vph=440;I0=8.9936;pfsc=0.2254;D=330;L=160;v=17.28;
D0=500;L=160;Pic=982.2;Pit=619.86;Wsr=6.5;Wcus=38.043;Wc=53.6;
Hsr=15.5;Sr=45;Dcrav=260.84;dcr=36.84;Dr=328.68;Wt=25.91;
Wcue=3.26;Wcur=7.42;            %Input Data
Pt=PnL+Pcus+Pcur;                          EFF=KW/(KW+Pt/1000)*100;
Rinp=KW*1000+Pfw+Pcur;                              SFL=Pcur/Rinp*100;
Tst=(Isc/Ir)^2*SFL/100;     Pmax=3*Vph*(Isc-I0)/2/(1+pfsc)*1e-3;
Acool1=(pi*D*(L*2.5)+2*pi*(D+50)*0.04)/1e6;
Acool2=Acool1*(1+0.1*v);
Acool3=pi*D0*L/1e6;    AcoolT=Acool2+Acool3;    Pst=Pcus+Pit+Pic;
Tr=0.03*Pst/      AcoolT;      Ars=Wsr*Hsr*Sr;       Dri=Dcrav-dcr;
Wri=(pi*(Dr^2-Dri^2)/4-Ars)*L*7.8e-6;
Wtot=1.01*(Wcus+Wt+Wc+Wri+Wcur+Wcue);    KgPKw=Wtot/KW;
%--------------------End of Program------------------------->
```

7.1.9 Computer Program in "C" in MATLAB for Complete Design

```
%3ph,KW=30;V=440;P=6;f=50Sq.Cage IM with Sq.Cage rotor
f2=fopen('Total_30KW_Output.m','w');
%------------Standard Curves/Tables for%Data-------------->
SKW=[1 2 5 10 20 50 100 500];SBav=[0.35 0.38 0.42 0.46 0.48
0.50 0.51 0.53];
Sq=[16e3 19e3 23e3 25e3 26e3 29e3 31e3 33e3];
SKWa=[5 10 20 50 100 200 500];
SPF6P=[0.82 0.83 0.85 0.87 0.89 0.9 0.92];%Table for 1000RPM
SEFF6P=[0.83 0.85 0.87 0.89 0.91 0.92 0.93];%Table for 1000RPM
SPF4P=[0.85 0.86 0.88 0.9 0.91 .92 .93];%Table: for 1500RPM
SEFF4P=[.85 .87 .88 .9 .91 .93 .94];%Table for 1500RPM
SD=[0.1 0.15 0.2 0.3 0.4 0.5 0.75 1];
SCDSW=[4 3.8 3.6 3.5 3.5 3.5 3.5 3.5];
```

```
%-------------BH Curve for--0.5mm,LOHYS Quality--------->
BB=[.1 .2 .3 .4 .5 .6  .7  .8  .9   1  1.1 1.2 1.3 1.4 1.5  1.6
1.7  1.8   1.9   2];
H= [50 65 70 80 90 100 110 120 150 180 220 295 400 580 1000
2400 5000 8900 15000 24000];
%-------Carters Coefft for Air Gap--->
Ratio=[0  1   2   3   4   5  6   7   8   9   10  11  12];
CC=  [0 .18 .33 .45 .53 .6 .66 .71 .75 .79 .82 .86 .89];
CC1= [0 .14 .27 .37 .44 .5 .54 .58 .62 .65 .68 .69 .7];
%(1)<-----------Main Dimensions----------------------->
KW=30;V=440;f=50;%P=6;Specification/Input Data------>
insW=3.4;Hw=4;HL=1;insH=6;nvd=2;bvd=0.01;ki=0.92;Bc=1.35;Vph=V;
Zr=1;kwr=1;cdb=6;Tb=6;cde=6;dd=0.05;Brc=1.35;   %Assumptions
Bav=interp1(SKW,SBav,KW,'spline');
q=interp1(SKW,Sq,KW,'spline');
pf=interp1(SKWa,SPF6P,KW,'spline');
eff=interp1(SKWa,SEFF6P,KW,'spline');
Kw=0.955;nvd=2;bvd=10;ki=0.92;Tstrip=1.9;insS=0.5;%Assumptions
insW=3.4;Hw=4;HL=1;insH=6;nvd=2;ki=0.92;Bc=1.35;%Assumptions
SD=[0.1 0.15 0.2 0.3 0.4 0.5 0.75 1];SCDSW=[4 3.8 3.6 3.5 3.5
3.5 3.5 3.5];
B=[0.8 1.2 1.6 2 2.4];  WpKg=[7 15 24 34 50];
Vph=V; KWinp=KW/eff; Iph=KWinp*1e3/(3*Vph*pf);
P=6; spp=3; Tstrip=1.9; Zsw=3; % Assumption
pf=interp1(SKWa,SPF6P,KW,'spline');
eff=interp1(SKWa,SEFF6P,KW,'spline');
if P==4 pf=interp1(SKWa,SPF4P,KW,'spline');end;
if P==4 eff=interp1(SKWa,SEFF4P,KW,'spline');end;
Ns=120*f/P;ns=Ns/60;C0=11*Kw*Bav*q*eff*pf*1e-3;
DsqL=1/C0*(KW/ns);                    L1=sqrt(DsqL/(0.135*P)^2);
L=floor(L1*100)*10;  Ls=(L-nvd*bvd);Li= ki*Ls;
D1=sqrt(DsqL/(L/1000));D=ceil(D1*100)*10;PP=pi*D/P;LbyPP=L/PP;
if LbyPP <0.8||LbyPP >2 continue; end;
v=pi*D*ns/1000;if v >30 continue;end;
FI=pi*D/P*L*Bav/1e6;
```

```
%(2)<--------------Stator Slots and Winding----->
Tphi=Vph/(4.44*f*FI*Kw);CDSW=interp1(SD,SCDSW,D/1000,'spline');
As1=Iph/CDSW;S=spp*P*3;SPitch=pi*D/S;
if SPitch <18||SPitch>=25 continue;end;
Zphi=2*Tphi; sph=S/3; Zs1=Zphi/sph;Zs=ceil(Zs1); Tph=Zs*sph/2;
FI=Vph/(4.44*f*Tph*Kw);                    Hstrip1=As1/Tstrip;
Hstrip=ceil(Hstrip1*2)/2; WbyT=Hstrip/Tstrip;
if WbyT <2.5||WbyT >=3.5 continue; end; As=0.967*Hstrip*Tstrip;
Ws=(Zsw*(Tstrip+insS)+insW);                             Zsh=Zs/Zsw;
Hs=(Zsh*(Hstrip+insS)+Hw+HL+insH+2);D13=D+2/3*Hs;sp13=pi*D13/S;
Wt13=sp13Ws;B13=FI*P*1e6/(Li*Wt13*S);Btmax=1.5*B13;
Lmt=(2*L+2.3*PP+240)/1000;  Rph=0.021*Lmt*Tph/As;
Pcus=3*Iph^2*Rph;       Wcus=Lmt*Tph*3*As*8.9e3;       FIc=FI/2;
Ac=FIc*1e6/Bc;Hc=Ac/Li; D01=D+2*(Hs+Hc);     D0=ceil(D01/10)*10;
Hc=(D0-D)/2-Hs;
PitpKg=interp1(B,WpKg,Btmax,'spline');
PicpKg=interp1(B,WpKg,Bc,'spline');
Wt=Li*Wt13*S*Hs*7.8e-6;  Dmcs=D+2*Hs+Hc; Wc=Ac*pi*Dmcs*7.8e-6;
Pit=PitpKg*Wt;Pic=PicpKg*Wc;
%-----------------------ROTOR----------------------->
Zr=1;kwr=1;cdb=6;Tb=6;cde=6;dd=50;Brc=1.35;   %Assumptions
kws=Kw;Ss=S;Lg1=0.2+2*sqrt(D*L/1e6);Lg=ceil(Lg1*100)/100;
Dr=D-2*Lg;d1=Ss-3*P;d2=Ss-P;d3=Ss-2*P; d4=Ss-5*P;d5=Ss-1;
d6=Ss-2;d7=Ss-7;d8=Ss-8;Sr=Ss-9;    sp2=pi*Dr/Sr;    Ir=0.85*Iph;
Ib=Ir*kws*Ss*Zs/(kwr*Sr*Zr);Abi=Ib/cdb;          Wb=ceil(Abi/Tb);
Ab=Tb*Wb*0.98;Wsr=Tb+0.5;Hsr=Wb+0.5;Lb=L+50;Rb=0.021*Lb/1e3/Ab;
Pcub=Ib^2*Rb*Sr;  Ie=Ib*Sr/P/pi;  Ae=Ie/cde;   Dme=Drdd;
Lme=pi*Dme/1000;Re=0.021*Lme/Ae;
Pcue=2*Ie^2*Re;Pcur=Pcub+Pcue;Rr=Pcur/(3*Ir^2);
Dr13=Dr2*2/3*Hsr;spr13=pi*Dr13/Sr;
Wtr13=spr13Wsr;Atr=Wtr13*Li*Sr/P;Brt=FI*1e6/Atr;
Brtmax=Brt*1.5;Ac=FI*1e6/2/Brc;dcr=Ac/Li;
Pfw=0.01*KW*1e3;PnL=Pit+Pic+Pfw;Iw=PnL/3/V;
Wcur=Lb*Sr*Ab*8.9e-6;Wcue=Lme*2*Ae*8.9e-3;
%(4)<----AmpTurns and Magnetizing-Current---------------------
>
```

```
%-------------BH Curve for--0.5mm,LOHYS Quality--------->
BB=[.1 .2 .3 .4 .5 .6  .7  .8  .9   1  1.1 1.2 1.3 1.4 1.5  1.6
1.7  1.8   1.9   2];
H= [50 65 70 80 90 100 110 120 150 180 220 295 400 580 1000
2400 5000 8900 15000 24000];
%-------Carters Coefft for Air Gap--->
Ratio=[0  1   2   3   4   5  6   7   8   9   10   11   12 ];
CC= [0 .18 .33 .45 .53 .6 .66 .71 .75 .79 .82 .86 .89 ];
CC1=  [0 .14 .27 .37 .44 .5 .54 .58 .62 .65 .68 .69 .7 ];
Wss0=4;Wsr0=2;   %Assumptionsatsc=interp1(BB,H,Bc,'spline');
Dcav=D+2*Hs+Hc;ATSC=pi*Dcav/P/3*atsc/1e3;         Bt30=B13*1.36;
atst=interp1(BB,H,Bt30,'spline');ATST=atst*Hs/1000;
ATS=ATSC+ATST;rat1=Wss0/Lg;
k01=interp1(Ratio,CC,rat1,'spline');
kgs=SPitch/(SPitch-Wss0*k01);    rat2=Wsr0/Lg;
k02=interp1(Ratio,CC,rat2,'spline'); spr0=pi*Dr/Sr;
kgr=spr0/(spr0-Wsr0*k02);  kg=kgs*kgr;  Lgd=Lg*kg;
rat3=bvd/Lg; kv=interp1(Ratio,CC1,rat3,'spline');
if rat3 >=12 kv=0.7;end;
Ld=L-kv*nvd*bvd;Aag=pi*D/P*Ld;Bg=FI*1e6/Aag;B30d=1.36*Bg;
ATg= 0.796*B30d*Lgd*1e3;
Btr30=Brt*1.36;atrt=interp1(BB,H,Btr30,'spline');
ATRT=atrt*Hsr/1e3;                             Dcrav=Dr-2*Hsr-dcr;
atrc=interp1(BB,H,Brc,'spline');ATRC=pi*Dcrav/1e3/P/3*atrc;
ATR=ATRC+ATRT;      ATT=ATS+ATR+ATg;    Im=P/2*ATT/(1.17*Kw*Tph);
I0=sqrt(Iw^2+Im^2); pf0=Iw/I0; I0byI=I0/Iph;
%(5)<-----------Short-Circuit-Current-------------------->');
h2=1.6;h2r=0;h3r=0;h4r=0.5;ks=1; %Assumptions
h1=Zsh*(Hstrip+insS);h3=Hw;h4=HL;bs=Ws;b0=Wss0;
Lmdss=h1/3/bs  +h2/bs  +2*h3/(bs+b0)  +h4/b0;  h1r=Wb;  br=Wsr;
br0=Wsr0;          Lmdsr=h1r/3/br+h2r/br+2*h3r/(br+br0)+h4r/br0;
Lmddsr=Kw^2*S/Sr*Lmdsr;  ssp=Lmdss+Lmddsr;    gd=S/P/3;  p=P/2;
Xs=15.8*f*L*ssp*Tph^2/(p*gd)*1e9;L0Lmd0=ks*PP^2/pi/SPitch/1000;
X0=15.8*f*L0Lmd0*Tph^2/(p*gd)*1e6;gs=S/P;gr=Sr/P;Xm=Vph/Im;
Xz=5/6*Xm*(1/gs^2+1/gr^2);          X=Xs+X0+Xz;           R=Rph+Rr;
Z=sqrt(R^2+X^2); Isc=Vph/Z; pfsc=R/Z;   RAT=Isc/Iph;
%(6)<--------------Performance-------------------->');
```

```
Pt=PnL+Pcus+Pcur;EFF=KW/(KW+Pt/1000)*100;Rinp=KW*1000+Pfw+Pcur;
SFL=Pcur/Rinp*100; Tst=(Isc/Ir)^2*SFL/100;

Pmax=3*Vph*(Isc-I0)/2/(1+pfsc)*1e-3;

Acool1=(pi*D*(L*2.5)+2*pi*(D+50)*0.04)/1e6;
Acool2=Acool1*(1+0.1*v);Acool3=pi*D0*L/1e6;

AcoolT=Acool2+Acool3;Pst=Pcus+Pit+Pic;   Tr=0.03*Pst/AcoolT;

Ars=Wsr*Hsr*Sr;Dri=Dcrav-dcr;

Wri=(pi*(Dr^2-Dri^2)/4-Ars)*L*7. 8e-6;

Wtot=1.01*(Wcus+Wt+Wc+Wri+Wcur+Wcue);KgPKw=Wtot/KW;

%--------------------End of Program------------------------->

fprintf(f2,'Design   of   30KW,440V,50HZ,3-Ph   Sq.Cage   Ind
Motor\n');

fprintf(f2,'*********************************************');

fprintf(f2,'\nInput Data:');

fprintf(f2,'\n-----------');

fprintf(f2,'\nParamter                    VALUES');

fprintf(f2,'\n--------                    ------');

fprintf(f2,'\nRating(KW)                  %5.1f',KW);

fprintf(f2,'\nVolts                       %5.0f',V);

fprintf(f2,'\nPoles                       %5.0f',P);

fprintf(f2,'\nHz                          %5.0f',f);

fprintf(f2,'\nInterpolated     values     from     curves:
Bav=%5.3f,q=%5.0f,eff=%5.3f,pf=%5.3f',Bav,q,pf,eff);

fprintf(f2,'\nOutput Results:');

fprintf(f2,'\n===============');

fprintf(f2,'\nParamter                    VALUES');

fprintf(f2,'\n--------                    ------');

fprintf(f2,'\nOutput Coefft(C0)           %6.2f',C0);

fprintf(f2,'\nSync.Speed(rps)             %5.2f',ns);

fprintf(f2,'\nDsqL                        %5.4f',DsqL);

fprintf(f2,'\nGross Length(mm)            %5.1f',L);

fprintf(f2,'\nNet iron Length(mm)         %5.1f',Li);

fprintf(f2,'\nStator Inner Dia(mm)        %5.1f',D);

fprintf(f2,'\nPeriphoral Speed(m/s)       %5.2f
(Max.Permissible:30)',v);
```

```
fprintf(f2,'\nPole-Pitch(mm)                      %5.1f',PP);
fprintf(f2,'\nLength   to   PP   ratio      %6.4f(Around   1   -->good)',LbyPP);
fprintf(f2,'\nSlots                               %5.0f',S);
fprintf(f2,'\nSlot-Pitch(mm)
%6.3f(Permissible:18to25mm)',SPitch);
fprintf(f2,'\nCond/Slot                           %5.0f',Zs);
fprintf(f2,'\nTurns/Ph                            %5.0f',Tph);
fprintf(f2,'\nFlux/Pole(Wb)                       %6.5f',FI);
fprintf(f2,'\nPhase Current(A)                    %5.2f',Iph);
fprintf(f2,'\nBare Strip  (w*t)mm
%5.2fX%4.2f',Hstrip,Tstrip);
fprintf(f2,'\nWidth to Thickness Ratio
%5.2f(Permissible:2.5to3.5)',WbyT);
fprintf(f2,'\nArea of CS cond(mm^2)               %7.3f',As);
fprintf(f2,'\nCt.density(A/mm^2)                  %7.3f',CDSW);
fprintf(f2,'\nInterpolated    values    of    W/Kg    of
St:Teeth=%5.2fandCore=%5.2f',PitpKg, PicpKg);
fprintf(f2,'\nNo       of       Strips(width*depth-wise)%4.0f
X%2.0f',Zsw,Zsh);
fprintf(f2,'\nSlot-Width (mm)                     %6.1f',Ws);
fprintf(f2,'\nSlot-Height(mm)Total               %6.1f',Hs);
fprintf(f2,'\d(1/3),SP(1/3),Wt(1/3)(m)
%6.4f,%6.4f,%6.4f',D13,sp13,Wt13);
fprintf(f2,'\nSt-Tooth-Flux-Dens(1/3)             %6.4f ',B13);
fprintf(f2,'\nSt-Tooth-Flux-Dens-Max(T)
%6.4f(Permissible:1.6to1.8)',B13*1.5);
fprintf(f2,'\nLength of mean-turn (m)             %6.3f',Lmt);
fprintf(f2,'\nResistance/Ph (ohm)                 %6.4f',Rph);
fprintf(f2,'\ndepth of St.Core(mm)                %6.2f',Hc);
fprintf(f2,'\nOuter Dia of St.Core(mm)            %6.1f',D0);
fprintf(f2,'\nSt.Cu.Loss(W)                       %6.1f',Pcus);
fprintf(f2,'\nWt of St-Teeth+Core(Kg)=           %6.2f+%6.2f=%
6.2f',Wt,Wc,Wt+Wc);
fprintf(f2,'\nIron      Loss=Teeth+Core(W)=        %5.1f+%5.1f=
%5.1f',Pit,Pic,Pit+Pic);
```

```
fprintf(f2,'\n-----------ROTOR----------------------------->');
fprintf(f2,'\nLength of Air-Gap(mm)        %6.4f',Lg);
fprintf(f2,'\Dia of Rotor(mm)              %6.1f',Dr);
fprintf(f2,'\nNo of Slots Should NE to:%2.0f,%2.0f,%2.0f,
%2.0f,%2.0f',d1,d2,d3,d4);
fprintf(f2,'\nShould NE to%2.0f,%2.0f,%2.0f,%2.0f,%2.0f',
d5,d6,d7,d8);
fprintf(f2,'\nNo of Rotor Slots Selected   %3.0f',Sr);
fprintf(f2,'\nRotor Slot-Pitch(mm)         %6.4f',sp2);
fprintf(f2,'\nEquivalent Rotor Ct(A)       %6.2f',Ir);
fprintf(f2,'\nRotor bar Ct(A)              %6.1f',Ib);
fprintf(f2,'\nRotor bar CS(mm^2)           %6.4f',Ab);
fprintf(f2,'\nBar (w*t)mm                  %5.1fX%4.1f',Wb,Tb);
fprintf(f2,'\nRotor Slot (w*h)mm         %5.1fX%4.1f',Wsr,Hsr);
fprintf(f2,'\nLength of Bar (m)            %5.3f',Lb);
fprintf(f2,'\nResistance/bar(m.ohm)        %6.4f',Rb*1e3);
fprintf(f2,'\nLosses in Rot.Bars (W)       %6.1f',Pcub);
fprintf(f2,'\nEnd ring Ct(A)               %6.1f',Ie);
fprintf(f2,'\nArea of end ring(mm^2)       %6.1f',Ae);
fprintf(f2,'\nResistance of end ring(m.ohm)%6.4f',Re*1e3);
fprintf(f2,'\nRot-Cu-
Loss=Bars+End.rings(W)=%5.1f+%5.1f=%5.1f',Pcub,Pcue, Pcur);
fprintf(f2,'\nEquivalent Rotor res(Ohm)   %6.3f',Rr);
fprintf(f2,'\nRotor(1/3)Slot-Pitch(mm)    %6.2f',spr13);
fprintf(f2,'\nRotor(1/3)tooth width(mm)   %6.2f',Wtr13);
fprintf(f2,'\nRt-Tooth-Flux-Dens(1/3)(T)  %6.4f ',Brt);
fprintf(f2,'\nRt-Tooth-Flux-Dens-Max(T)
%6.4f(Permissible:1.2to1.5)',Brt*1.5);
fprintf(f2,'\nDepth of Rotor core (mm)     %5.2f',dcr);
fprintf(f2,'\n----------------------No-Load-Losses-------------
--------->');
fprintf(f2,'\nN0-Load Losses (W)           %5.1f',PnL);
fprintf(f2,'\nN0-Load Wattful Current(A) %5.3f',Iw);
```

```
fprintf(f2,'\n-------------------Magnetizing-Current-----------
--------->');

fprintf(f2,'\nInterpolated values of at/m of
St:Core=%5.1fandTeeth=%5.1f',atsc,atst);

fprintf(f2,'\nInterpolated values of Caters
Coeffts:k01=%5.3f,k02=%5.3f,kv= %5.3f', k01,k02,kv);

fprintf(f2,'\nStator AT:Core+Teeth:
%5.1f+%5.1f=%5.1f',ATSC,ATST,ATS);

fprintf(f2,'\nRotor AT :Core+Teeth:
%5.1f+%5.1f=%5.1f',ATRC,ATRT,ATR);

fprintf(f2,'\nTotal AT :Stator+Rotor+Airgap: %5.1f+%5.1f+%5.1f=
%5.1f',ATS,ATR, ATg,ATT);

fprintf(f2,'\nN0-Load Current(A) Iw=%5.2f,Im=%5.2f and I0=%5.2f
at pf=%5.3f',Iw, Im,I0,pf0);

fprintf(f2,'\nI0/Iph ratio
%5.3f',I0byI);

fprintf(f2,'\n-------------------Short-Circuit-Current---------
----------->');

fprintf(f2,'\nSlot-Permeances: Stator=%5.3f,Rotor=%5.3f and
Rotor refered to stator= %5.3f',Lmdss,Lmdsr,Lmddsr);

fprintf(f2,'\nSpecific-Slot-Permeance=%5.3f',ssp);

fprintf(f2,'\nTotalReactance(X):Slot+Overhang+Zig-
Zag=%5.3f+%5.3f+%5.3f=% 5.3fohms',Xs,X0,Xz,X);

fprintf(f2,'\nShort-Circuit: R=%5.2f,Z=%5.2f and Isc=%5.1f at
pf=%5.3f;Isc/IFL=% 5.3f',R,Z,Isc,pfsc,RAT);

fprintf(f2,'\nTotalLosses(PnL+Pcus+Pcur)=%5.1f+%5.1f+%5.1f=%5.1
fWandEfficiency=%5.2f perc',PnL,Pcus,Pcur,Pt,EFF);

fprintf(f2,'\nSlip at FL        =%5.3f X Perc',SFL);

fprintf(f2,'\nStarting Tq       =%5.3f X FL-Tq',Tst);

fprintf(f2,'\nMax.Output(KW)=%5.1f',Pmax);              %D=0.15;
v=11.75;D0=0.24;L=0.09;

fprintf(f2,'\nTemp-Rise(deg-C)  =%5.1f',Tr);

fprintf(f2,'\nTotalWt(Kg)=Wcus+Wt+Wc+Wri+Wcur+Wcue=%5.1f+%5.1f+
%5.1f+%5.1f+%4.1f+%4.1f=%5.1f',Wcus,Wt,Wc,Wri,Wcur,Wcue,Wtot);

fprintf(f2,'\nTotal Wt(Kg)=%5.1f and Kg/KW=%5.2f',Wtot,KgPKw);

fprintf(f2,'\n<------------End of Output-------->');

fclose(f2);

%----------------------------------------------------
```

7.1.9 *(a) Computer Output Results for Complete Design*

Design of 30KW, 440V, 50HZ,3-Ph Sq.Cage Ind Motor

****

Input Data:

Parameter	VALUES
Rating(KW)	30.0
Volts	440
Poles	6
Hz	50

Interpolated values from curves: Bav = 0.484, q = 26802, eff = 0.862, pf = 0.880

Output Results:

================

Parameter	VALUES
Output Coefft (C0)	103.34
Sync.Speed (rps)	16.67
DsqL	0.0174
Gross Length (mm)	160.0
Net iron Length (mm)	128.8
Stator Inner Dia (mm)	330.0
Periphoral Speed (m/s)	17.28 (Max.Permissible:30)
Pole-Pitch (mm)	172.8
Length to PP ratio	0.9260 (Around 1 -->good)
Slots	54
Slot-Pitch (mm)	19.199 (Permissible: 18 to 25mm)
Cond/Slot	18
Turns/Ph	162
Flux/Pole (Wb)	0.01281
Phase Current (A)	29.96
Bare Strip (w*t) mm	5.00×1.90

Width to Thickness Ratio	2.63 (Permissible: 2.5 to 3.5)
Area of CS cond (mm^2)	9.186
Ct.density (A/mm^2)	3.500

Interpolated values of W/Kg of St:Teeth = 23.92 and Core = 18.32

No. of Strips (width*depth – wise)	3 × 6
Slot-Width (mm)	10.6
Slot-Height (mm) Total	46.0
St-Tooth-Flux-Dens (1/3)	1.0644
St-Tooth-Flux-Dens-Max (T)	1.5966 (Permissible: 1.6 to 1.8)
Length of mean-turn (m)	0.957
Resistance/Ph (ohm)	0.3546
Depth of St.Core (mm)	39.00
Outer Dia of St.Core (mm)	500.0
St.Cu.Loss (W)	954.6
Wt. of St – Teeth + Core(Kg)=	25.91 + 53.60 = 79.51
Iron Loss = Teeth + Core(W)=	619.9 + 982.2 = 1602.1

------------------------ROTOR--------------------------->

Length of Air-Gap (mm)	0.6600
No. of Slots Should NE to:	36, 48, 42, 24,
Should NE to	53, 52, 47, 46,
No. of Rotor Slots Selected	45
Rotor Slot-Pitch (mm)	22.9462
Equivalent Rotor Ct (A)	25.46
Rotor bar Ct (A)	525.3
Rotor bar CS (mm^2)	88.2000
Bar (w*t) mm	15.0 × 6.0
Rotor Slot (w*h) mm	6.5 × 15.5
Length of Bar (m)	210.000
Resistance/bar (m.ohm)	0.0500
Losses in Rot.Bars (W)	620.8
End ring Ct (A)	1254.0
Area of end ring (mm^2)	209.0
Resistance of end ring (m.ohm)	0.0880

Rot–Cu–Loss = Bars+End.rings (W) =620.8 + 276.7 = 897.5

Equivalent Rotor res (Ohm) 0.461

Rotor (1/3) Slot–Pitch (mm) 21.50

Rotor (1/3) tooth width (mm) 15.00

Rt-Tooth-Flux-Dens (1/3) (T) 0.8839

Rt-Tooth-Flux-Dens-Max (T) 1.3259 (Permissible:1.2 to 1.5)

Depth of Rotor core (mm) 36.84

----------------------No-Load-Losses---------------------->

No-Load Losses (W) 1902.1

No-Load Wattful Current (A) 1.441

-------------------Magnetizing-Current-------------------->

Interpolated values of at/m of St:Core = 480.8 and Teeth = 706.5

Interpolated values of Caters Coeffts: k01 = 0.663, k02 = 0.453, kv = 0.700

Stator AT: Core + Teeth: 38.7 + 32.5 = 71.2

Rotor AT: Core + Teeth: 21.9 + 4.6 = 26.5

Total AT: Stator + Rotor + Airgap: 71.2 + 26.5 + 438.3 = 536.0

N0-Load Current(A) Iw = 1.44, Im = 8.88 and I0 = 9.00 at pf = 0.160

I0/Iph ratio 0.300

-------------------Short-Circuit-Current-------------------->

Slot-Permeances: Stator = 1.987, Rotor = 1.019 and Rotor referred to stator = 1.115

Specific-Slot-Permeance = 3.102

TotalReactance(X): Slot + Overhang + Zig – Zag = 1.143 + 1.140 + 1.243 = 3.527 ohms

Short-Circuit: R = 0.82, Z = 3.62 and Isc = 121.5 at pf = 0.225; Isc/IFL = 4.057

Total Losses (PnL + Pcus + Pcur) = 1902.1 + 954.6 + 897.5 = 3754.2 W and Efficiency = 88.88 perc

Slip at FL = 2.877 × Perc

Starting Tq = 0.655 × FL – Tq

Max.Output (KW) = 60.6

Temp-Rise (deg-C) = 55.5

Total Wt (Kg) = Wcus + Wt + Wc + Wri + Wcur + Wcue = 38.0 + 25.9 + 53.6 + 51.0 + 7.4 + 3.3 = 179.3

Total Wt (Kg) = 179.3 and Kg/KW = 5.98

<------------End of Output-------->

7.1.10 Modifications to be done in the above Program to get Optimal Design

1. Insert "for" Loops for the following parameters to iterate the total program between min and max permissible limits for selecting the feasible design variants:
   (a) No of Poles
   (b) St-Wdg Current density
   (c) Slots/pole/phase
   (d) Stator conductor thickness
   (e) No. of conductors width-wise.
2. Insert also minimum or maximum range of required objective functional values as constraint values, for example (a) Efficiency, (b) Kg/KW, (c) Temp-rise (d) I0/I ratio.
3. Run the program to get various possible design variants.

Note: From the feasible design variants printed in the output, select that particular design fulfilling the objective of optimal parameter for the design.

7.1.11 Computer Program in "C" in MATLAB for Optimal Design

```
%3ph,KW=30;V=440;P=6;f=50Sq.Cage IM wth Sq.Cage Rotor

f2=fopen('Optimal_30KW_Output.m','w');

%------------Standard Curves/Tables for%Data-------------->

SKW=[1 2 5 10 20 50 100 500];SBav=[0.35 0.38 0.42 0.46 0.48
0.50 0.51 0.53];

Sq=[16e3 19e3 23e3 25e3 26e3 29e3 31e3 33e3];

SKWa=[5 10 20 50 100 200 500];

SPF6P=[0.82 0.83 0.85 0.87 0.89 0.9 0.92];%Table for 1000RPM

SEFF6P=[0.83 0.85 0.87 0.89 0.91 0.92 0.93];

%Table for 1000RPM

SPF4P=[0.85 0.86 0.88 0.9 0.91 .92 .93];%Table for  1500RPM

SEFF4P=[.85 .87 .88 .9 .91 .93 .94];%Table for 1500RPM

SD=[0.1 0.15 0.2 0.3 0.4 0.5 0.75 1];

SCDSW=[4 3.8 3.6 3.5 3.5 3.5 3.5 3.5];

%-------------BH Curve for--0.5mm,LOHYS Quality--------->
```

```
BB=[.1 .2 .3 .4 .5 .6  .7  .8  .9   1  1.1 1.2 1.3 1.4 1.5
1.6  1.7  1.8   1.9   2];

H= [50 65 70 80 90 100 110 120 150 180 220 295 400 580 1000
2400 5000 8900 15000 24000];

%-------Carters Coefft for Air Gap--->

Ratio=[0  1   2   3   4   5  6   7   8   9   10  11  12];

CC= [0 .18 .33 .45 .53 .6 .66 .71 .75 .79 .82 .86 .89];

CC1= [0 .14 .27 .37 .44 .5 .54 .58 .62 .65 .68 .69 .7];

%(1)<-----------Main Dimensions----------------------->

KW=30;V=440;f=50;%P=6;Specification/Input Data------>

fprintf(f2,'Optimal  Design  of%4d  KW,%4d  V,%3d  HZ  3-Ph
Sq.Cage Ind Motor\n',KW,V,f);

fprintf(f2,'Sn  P Zs  D m/s  L   S wcaXhca CDSW WsXHs  Btmax
Tr  D0  Eff   Wtot KgPkw I0/I pf0 pfsc Slip\n');

fprintf(f2,'-- -- -- ------ --- -- ------- ---- ------ -----
-- ---- ----- ---- ----- ---- --- ---- ----');

insW=3.4;Hw=4;HL=1;insH=6;nvd=2;bvd=0.01;ki=0.92;Bc=1.35;

Vph=V; Zr=1;kwr=1;cdb=6;Tb=6;cde=6;dd=0.05;Brc=1.35;

sn=0;M1=0;M2=0;M3=0;M4=0;EFFmax=90;minKgPkw=9;minTr=75;

minI0byI=5; %Assumptions

Bav=interp1(SKW,SBav,KW,'spline');

q=interp1(SKW,Sq,KW,'spline');

Vph=V;Kw=0.955;nvd=2;bvd=10;ki=0.92;Tstrip=1.9;insS=0.5;

insW=3.4;Hw=4;HL=1;insH=6;nvd=2;ki=0.92;Bc=1.35;%Assumptions

SD=[0.1 0.15 0.2 0.3 0.4 0.5 0.75 1];SCDSW=[4 3.8 3.6 3.5
3.5 3.5 3.5 3.5];

B=[0.8 1.2 1.6 2 2.4];       WpKg=[7 15 24 34 50];

for P=4:2:10;

pf=interp1(SKWa,SPF6P,KW,'spline');

eff=interp1(SKWa,SEFF6P,KW,'spline');

if P==4 pf=interp1(SKWa,SPF4P,KW,'spline');end;
```

```
if P==4 eff=interp1(SKWa,SEFF4P,KW,'spline');end;

KWinp=KW/eff;Iph=KWinp*1e3/(3*Vph*pf);

Ns=120*f/P;ns=Ns/60;C0=11*Kw*Bav*q*eff*pf*1e-3;
DsqL=1/C0*(KW/ns);

L1=sqrt(DsqL/(0.135*P)^2);L=floor(L1*100)*10;
Ls=(L-nvd*bvd);Li=ki*Ls;
D1=sqrt(DsqL/(L/1000));D=ceil(D1*100)*10;PP=pi*D/P;

LbyPP=L/PP;  if  LbyPP  <0.8||LbyPP  >2  continue;  end;
v=pi*D*ns/1000;  if v >30 continue;end;

FI=pi*D/P*L*Bav/1e6;Tphi=Vph/(4.44*f*FI*Kw);

%CDSW=interp1(SD,SCDSW,D,'spline');

for CDSW=3:0.5:5;  As1=Iph/CDSW;

for spp=3:1:5;    S=spp*P*3; SPitch=pi*D/S;

if SPitch <18||SPitch>=25 continue; end;

Zphi=2*Tphi;sph=S/3;Zs1=Zphi/sph;Zs=ceil(Zs1);Tph=Zs*sph/2;

FI=Vph/(4.44*f*Tph*Kw); for Tstrip=1:0.1:2;

Hstrip1=As1/Tstrip;Hstrip=ceil(Hstrip1*2)/2;

WbyT=Hstrip/Tstrip;

If WbyT <2.5||WbyT >=3.5 continue;end;
As=0.967*Hstrip*Tstrip;

for Zsw=1:1:4;

Ws=(Zsw*(Tstrip+insS)+insW);Zsh=Zs/Zsw;

Hs=(Zsh*(Hstrip+insS)+Hw+HL+insH+2);

D13=D+2/3*Hs;sp13=pi*D13/S;Wt13=sp13-Ws;At13=Li*Wt13;

B13= FI*P*1e6/(At13*S);

Btmax=1.5*B13;if Btmax >1.8||Btmax <1.4 continue;end;

Lmt=(2*L+2.3*PP+240)/1000;Rph=0.021*Lmt*Tph/As;

Pcus=3*Iph^2*Rph;

Wcus=Lmt*Tph*3*As*8.9e-3;FIc=FI/2;Ac=FIc*1e6/Bc; Hc=Ac/Li;

D01=D+2*(Hs+Hc);D0=ceil(D01/10)*10;Hc=(D0-D)/2-Hs;

Wt=At13*S*Hs*7.8e-6;Dmcs=D+2*Hs+Hc;Wc=Ac*pi*Dmcs*7.8e-6;
```

```
PitpKg=interp1(B,WpKg,Btmax,'spline');

PicpKg=interp1(B,WpKg,Bc,'spline');

Pit=PitpKg*Wt;Pic=PicpKg*Wc;

%-----------------------ROTOR----------------------->

Zr=1;kwr=1;cdb=6;Tb=6;cde=6;dd=50;Brc=1.35;   %Assumptions

kws=Kw;Ss=S;Lg1=0.2+2*sqrt(D*L/1e6);Lg=ceil(Lg1*100)/100;
Dr=D-2*Lg;d1=Ss-3*P;d2=Ss-P;d3=Ss-2*P;d4=Ss-5*P;

d5=Ss-1;d6=Ss-2;d7=Ss-7;d8=Ss-8;Sr=Ss-9;sp2=pi*Dr/Sr;
Ir=0.85*Iph;Ib=Ir*kws*Ss*Zs/(kwr*Sr*Zr);          Abi=Ib/cdb;
Wb=ceil(Abi/Tb);Ab=Tb*Wb*0.98;Wsr=Tb+0.5;Hsr=Wb+0.5;Lb=L+50;
Rb=0.021*Lb/1e3/Ab;      Pcub=Ib^2*Rb*Sr;      Ie=Ib*Sr/P/pi;
Ae=Ie/cde;Dme=Dr-dd;Lme=pi* Dme/1000;Re= 0.021*Lme/Ae;

Pcue=2*Ie^2*Re;Pcur=Pcub+Pcue;Rr=Pcur/(3*Ir^2);

Dr13=Dr-2*2/3*Hsr;spr13= pi*Dr13/Sr;

Wtr13=spr13-Wsr;Atr=Wtr13*Li*Sr/P;Brt=FI*1e6/Atr;Brtmax=
Brt*1.5;if Brtmax >1.5 continue;end;

Ac=FI*1e6/2/Brc;dcr=Ac/Li;Pfw=0.01*KW*1e3;

 PnL=Pit+Pic+Pfw; Iw=PnL/3/V;

Wcur=Lb*Sr*Ab*8.9e-6; Wcue=Lme*2*Ae*8.9e-3;

%(4)<--AmpTurns and Magnetizing-Current---------------------
->

%-------------BH Curve for--0.5mm,LOHYS Quality--------->

BB=[.1 .2 .3 .4 .5 .6  .7  .8  .9   1  1.1 1.2 1.3 1.4 1.5
1.6  1.7  1.8   1.9   2];

H= [50 65 70 80 90 100 110 120 150 180 220 295 400 580 1000
2400 5000 8900 15000 24000];

%-------Carters Coefft for Air Gap--->

Ratio=[0  1   2   3   4   5  6   7   8   9   10  11  12 ];

CC=   [0 .18 .33 .45 .53 .6 .66 .71 .75 .79 .82 .86 .89 ];

CC1=  [0 .14 .27 .37 .44 .5 .54 .58 .62 .65 .68 .69 .7 ];

Wss0=4;Wsr0=2; %Assumptions

atsc=interp1(BB,H,Bc,'spline');Dcav=D+2*Hs+Hc;
```

```
ATSC=pi*Dcav/P/3*atsc/1e3;

Bt30=B13*1.36;atst=interp1(BB,H,Bt30,'spline');

ATST=atst*Hs/1000;ATS=ATSC+ATST;

rat1=Wss0/Lg;k01=interp1(Ratio,CC,rat1,'spline');

kgs=SPitch/(SPitch-Wss0*k01);                    rat2=Wsr0/Lg;
k02=interp1(Ratio,CC,rat2,'spline');spr0=pi*Dr/Sr;
kgr=spr0/(spr0-Wsr0*k02); kg=kgs*kgr; Lgd=Lg*kg;rat3=bvd/Lg;
kv=interp1(Ratio,CC1,rat3,'spline');

if  rat3  >=12  kv=0.7;end;  Ld=L-kv*nvd*bvd;  Aag=pi*D/P*Ld;
Bg=FI*1e6/Aag;    B30d=1.36*Bg;    ATg=0.796    *B30d*Lgd*1e3;
Btr30=Brt*1.36;          atrt=interp1(BB,H,Btr30,'spline');
ATRT=atrt*Hsr/1e3;                         Dcrav=Dr-2*Hsr-dcr;
atrc=interp1(BB,H,Brc,'spline');           ATRC=pi*Dcrav/1e3/P
/3*atrc;  ATR=ATRC+ATRT;ATT=ATS+ATR+ATg;

Im=P/2*ATT/(1.17*Kw*Tph);I0=sqrt(Iw^2+Im^2);

pf0=Iw/I0;I0byI=I0/Iph;

%(5)<----------Short-Circuit-Current-------------------->');

h2=1.6;h2r=0;h3r=0;h4r=0.5;ks=1; %Assumptions

h1=Zsh*(Hstrip+insS);h3=Hw;h4=HL;bs=Ws;b0=Wss0;

Lmdss=h1/3/bs   +h2/bs   +2*h3/(bs+b0)   +h4/b0;   h1r=Wb;
br=Wsr; br0=Wsr0;

Lmdsr=h1r/3/br +h2r/br +2*h3r/(br+br0) +h4r/br0;
Lmddsr=Kw^2*S/Sr*Lmdsr; ssp = Lmdss+Lmddsr;

gd=S/P/3;p=P/2;Xs=15.8*f*L*ssp*Tph^2/(p*gd)*1e-9;
L0Lmd0=ks*PP^2/pi/ SPitch/1000;

X0=15.8*f*L0Lmd0*Tph^2/(p*gd)*1e-6;      gs=S/P;      gr=Sr/P;
Xm=Vph/Im;   Xz=5/6  *Xm*(1/gs^2+1/gr^2);X=Xs+X0+Xz;  R=Rph+Rr;
Z=sqrt(R^2+X^2); Isc=Vph/Z; pfsc=R/Z; RAT=Isc/Iph;

%(6)<--------------Performance-------------------->');

Pt=PnL+Pcus+Pcur;EFF=KW/(KW+Pt/1000)*100;
if    EFF    <88.8continue;    end;    Rinp=KW*1000+Pfw+Pcur;
SFL=Pcur/Rinp*100; Tst=(Isc/Ir)^2*SFL/100;

Pmax=3*Vph*(Isc-I0)/2/(1+pfsc)*1e-3;
```

```
Acool1=(pi*D*(L*2.5)+2*pi*(D+50)*0.04)/1e6;
Acool2=Acool1*(1+0.1*v);Acool3=pi*D0*L/1e6;
AcoolT=Acool2+Acool3;   Pst=Pcus+Pit+Pic;
Tr=0.03*Pst/AcoolT;  if Tr>56 continue;end; Ars=Wsr*Hsr*Sr;
Dri=Dcrav-dcr; Wri=(pi*(Dr^2-Dri^2)/4-Ars)*L*7.8e-6;
Wtot=1.01*(Wcus+Wt+Wc+Wri+Wcur+Wcue);KgPkw=Wtot/KW;
if KgPkw >6.1 continue;end;
%------------------End of Program-------------------------->
sn=sn+1;              if EFF >=EFFmax  EFFmax=EFF;  end;
if abs(EFF-EFFmax)<=2e-3 M2=sn; end;
if KgPkw <=minKgPkw  minKgPkw=KgPkw;end;
if abs(KgPkw-minKgPkw)<=0.001 M1=sn; end;
if   Tr<=minTr   minTr=Tr;end;ifabs(Tr-minTr)<=0.0001   M3=sn;
end;
if I0byI <=minI0byI  minI0byI=I0byI;end;
if abs(I0byI-minI0byI)<=0.0001 M4=sn; end;
fprintf(f2,'\n%2d%3d%3d%4.0f%3.0f%4.0f%3d%4.1fX%3.1f%4.1f',s
n,P,Zs,D,v,L,S,Tstrip,Hstrip,CDSW);
fprintf(f2,'%5.1fX%2.0f%6.3f%5.1f%4.0f%6.2f%5.0f%6.2f%5.2f%5
.2f%5.2f%4.1f',Ws,Hs,Btmax,Tr,D0,EFF,Wtot,KgPkw,I0byI,pf0,pf
sc,SFL);end;end;end;end;end;
fprintf(f2,'\n--- ----- ----- ----  ---- --- ---- -- ---- --
------ --------- --- --- ----- ----  ---- \n');
fprintf(f2,'Selection   of   Design   Variant   based   on
Optimization Criteria:');
fprintf(f2,'\nIf Maximum Efficiency is Required   , Select
Variant(Sn)=%3d  (%5.2f perc)',M2,EFFmax);
fprintf(f2,'\nIf Minimum   Kg/KW       is Required, Select
Variant(Sn)=%3d  (%5.2f)', M1,minKgPkw);
fprintf(f2,'\nIf  Minimum       Temp-Rise  is  Reqd,  Select
Variant(Sn)=%3d  (%4.2f)', M3,minTr);
fprintf(f2,'\nIf Minimum  ratio of I0/I   is Reqd, Select
Variant(Sn)=%3d  (%4.2f)',M4,minI0byI);fclose(f2);sn,Hs,Hc,
```

7.1.11 (a) Computer Output Results for Optimal Design

Optimal Design of 30 KW, 440 V, 50 HZ 3 -Ph Sq.Cage Ind Motor

Sn	P	Zs	D	m/s	L	S	wca×hca	CDSW	Ws×Hs	Btmax	Tr	D0	Eff	Wtot	KgPkw	I0/I	pf0	pfsc	Slip
1	4	20	250	20	190	36	1.8×5.5	3.0	12.6×43	1.573	51.0	420	90.05	172	5.74	0.27	0.19	0.17	2.3
2	4	20	250	20	190	36	1.9×5.5	3.0	13.0×43	1.628	50.4	420	90.13	173	5.78	0.27	0.18	0.17	2.3
3	4	20	250	20	190	36	2.0×5.0	3.0	13.4×41	1.711	50.5	420	90.11	169	5.64	0.29	0.17	0.18	2.3
4	4	20	250	20	190	36	1.6×5.5	3.5	11.8×43	1.472	52.5	420	89.87	170	5.67	0.26	0.19	0.18	2.3
5	4	20	250	20	190	36	1.7×5.0	3.5	12.2×41	1.539	52.6	420	89.86	166	5.54	0.27	0.18	0.19	2.3
6	4	20	250	20	190	36	1.8×5.0	3.5	12.6×41	1.593	51.8	420	89.95	167	5.57	0.27	0.18	0.18	2.3
7	4	20	250	20	190	36	1.5×5.0	4.0	11.4×41	1.443	54.6	420	89.61	164	5.48	0.26	0.18	0.19	2.3
8	4	20	250	20	190	36	1.6×4.5	4.0	11.8×38	1.507	54.8	410	89.61	160	5.33	0.26	0.18	0.20	2.3
9	4	20	250	20	190	36	1.7×4.5	4.0	12.2×38	1.558	53.8	410	89.74	161	5.36	0.27	0.18	0.20	2.3
10	6	18	330	17	160	54	1.9×5.5	3.0	10.6×49	1.579	54.6	510	88.97	188	6.26	0.30	0.16	0.21	2.9
11	6	18	330	17	160	54	2.0×5.0	3.0	10.9×46	1.644	54.6	500	88.98	182	6.08	0.31	0.16	0.22	2.9
12	6	18	330	17	160	54	1.9×5.0	3.5	10.6×46	1.597	55.5	500	88.88	181	6.04	0.30	0.16	0.23	2.9

:

Selection of Design Variant based on Optimization Criteria

If Maximum Efficiency is Required, Select Variant (Sn) = 2 (90.13 perc)

If Minimum Kg/KW is Required, Select Variant (Sn) = 8 (5.33)

If Minimum Temp-Rise is Required, Select Variant (Sn) = 2 (50.37)

If Minimum ratio of I0/I is Required, Select Variant (Sn) = 7 (0.26).

7.2 Slip-Ring Type Induction Motor

Problem: Design a 30 KW, 440V, 50 Hz, delta connected, 3 ph Slip-ring type Ind.Motor.

The stator design is the same as that of Squirrel Cage type. Only Rotor design is made for 3ph-wound rotor with slip-rings. It means only chapter 7.1.5 (part-3) is replaced with 7.2.2 along with required corrections in subsequent chapters accordingly.

Stator Data taken from that of Sq.Cage Design

KW rating (KW) = 30, Voltage Rating (V = Vph) = 440,

Phase Current (Iph) = 29.96 A

Inner dia (D) = 300mm, Total length (L) = 160 mm , Poles = 6,

Turns/pf (Tph) = 162, Slots = 54, Cond/slot (Zs) = 18, Pole Pitch (PP) = 172.8 mm,

Net Iron length (Li) = 128.8 mm, Flux/Pole (FI) = 0.01281 Wb,

Winding factor (Kw) = 0.955,

Iron Loss in tooth (Pit) = 619.86 W, Iron Loss in Core (Pic) = 982.2 W.

Using this Data Rotor Design is made as Follows

7.2.1 Sequential Steps for Design of Each Part and Programming Simultaneously

(a) Calculate Air gap length, Rot dia, no of Rotor Slots, Cond size, Check for Slot Balance. Calculate Copper Losses, Flux densities in tooth and Core, Weight of rotor copper.

(b) Calculate Carter Coefft and Ampere-turns for Air gap, Stator tooth, Stator core, Rotor tooth, Rotor core and Total No-load AT, Magnetizing current, No-Load PF.

(c) Calculation of Reactance, Short-circuit current, and Short-ckt PF.

(d) Calculation of total losses, Efficiency, Slip, Starting torque, Temp-rise, Total Weight and Kg/KW.

Note: By adding programs established for each part sequentially we get the Program for complete design.

7.2.2 Design of Slip-Ring Rotor (Part-3)

Air-gap length (Lg) = $0.2 + 2 \times \sqrt{D \times L / 10^6} = 0.2 + 2 \times \sqrt{330 \times 160 / 10^6} = 0.66$ mm

Rotor dia(Dr) = D – 2 × Lg = 330 – 2 × 0.66 = 328.68 mm

Assuming Slots/Pole/Ph for rotor = 3.5, No. of rotor slots (Sr) = 3.5 × 3 × 6 = 63

$$\text{Slot Pitch (sp2)} = \frac{\pi \times Dr}{Sr} = \frac{\pi \times 328.68}{63} = 16.39 \text{ mm}$$

Assuming Rotor winding is of Star connected and

No load Line to Line voltage = 100,

$$\text{Turns/ph (TphR)} = \frac{Er \times kws \times Tph}{\sqrt{3} \times Vph \times kwr} = \frac{100 \times 0.955 \times 162}{\sqrt{3} \times 440 \times 0.955} = 21.3$$

$$\text{Conductors/Slot (Zr)}\ \frac{\text{TphR} \times 3 \times 2}{\text{Sr}} = \frac{21.3 \times 3 \times 2}{63}$$

$$= 2.02 \approx 2 \text{ (rounded off to nearest even integer)}$$

$$\text{Corrected Turns/ph (TphR)} = \frac{\text{Zr} \times \text{Sr}}{2 \times 3} = \frac{2 \times 63}{6} = 21$$

Equivalent Rotor Current (Ird) = 0.85 × Iph = 0.85 × 29.96 = 25.466 A

$$\text{Rotor current (Ir)} = \text{Ird} \times \frac{\text{kws} \times \text{Ss} \times \text{Zs}}{\text{kwr} \times \text{Sr} \times \text{Zr}} = 25.446 \times \frac{0.955 \times 54 \times 18}{0.955 \times 63 \times 2} = 196.45 \text{ A}$$

Assuming Current density (cdr) = 5 A/mm^2,

CS of conductor (Acr) = Ir/cdr = 196.45/5 = 39.29 mm^2,

Assuming width of conductor (wcR) = 4 mm,

Height of cond (hcR) = Acr/wcR = 39.29/4 = 9.83≈ 10 mm

Taking 0.98 factor for edges rounding off,

corrected (Acr) = 0.98 × wcR × hcR = 0.98 × 4 × 10 = 39.2 mm^2,

Width of slot (Wsr) = wcR + 2.5 = 4 + 2.5 = 6.5 mm

Assuming wedge ht (HwR) = 4 mm and Lip ht (HLR) = 1 mm,

Height of slot (Hsr) = Zr × (hcR + 0.5) + 3 + HwR + HLR

= 2 × (10 + 0.5) + 3 + 4 + 1 = 29 mm

Cross-Section of Rotor Slot (dimensions in mm)

Height-Wise			
Lip			1.0
Wedge			4.0
Ins.Under Wedge			0.5
Cu.Conductor	bare + insulation	10 + 0.5 = 10.5	
Top-Layer	for 1 conductor	1 × 10.5 = 10.5	10.5
Ins.between Layers			0.5
Bottom- Layer	for 1 conductor	1 × 10.5 = 10.5	10.5
Bottom Insulation			0.5
Slack			1.5
Total height			**29.0**
Width-Wise			
Cu.Conductor	bare + insulation	4 + 0.5 = 4.5	
	for 1 conductor	1 × 4.5 = 4.5	4.5
Insulation on sides	2 × 0.6 = 1.2		1.2
Slack			0.8
Total width			**6.5**

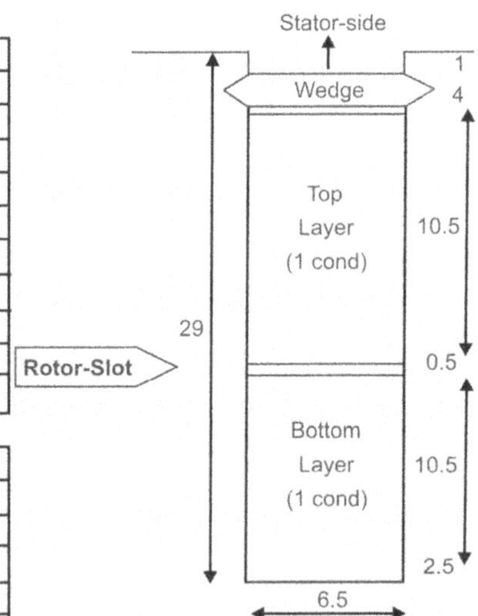

Length of mean turn (LmtR) = (2 × L + 2.3 × PP + 80)/1000

= (2 × 160 + 2.3 × 172.8 + 80)/1000 = 0.7974 m

Resistance/ph (RphR) = 0.021 × LmtR × TphR/Acr

= 0.021 × 0.7974 × 21/39.2 = 8.9712 mΩ

Rotor Cu.Loss (Pcur) = $3 \times r^2 \times RphR = 3 \times 196.45^2 \times 8.9712/1000 = 1038.7$ W

Dia of rotor at ⅓tooth ht from tip (Dr13) = $Dr-2 \times ⅔ \times Hsr$

$= 328.68 - 2 \times ⅔ \times 29 = 290.01$

Rotor slot pitch at Dr13 (spr13) = $\dfrac{\pi \times Dr13}{Sr} = \dfrac{\pi \times 290.01}{63} = 14.462$ mm

Width of tooth at Dr13 (Wtr13) = spr13 – Wsr = 14.4620 – 6.5 = 7.962 mm

Area of tooth at Dr13 (Atr) = Wtr13 × Li × Sr/P = 7.962 × 128.8 × 63/6 = 10768 mm^2

Flux density in tooth (Brt) = FI × 10^6/Atr = 0.01281 × 10^6/10768 = 1.1897 T

Max Flux density in tooth (Brtmax) = Brt × 1.5 = 1.1897 × 1.5 = 1.7845 T (<1.8 and hence OK)

Assuming Flux density in Rotor Core (Brc) = 1.35 T,

Area of Core (Ac) = $\dfrac{FI \times 10^6}{2 \times Brc} = \dfrac{0.01281 \times 10^6}{2 \times 1.35} = 4744.8\ m^2$

Depth of core (dcr) = $\dfrac{Ac}{Li} = \dfrac{4744.8}{128.8} = 36.84$ mm

Assuming Friction and Windage Loss (Pfw) = 1%

= 0.01 × KW × 1000 = 0.01 × 30 × 1000 = 300 W

No load Loss (PnL) = Pit + Pic + Pfw = 619.9 + 982.2 + 300 = 1902.1 W

Active/Wattful Component of No-load Current (Iw) = $\dfrac{PnL}{3 \times V} = \dfrac{1902.1}{3 \times 440} = 1.441$ A

Wt. of Rotor Copper (Wcur) = LmtR × TphR × 3 × Acr × 8.9 × 10^{-6}

= 0.7974 × 21 × 3 × 39.2 × 8.9 × 10^{-6} = 17.527 Kg

7.2.2 (a) Computer Program in "C" in MATLAB for Part-3

```
%--------------------Slip-Ring-ROTOR------------------------>
D=330;L=160;P=6;Tph=162;S=54;Kw=0.955;Vph=440;Iph=29.96;Zs=18;
PP=172.8;Li=128.8;FI=0.01281;KW=30;V=440;Pit=619.86;Pic=982.2;
sppR=3.5;Ss=S;kws=Kw;    kwr=kws;Er=100;cdr=5;wcR=4;HwR=4;
HLR=1; Brc=1.35; %Assumtions
Lg1=0.2 + 2*sqrt(D*L/1e6);Lg=ceil(Lg1*100)/100; Dr=D-2*Lg;
Sr=sppR*3*P; sp2=pi*Dr/Sr;     TphR1=Er/sqrt(3)/Vph*kws/kwr*Tph;
Zr1=TphR1*3*2/Sr;     Zr=floor(Zr1);          TphR=Zr*Sr/2/3;
Ird=0.85*Iph;  Ir=Ird*(kws*Ss*Zs)/(kwr*Sr*Zr);       Acr1=Ir/cdr;
hcR1=Acr1/wcR;  hcR=ceil(hcR1);  Acr=wcR*hcR*0.98;  Wsr=wcR+2.5;
```

```
Hsr=Zr*(hcR+0.5)+3+HwR+HLR;            LmtR=(2*L+2.3*PP+80)/1000;
RphR=0.021*LmtR*TphR/Acr; Pcur=3 *Ir^2*RphR;
Dr13=Dr-2*2/3*Hsr;        spr13=pi*Dr13/Sr;       Wtr13=spr13-Wsr;
Atr=Wtr13*Li*Sr/P; Brt=FI*1e6/Atr;                   Brtmax=Brt*1.5;
Ac=FI*1e6/2/Brc; dcr=Ac/Li; Pfw=0.01*KW*1e3;     PnL=Pit+Pic+Pfw;
Iw=PnL/3/V; Rr=RphR/(TphR/Tph)^2;   Wcur=LmtR*TphR*3*Acr*8.9e-3;
%-------------------------------------------------------
```

Other Changes to be Made are as Follows

h1r = wcR; h2r = 1.6 ; h3r = HwR ; h4r = HLR;

br = Wsr → in 5th part(Short-Circuit Current)

For Short circuit Torque (Tst) = (Isc/Ird)^2 × SFL/100 and

Wcue = 0; → in 6th part(Performance)

Remove Print statements whereever "bar" and "end rings" appear

Incorporating these Changes, The Modified Total Program is Given Below

7.2.3 Computer Program in "C" in MATLAB for Complete Design

```
%3ph,KW=30;V=440;P=6;f=50 Slip-Ring IM
f2=fopen('Total_30KW_SR_Output.m','w');
%------------Standard Curves/Tables for%Data-------------->
SKW=[1  2  5  10  20  50  100  500];SBav=[0.35  0.38  0.42  0.46  0.48
0.50 0.51 0.53];
Sq=[16e3 19e3 23e3 25e3 26e3 29e3 31e3 33e3];
SKWa=[5 10 20 50 100 200 500];
SPF6P=[0.82 0.83 0.85 0.87 0.89 0.9 0.92];%Table for 1000RPM
SEFF6P=[0.83 0.85 0.87 0.89 0.91 0.92 0.93];%Table  for 1000RPM
SPF4P=[0.85 0.86 0.88 0.9 0.91 .92 .93];%Table for 1500RPM
SEFF4P=[.85 .87 .88 .9 .91 .93 .94];%Table  for 1500RPM
SD=[0.1 0.15 0.2 0.3 0.4 0.5 0.75 1];
SCDSW=[4 3.8 3.6 3.5 3.5 3.5 3.5 3.5];
%-------------BH Curve for--0.5mm,LOHYS Quality--------->
BB=[.1 .2 .3 .4 .5 .6  .7  .8  .9   1  1.1 1.2 1.3 1.4 1.5  1.6
1.7  1.8   1.9   2];
H= [50  65  70  80  90  100  110  120  150  180  220  295  400  580  1000
2400 5000 8900 15000 24000];
%-------Carters Coefft for Air Gap--->
Ratio=[0  1   2   3   4   5  6   7   8   9   10  11  12];
```

```
CC= [0 .18 .33 .45 .53 .6 .66 .71 .75 .79 .82 .86 .89];
CC1=[0 .14 .27 .37 .44 .5 .54 .58 .62 .65 .68 .69 .7];
%(1)<-----------Main Dimensions----------------------->
KW=30;V=440;f=50;%P=6;Specification/Input Data------>
insW=3.4;Hw=4;HL=1;insH=6;nvd=2;bvd=0.01;ki=0.92;Bc=1.35;Vph=V;
Zr=1;kwr=1;cdb=6;Tb=6;cde=6;dd=0.05;Brc=1.35;   %Assumptions
Bav=interp1(SKW,SBav,KW,'spline');
q=interp1(SKW,Sq,KW,'splin e');
pf=interp1(SKWa,SPF6P,KW,'spline');
eff=interp1(SKWa,SEFF6P,KW,'spline');
Kw=0.955;nvd=2;bvd=10;ki=0.92; Tstrip=1.9;insS=0.5;%Assumptions
insW=3.4;Hw=4;HL=1;insH=6;nvd=2;ki=0.92;Bc=1.35;%Assumptions
SD=[0.1 0.15 0.2 0.3 0.4 0.5 0.75 1];
SCDSW=[4 3.8 3.6 3.5 3.5 3.5 3.5 3.5];
B=[0.8 1.2 1.6 2 2.4];WpKg=[7 15 24 34 50];
Vph=V;KWinp=KW/eff;Iph=KWinp*1e3/(3*Vph*pf);
P=6;spp=3;Tstrip=1.9;Zsw=3; % Assumption
pf=interp1(SKWa,SPF6P,KW,'spline');
eff=interp1(SKWa,SEFF6P,KW,'spline');
if P==4 pf=interp1(SKWa,SPF4P,KW,'spline');end;
if P==4 eff=interp1(SKWa,SEFF4P,KW,'spline');end;Ns=120*f/P;
ns=Ns/60; C0=11*Kw*Bav*q*eff*pf*1e-3; DsqL=1/C0*(KW/ ns);
L1=sqrt(DsqL/(0.135*P)^2); L=floor(L1*100)*10;Ls=(L-nvd*bvd);
Li= ki*Ls; D1=sqrt(DsqL/(L/1000));
D=ceil(D1*100)*10; PP=pi*D/P; LbyPP=L/PP;
if LbyPP <0.8||LbyPP >2 continue;end;
v=pi*D*ns/1000;if v >30 continue;end;
FI=pi*D/P*L*Bav/1e6;
%(2)<--------------Stator Slots and Winding----->
Tphi=Vph/(4.44*f*FI*Kw);CDSW=interp1(SD,SCDSW,D/1000,'spline');
As1=Iph/CDSW;  S=spp*P*3;  SPitch=pi*D/S;
if SPitch <18||SPitch>=25 continue;end;
```

```
Zphi=2*Tphi;sph=S/3;Zs1=Zphi/sph;Zs=ceil(Zs1);Tph=Zs*sph/2;
FI=Vph/(4.44*f*Tph*Kw);    Hstrip1=As1/Tstrip;
Hstrip=ceil(Hstrip1*2)/2; WbyT=Hstrip/Tstrip;

if WbyT <2.5||WbyT >=3.5 continue;end; As=0.967*Hstrip*Tstrip;

Ws=(Zsw*(Tstrip+insS)+insW);Zsh=Zs/Zsw;

Hs=(Zsh*(Hstrip+insS)+Hw+HL+insH+2);

D13=D+2/3*Hs;sp13=pi*D13/S;Wt13=sp13-Ws;
B13=FI*P*1e6/(Li*Wt13*S); Btmax=1.5*B13;

Lmt=(2*L+2.3*PP+240)/1000;Rph=0.021*Lmt*Tph/As;

Pcus=3*Iph^2*Rph;              Wcus=Lmt*Tph*3*As*8.9e-3;FIc=FI/2;
Ac=FIc*1e6/Bc; Hc = Ac/Li; D01=D+2*(Hs+Hc); D0=ceil(D01/10)*10;
Hc=(D0-D)/2-Hs;            PitpKg=interp1(B,WpKg,Btmax,'spline');
PicpKg=interp1(B,WpKg,Bc,'spline');

Wt=Li*Wt13*S*Hs*7.8e-6; Dmcs=D+2*Hs+Hc; Wc=Ac*pi*Dmcs*7.8e-6;

Pit=PitpKg*Wt;Pic=PicpKg*Wc;

%--------------------Slip-Ring-ROTOR------------------------->

sppR=3.5;Ss=S;kws=Kw;kwr=kws;Er=100;cdr=5;wcR=4;HwR=4;HLR=1;

Brc=1.35; %Assumtions

Lg1=0.2 + 2*sqrt(D*L/1e6);Lg=ceil(Lg1*100)/100;Dr=D-2*Lg;

Sr=sppR*3*P;sp2=pi*Dr/Sr;TphR1=Er/sqrt(3)/Vph*kws/kwr*Tph;

Zr1=TphR1*3*2/Sr;Zr=floor(Zr1);TphR=Zr*Sr/2/3;Er=TphR/TphR1*Er;


Ird=0.85*Iph;Ir=Ird*(kws*Ss*Zs)/(kwr*Sr*Zr);Acr1=Ir/cdr;
hcR1=Acr1/wcR;   hcR=ceil(hcR1);  Acr=wcR*hcR*0.98;  Wsr=wcR+2.5;
Hsr=Zr*(hcR+0.5)+3+HwR+HLR;

LmtR=(2*L+2.3*PP+80)/1000;             RphR=0.021*LmtR*TphR/Acr;
Pcur=3*Ir^2*RphR;      Dr13=Dr-2*2/3*Hsr;      spr13=pi*Dr13/Sr;
Wtr13=spr13-Wsr; Atr=Wtr13*Li *Sr/P;     Brt=FI*1e6/Atr;
Brtmax=Brt*1.5;Ac=FI*1e6/2/Brc;dcr=Ac/Li;

Pfw=0.01*KW*1e3; PnL=Pit+Pic+Pfw; Iw=PnL/3/V;

Rr=RphR/(TphR/ Tph)^2;     Wcur=LmtR*TphR*3*Acr*8.9e-3;

%(4)<------------AmpTurns and Magnetizing-Current--------------
-------->

%-------------BH Curve for--0.5mm,LOHYS Quality--------->

BB=[.1 .2 .3 .4 .5 .6  .7  .8  .9   1  1.1 1.2 1.3 1.4 1.5  1.6
1.7  1.8   1.9   2];
```

```
H= [50 65 70 80 90 100 110 120 150 180 220 295 400 580 1000
2400 5000 8900 15000 24000];

%-------Carters Coefft for Air Gap--->

Ratio=[0  1   2   3   4   5  6   7   8   9   10  11  12 ];

CC=   [0 .18 .33 .45 .53 .6 .66 .71 .75 .79 .82 .86 .89 ];

CC1=  [0 .14 .27 .37 .44 .5 .54 .58 .62 .65 .68 .69 .7 ];

Wss0=4;Wsr0=2; %Assumptions

atsc=interp1(BB,H,Bc,'spline');

Dcav=D+2*Hs+Hc;ATSC=pi*Dcav/P/3*atsc/1e3;

Bt30=B13*1.36;atst=interp1(BB,H,Bt30,'spline');

ATST=atst*Hs/1000;ATS=ATSC+ATST;

rat1=Wss0/Lg;k01=interp1(Ratio,CC,rat1,'spline');

kgs=SPitch/(SPitch-Wss0*k01);

rat2=Wsr0/Lg;k02=interp1(Ratio,CC,rat2,'spline');spr0=pi*Dr/Sr;

kgr=spr0/(spr0-Wsr0*k02);kg=kgs*kgr;Lgd=Lg*kg;

rat3=bvd/Lg;kv=interp1(Ratio,CC1,rat3,'spline');

if  rat3  >=12  kv=0.7;end;  Ld=L-kv*nvd*bvd;  Aag=pi*D/P*Ld;
Bg=FI*1e6/Aag;B30d=1.36*Bg;ATg= 0.796 *B30d*Lgd*1e3;

Btr30=Brt*1.36;              atrt=interp1(BB,H,Btr30,'spline');
ATRT=atrt*Hsr/1e3;                                Dcrav=Dr-2*Hsr-dcr;
atrc=interp1(BB,H,Brc,'spline'); ATRC=pi *Dcrav/1e3/P/3*atrc;

ATR=ATRC+ATRT; ATT=ATS+ATR+ATg; Im=P/2*ATT/(1.17*Kw*Tph);

 I0=sqrt(Iw^2+Im^2); pf0=Iw/I0;I0byI=I0/Iph;

%(5)<-------------Short-Circuit-Current------------------->');

h2=1.6;h2r=1.6;h3r=HwR;h4r=HLR;ks=1; %Assumptions

h1=Zsh*(Hstrip+insS);h3=Hw;h4=HL;bs=Ws;b0=Wss0;

Lmdss=h1/3/bs   +h2/bs   +2*h3/(bs+b0)   +h4/b0;h1r=wcR;br=Wsr;
br0= Wsr0;    Lmdsr=h1r/3/br +h2r/br +2*h3r/(br+br0) +h4r/br0;

Lmddsr=Kw^2*S/Sr*Lmdsr;ssp=Lmdss+Lmddsr;

gd=S/P/3;p=P/2;Xs=15.8*f*L*ssp*Tph^2/(p*gd)*1e-9;

L0Lmd0= ks*PP^2/pi/SPitch/1000;

X0=15.8*f*L0Lmd0*Tph^2/(p*gd)*1e-6;gs=S/P;gr=Sr/P;
Xm=Vph/Im;Xz=5/6*Xm        *(1/gs^2+1/gr^2);        X=Xs+X0+Xz;
R=Rph+Rr;Z=sqrt(R^2+X^2);Isc=Vph/Z;pfsc=R/Z;RAT=Isc/Iph;

%(6)<--------------Performance-------------------->');
```

```
Pt=PnL+Pcus+Pcur;EFF=KW/(KW+Pt/1000)*100;
Rinp=KW*1000+Pfw+Pcur;SFL = Pcur/Rinp*100;
Tst=(Isc/Ird)^2*SFL/100;Pmax=3*Vph*(Isc-I0)/2/(1+pfsc)*1e-3;
Acool1=(pi*D*(L*2.5)+2*pi*(D+50)*0.04)/1e6;
Acool2=Acool1*(1+0.1*v);Acool3=pi*D0*L/1e6;
AcoolT=Acool2+Acool3;Pst=Pcus+Pit+Pic;Tr=0.03*Pst/AcoolT;
Ars=Wsr*Hsr*Sr;Dri=Dcrav-dcr;
Wri=(pi*(Dr^2-Dri^2)/4-Ars)*L*7.8e-6;
Wtot=1.01*(Wcus+Wt+Wc+Wri+Wcur);KgPKw=Wtot/KW;
%--------------------End of Program------------------------->
fprintf(f2,'Design  of  30KW,440V,50HZ,3-Ph  Slip-Ring   Ind
Motor\n');
fprintf(f2,'*********************************************');
fprintf(f2,'\nInput Data:');
fprintf(f2,'\n----------');
fprintf(f2,'\nParamter                          VALUES');
fprintf(f2,'\n--------                          ------');
fprintf(f2,'\nRating(KW)                        %5.1f',KW);
fprintf(f2,'\nVolts                             %5.0f',V);
fprintf(f2,'\nPoles                             %5.0f',P);
fprintf(f2,'\nHz                                %5.0f',f);
fprintf(f2,'\nInterpolated values from curves:
Bav=%5.3f,q=%5.0f,eff=%5.3f,pf=%5.3f',Bav,q,pf,eff);
fprintf(f2,'\nOutput Results:');
fprintf(f2,'\n==============');
fprintf(f2,'\nParamter                          VALUES');
fprintf(f2,'\n--------                          ------');
fprintf(f2,'\nOutput Coefft(C0)                 %6.2f',C0);
fprintf(f2,'\nSync.Speed(rps)                   %5.2f',ns);
fprintf(f2,'\nDsqL                              %5.4f',DsqL);
fprintf(f2,'\nGross Length(mm)                  %5.1f',L);
fprintf(f2,'\nNet iron Length(mm)               %5.1f',Li);
fprintf(f2,'\nStator Inner Dia(mm)              %5.1f',D);
```

```
fprintf(f2,'\nPeriphoral Speed(m/s)          %5.2f
(Max.Permissible:30)',v);
fprintf(f2,'\nPole-Pitch(mm)                 %5.1f',PP);
fprintf(f2,'\nLength to PP ratio %6.4f(Around 1 -->good)',
LbyPP);
fprintf(f2,'\nSlots                          %5.0f',S);
fprintf(f2,'\nSlot-Pitch(mm)
%6.3f(Permissible:18to25mm)',SPitch);
fprintf(f2,'\nCond/Slot                      %5.0f',Zs);
fprintf(f2,'\nTurns/Ph                       %5.0f',Tph);
fprintf(f2,'\nFlux/Pole(Wb)                  %6.5f',FI);
fprintf(f2,'\nPhase Current(A)               %5.2f',Iph);
fprintf(f2,'\nBare Strip (w*t)mm             %5.2fX%4.2f',
Hstrip,Tstrip);
fprintf(f2,'\nWidth to Thickness Ratio %5.2f(Permissible:
2.5to3.5)',WbyT);
fprintf(f2,'\nArea of CS cond(mm^2)          %7.3f',As);
fprintf(f2,'\nCt.density(A/mm^2)             %7.3f',CDSW);
fprintf(f2,'\nInterpolated values of W/Kg of
St:Teeth=%5.2fandCore=%5.2f',PitpKg, PicpKg);
fprintf(f2,'\nNo of Strips(width*depth-wise)%4.0f
X%2.0f',Zsw,Zsh);
fprintf(f2,'\nSlot-Width (mm)                %6.1f',Ws);
fprintf(f2,'\nSlot-Height(mm)Total           %6.1f',Hs);
fprintf(f2,'\d(1/3),SP(1/3),Wt(1/3)(m)%6.4f,%6.4f,%6.4f',D13,sp
13,Wt13);
fprintf(f2,'\nSt-Tooth-Flux-Dens(1/3)        %6.4f ',B13);
fprintf(f2,'\nSt-Tooth-Flux-Dens-Max(T)    %64f(Permissible:1.
6to1.8)',B13*1.5);
fprintf(f2,'\nLength of mean-turn (m)        %6.3f',Lmt);
fprintf(f2,'\nResistance/Ph (ohm)            %6.4f',Rph);
fprintf(f2,'\ndepth of St.Core(mm)           %6.2f',Hc);
fprintf(f2,'\nOuter Dia of St.Core(mm)       %6.1f',D0);
fprintf(f2,'\nSt.Cu.Loss(W)                  %6.1f',Pcus);
fprintf(f2,'\nWt of St-Teeth+Core(Kg)= %6.2f+%6.2f= %6.2f',Wt,
Wc,Wt+Wc);
```

```
fprintf(f2,'\nIron Loss=Teeth+Core(W)=        %5.1f+%5.1f=%
5.1f',Pit,Pic,Pit+Pic);
fprintf(f2,'\n-----------------Slip-Ring-ROTOR-----------------
---->');
fprintf(f2,'\nLength of Air-Gap(mm)              %6.4f',Lg);
fprintf(f2,'\Dia of Rotor(mm)                   %6.1f',Dr);
fprintf(f2,'\nNo of Rotor Slots Selected         %3.0f',Sr);
fprintf(f2,'\nRotor Slot-Pitch(mm)               %6.2f',sp2);
fprintf(f2,'\nEquivalent Rotor Ct(A)             %6.2f',Ir);
fprintf(f2,'\nRotor Cond CS(mm^2)                %6.4f',Acr);
fprintf(f2,'\nRot Cond (w*h)mm %5.1fX%4.1f',wcR,hcR);
fprintf(f2,'\nRotor Slot (w*h)mm %5.1fX%4.1f',Wsr,Hsr);
fprintf(f2,'\nRotor Cond/slot                    %6.0f',Zr);
fprintf(f2,'\nRotor Turns/Ph                     %6.0f',TphR);
fprintf(f2,'\nMean Length of rot-turn(m)         %5.3f',LmtR);
fprintf(f2,'\nResistance/phase(m.ohm)          %6.4f',RphR*1e3);
fprintf(f2,'\nRotor Cu Loss       (W)            %6.1f',Pcur);
fprintf(f2,'\nRotor res refered to St(Ohm)       %6.3f',Rr);
fprintf(f2,'\nRotor(1/3)Slot-Pitch(mm)           %6.2f',spr13);
fprintf(f2,'\nRotor(1/3)tooth width(mm)          %6.2f',Wtr13);
fprintf(f2,'\nRt-Tooth-Flux-Dens(1/3)(T)         %6.4f ',Brt);
fprintf(f2,'\nRt-Tooth-Flux-Dens-Max(T)        %6.4f(Max:1.8T)',
Brt*1.5);
fprintf(f2,'\nDepth of Rotor core (mm)           %5.2f',dcr);
fprintf(f2,'\n---------------------No-Load-Losses-------------
--->');
fprintf(f2,'\nN0-Load Losses (W)                 %5.1f',PnL);
fprintf(f2,'\nN0-Load Wattful Current(A)         %5.3f',Iw);
fprintf(f2,'\n------------------Magnetizing-Current-----------
--->');
fprintf(f2,'\nInterpolated     values     of     at/m     of
St:Core=%5.1fandTeeth=%5.1f',atsc,atst);
fprintf(f2,'\nInterpolated       values       of       Caters
Coeffts:k01=%5.3f,k02=%5.3f,kv=%5.3f', k01,k02,kv);
```

```
fprintf(f2,'\nStator     AT:Core+Teeth:     %5.1f+%5.1f=%5.1f',
ATSC,ATST,ATS);

fprintf(f2,'\nRotor AT :Core+Teeth:     %5.1f+%5.1f=%5.1f',
ATRC,ATRT,ATR);

fprintf(f2,'\nTotal
AT:Stator+Rotor+Airgap:%5.1f+%5.1f+%5.1f=%5.1f',ATS,ATR,
ATg,ATT);

fprintf(f2,'\nN0-Load Current(A) Iw=%5.2f,Im=%5.2f and I0=%5.2f
at pf= %5.3f', Iw,Im,I0,pf0);

fprintf(f2,'\nI0/Iph ratio                    %5.3f',I0byI);

fprintf(f2,'\n----------------Short-Circuit-Current------------
--->');

fprintf(f2,'\nSlot-Permeances:   Stator=%5.3f,Rotor=%5.3f   and
Rotor refered to stator=%5.3f',Lmdss,Lmdsr,Lmddsr);

fprintf(f2,'\nSpecific-Slot-Permeance=          %5.3f',ssp);

fprintf(f2,'\nTotalReactance(X):Slot+Overhang+Zig-
Zag=%5.3f+%5.3f+%5.3f= %5.3f ohms',Xs,X0,Xz,X);

fprintf(f2,'\nShort-Circuit: R=%5.2f,Z=%5.2f and Isc=%5.1f at
pf=%5.3f;Isc/IFL= %5.3f',R,Z,Isc,pfsc,RAT);

fprintf(f2,'\nTotalLosses(PnL+Pcus+Pcur)=%5.1f+%5.1f+%5.1f=%5.1
fWandEfficiency=%5.2f perc',PnL,Pcus,Pcur,Pt,EFF);

fprintf(f2,'\nSlip at FL=                 %5.3f X Perc',SFL);

fprintf(f2,'\nStarting Tq =               %5.3f X FL-Tq',Tst);

fprintf(f2,'\nMax.Output(KW) =            %5.1f',Pmax);

fprintf(f2,'\nTemp-Rise(deg-C) =          %5.1f',Tr);

fprintf(f2,'\nTotalWt(Kg)=1.01*(Wcus+Wt+Wc+Wri+Wcur)=1.01*(%5.1
f+%5.1f+%5.1f+%5.1f+%4.1f)=%5.1f',Wcus,Wt,Wc,Wri,Wcur,Wtot);

fprintf(f2,'\nTotal        Wt(Kg)=%5.1f        and        Kg/KW=
%5.2f',Wtot,KgPKw);

fprintf(f2,'\n<-----------End of Output-------->');

fclose(f2);
```

7.2.3 *(a) Computer Output Results for Complete Design*

Design of 30KW, 440V, 50HZ, 3-Ph Slp-Ring Ind Motor

****

Input Data

\-\-\-\-\-\-\-\-\-\-

Parameter VALUES

--------	------
Rating(KW)	30.0
Volts	440
Poles	6
Hz	50

Interpolated values from curves: Bav=0.484,q=26802,eff=0.862,pf=0.880

Output Results:

================

Paramter	VALUES
--------	------
Output Coefft(C0)	103.34
Sync.Speed(rps)	16.67
DsqL	0.0174
Gross Length (mm)	160.0
Net iron Length (mm)	128.8
Stator Inner Dia(mm)	330.0
Periphoral Speed(m/s)	17.28 (Max.Permissible:30)
Pole-Pitch(mm)	172.8
Length to PP ratio	0.9260(Around 1 -->good)
Slots	54
Slot-Pitch(mm)	19.199(Permissible:18to25mm)
Cond/Slot	18
Turns/Ph	162
Flux/Pole(Wb)	0.01281
Phase Current(A)	29.96
Bare Strip (w*t)mm	5.00 × 1.90
Width to Thickness Ratio	2.63(Permissible:2.5to3.5)
Area of CS cond(mm^2)	9.186
Ct.density(A/mm^2)	3.500

Interpolated values of W/Kg of St:Teeth = 23.92andCore=18.32

No. of Strips(width*depth-wise)	3×6
Slot-Width (mm)	10.6
Slot-Height(mm)Total	46.0

St-Tooth-Flux-Dens(1/3)	1.0644
St-Tooth-Flux-Dens-Max(T)	1.5966(Permissible:1.6to1.8)
Length of mean-turn (m)	0.957
Resistance/Ph (ohm)	0.3546
depth of St.Core(mm)	39.00
Outer Dia of St.Core(mm)	500.0
St.Cu.Loss(W)	954.6
Wt of St-Teeth+Core(Kg)=	25.91+ 53.60= 79.51
Iron Loss=Teeth+Core(W)=	619.9+982.2=1602.1

----------------Slip-Ring-ROTOR-------------------->

Length of Air-Gap(mm)	0.6600
No. of Rotor Slots Selected	63
Rotor Slot-Pitch(mm)	16.39
Equivalent Rotor Ct(A)	196.44
Rotor Cond CS(mm^2)	39.2000
Rot Cond (w*h)mm	4.0×10.0
Rotor Slot (w*h)mm	6.5×29.0
Rotor Cond/slot	2
Rotor Turns/Ph	21
Mean Length of rot-turn(m)	0.797
Resistance/phase(m.ohm)	8.9709
Rotor Cu Loss (W)	1038.5
Rotor res refered to St(Ohm)	0.534
Rotor(1/3)Slot-Pitch(mm)	14.46
Rotor(1/3)tooth width(mm)	7.96
Rt-Tooth-Flux-Dens(1/3)(T)	1.1898
Rt-Tooth-Flux-Dens-Max(T)	1.7846(Max:1.8T)
Depth of Rotor core (mm)	36.84

---------------------No-Load-Losses--------------->

N0-Load Losses (W)	1902.1
N0-Load Wattful Current(A)	1.441

------------------Magnetizing-Current------------->

Interpolated values of at/m of St:Core=480.8andTeeth=706.5

Interpolated values of Caters Coeffts: k01=0.663,k02=0.453,kv=0.700

Stator AT:Core+Teeth: 38.7+ 32.5= 71.2

Rotor AT :Core+Teeth: 19.6+ 80.7=100.3

Total AT :Stator+Rotor+Airgap: 71.2+100.3+445.7=617.2

N-Load Current(A) Iw= 1.44,Im=10.23 and I0=10.33 at pf=0.139

I0/Iph ratio 0.345

----------------Short-Circuit-Current-------------->

Slot-Permeances: Stator=1.987, Rotor=1.892 and Rotor refered to stator=1.479

Specific-Slot-Permeance=3.466

TotalReactance(X): Slot+Overhang+Zig-Zag=1.278+1.140+0.768=3.186 ohms

Short-Circuit: R= 0.89,Z= 3.31 and Isc=133.0 at pf=0.269;Isc/IFL=4.441

Total Losses(PnL+Pcus+Pcur)=1902.1+954.6+1038.5=3895.2 WandEfficiency=88.51 perc

Slip at FL = 3.314 X Perc

Starting Tq = 0.905 X FL-Tq

Max.Output(KW) = 63.8

Temp-Rise(deg-C) = 55.5

Total Wt(Kg) = 1.01 × (Wcus + Wt + Wc + Wri + Wcur)

= 1.01 × (38.0 + 25.9 + 53.6 + 53.0 + 17.5) = 190

Total Wt(Kg)=190and Kg/KW= 6.333 <------------End of Output-------->

7.2.4 Modifications to be Done in the above Program to get Optimal Design

1. Insert “for” Loops for the following parameters to iterate the total program between min and max permissible limits for selecting the feasible design variants:
   (a) No of Poles
   (b) St-Wdg Current density
   (c) Slots/pole/phase
   (d) Stator conductor thickness
   (e) No. of conductors width-wise
   (f) Rotor winding Current density.
2. Insert also minimum or maximum range of required objective functional values as constraint values, for example (a) Efficiency, (b) Kg/KW, (c) Temp-rise (d) I0/I ratio.
3. Run the program to get various possible design variants.

Note: From the feasible design variants printed in the output, select that particular design fulfilling the objective of optimal parameter for the design.

7.2.5 Computer Program in "C" in MATLAB for Optimal Design

```
%3ph,KW=30;V=440;P=6;f=50 Slip-Ring IM
f2=fopen('Optimal_30KW_SR_Output.m','w');
%------------Standard Curves/Tables for%Data-------------->
SKW=[1 2 5 10 20 50 100 500];
SBav=[0.35 0.38 0.42 0.46 0.48 0.50 0.51 0.53];
Sq=[16e3 19e3 23e3 25e3 26e3 29e3 31e3 33e3];
SKWa=[5 10 20 50 100 200 500];
SPF6P=[0.82 0.83 0.85 0.87 0.89 0.9 0.92]; Table for 1000RPM
SEFF6P=[0.83 0.85 0.87 0.89 0.91 0.92 0.93];%Table for 1000RPM
SPF4P=[0.85 0.86 0.88 0.9 0.91 .92 .93];%Table for 1500RPM
SEFF4P=[.85 .87 .88 .9 .91 .93 .94];%Table for 1500RPM
SD=[0.1 0.15 0.2 0.3 0.4 0.5 0.75 1];
SCDSW=[4 3.8 3.6 3.5 3.5 3.5 3.5 3.5];
%-------------BH Curve for--0.5mm,LOHYS Quality--------->
BB=[.1 .2 .3 .4 .5 .6  .7  .8  .9   1  1.1 1.2 1.3 1.4 1.5  1.6
1.7  1.8   1.9   2];
H= [50 65 70 80 90 100 110 120 150 180 220 295 400 580 1000
2400 5000 8900 15000 24000];
%-------Carters Coefft for Air Gap--->
Ratio= [0  1   2   3   4   5  6   7   8   9   10  11  12];
CC=   [0 .18 .33 .45 .53 .6 .66 .71 .75 .79 .82 .86 .89];
CC1=  [0 .14 .27 .37 .44 .5 .54 .58 .62 .65 .68 .69 .7];
%(1)<-----------Main Dimensions----------------------->
KW=30;V=440;f=50;           %P=6;Specification/Input Data------>
fprintf(f2,'Optimal Design of%4d KW,%4d V,%3d HZ 3-Ph Slip-ring
Ind Motor\n',KW,V,f);
fprintf(f2,'Sn   P Zs   D m/s  L    S wcaXhca CDSW WsXHs  Btmax
Tr  D0  Eff   KgPkw I0/I pf0 pfsc Slip Zr hcR cdr Brtmx \n');
fprintf(f2,'-- -- -- ------ --- -- ------- ---- ------ -----  -
- ---- ----- ----- ---- --- ---- ---- -- --- --- -----');
insW=3.4;Hw=4;HL=1;insH=6;nvd=2;bvd=0.01;ki=0.92;Bc=1.35;Vph=V;
%Assumptions
Zr=1;kwr=1;cdb=6;Tb=6;cde=6;dd=0.05;Brc=1.35;   %Assumptions
```

```
sn=0;M1=0;M2=0;M3=0;M4=0;EFFmax=80;minKgPkw=9;minTr=75;
minI0byI=5; %Assumptions
Bav=interp1(SKW,SBav,KW,'spline');
q=interp1(SKW,Sq,KW,'spline');
Vph=V;Kw=0.955;nvd=2;bvd=10;ki=0.92;Tstrip=1.9;insS=0.5;
insW=3.4;Hw=4;HL=1;insH=6;nvd=2;ki=0.92;Bc=1.35;%Assumptions
SD=[0.1 0.15 0.2 0.3 0.4 0.5 0.75 1];
SCDSW=[4 3.8 3.6 3.5 3.5 3.5 3.5 3.5];
B=[0.8 1.2 1.6 2 2.4];WpKg=[7 15 24 34 50];
for P=4:2:10;
pf=interp1(SKWa,SPF6P,KW,'spline');
eff=interp1(SKWa,SEFF6P,KW,'spline');
if P==4 pf=interp1(SKWa,SPF4P,KW,'spline');end;
if P==4 eff=interp1(SKWa,SEFF4P,KW,'spline');end;
KWinp=KW/eff;Iph=KWinp*1e3/(3*Vph*pf);
Ns=120*f/P;ns=Ns/60;C0=11*Kw*Bav*q*eff*pf*1e-3;
DsqL=1/C0*(KW/ns);                     L1=sqrt(DsqL/(0.135*P)^2);
L=floor(L1*100)*10;          Ls=(L-nvd*bvd);            Li=ki*Ls;
D1=sqrt(DsqL/(L/1000));D=ceil(D1*100)*10; PP=pi*D/P;LbyPP=L/PP;
if LbyPP <0.8||LbyPP >2 continue; end; v=pi*D*ns/1000;
if v >30 continue;end;  FI=pi*D/P*L*Bav/1e6;
Tphi=Vph/(4.44*f*FI*Kw); %CDSW=interp1(SD,SCDSW,D,'spline');
for CDSW=3:0.5:5; As1=Iph/CDSW;
for spp=3:1:5;
S=spp*P*3;SPitch=pi*D/S;if SPitch <18||SPitch>=25 continue;end;
Zphi=2*Tphi;sph=S/3;Zs1=Zphi/sph;Zs=ceil(Zs1);Tph=Zs*sph/2;
FI=Vph/(4.44*f*Tph*Kw);
for Tstrip=1:0.1:2;
Hstrip1=As1/Tstrip;Hstrip=ceil(Hstrip1*2)/2;WbyT=Hstrip/Tstrip;
if WbyT <2.5||WbyT >=3.5 continue;end; As=0.967*Hstrip*Tstrip;
for Zsw=1:1:4;
Ws=(Zsw*(Tstrip+insS)+insW);                          Zsh=Zs/Zsw;
Hs=(Zsh*(Hstrip+insS)+Hw+HL+insH+2);
```

```
D13=D+2/3*Hs;sp13=pi*D13/S;Wt13=sp13-Ws;
At13=Li*Wt13;B13=FI*P*1e6/ (At13*S);

Btmax=1.5*B13;if Btmax >1.8||Btmax <1.4 continue;end;

Lmt=(2*L+2.3*PP+240)/1000; Rph=0.021*Lmt*Tph/As;

Pcus=3*Iph^2*Rph;             Wcus=Lmt*Tph*3*As*8.9e-3;FIc=FI/2;
Ac=FIc*1e6/Bc; Hc=Ac /Li; D01=D+2*(Hs+Hc);
D0=ceil(D01/10)*10;Hc=(D0-D)/2-Hs;

Wt=At13*S*Hs*7.8e-6; Dmcs=D+2*Hs+Hc; Wc=Ac*pi*Dmcs*7.8e-6;

PitpKg=interp1(B,WpKg,Btmax,'spline');
PicpKg=interp1(B,WpKg,Bc,'spline');

Pit=PitpKg*Wt;Pic=PicpKg*Wc;

%--------------------Slip-Ring-ROTOR------------------------>

sppR=3.5;Ss=S;kws=Kw;kwr=kws;Er=100;cdr=5;wcR=4;HwR=4;HLR=1;
Brc=1.35; %Assumtions

Lg1=0.2 + 2*sqrt(D*L/1e6);Lg=ceil(Lg1*100)/100;Dr=D-2*Lg;

Sr=sppR*3*P;sp2=pi*Dr/Sr;TphR1=Er/sqrt(3)/Vph*kws/kwr*Tph;
Zr1=TphR1*3*2/Sr;Zr=floor(Zr1);TphR=Zr*Sr/2/3;Er=TphR/TphR1*Er;

Ird=0.85*Iph;Ir=Ird*(kws*Ss*Zs)/(kwr*Sr*Zr);
for cdr=4:0.5:5.5;Acr1=Ir/cdr; hcR1=Acr1/wcR;

hcR=ceil(hcR1);Acr=wcR*hcR*0.98;Wsr=wcR+2.5;
Hsr=Zr*(hcR+0.5)+3+HwR+HLR;           LmtR=(2*L+2.3*PP+80)/1000;
RphR=0.021*LmtR*TphR/Acr;  Pcur=3*Ir^2*RphR;

Dr13=Dr-2*2/3*Hsr;spr13=pi*Dr13/Sr;           Wtr13=spr13-Wsr;
Atr=Wtr13 *Li*Sr/P;  Brt=FI*1e6/Atr; Brtmax=Brt*1.5;

if Brtmax >1.8 continue;end;     Ac=FI*1e6/2/Brc;dcr=Ac/Li;

Pfw=0.01*KW*1e3;           PnL=Pit+Pic+Pfw;           Iw=PnL/3/V;
Rr=RphR/(TphR/Tph)^2;Wcur=LmtR*TphR*3*Acr*8.9e-3;

%(4)<-----AmpTurns and Magnetizing-Current---------------------
->

%-------------BH Curve for--0.5mm,LOHYS Quality--------->

BB=[.1 .2 .3 .4 .5 .6  .7  .8  .9   1  1.1 1.2 1.3 1.4 1.5  1.6
1.7  1.8   1.9   2];

H= [50 65 70 80 90 100 110 120 150 180 220 295 400 580 1000
2400 5000 8900 15000 24000];

%-------Carters Coefft for Air Gap--->

Ratio=[0  1   2   3   4   5  6   7   8   9   10  11  12 ];

CC=   [0 .18 .33 .45 .53 .6 .66 .71 .75 .79 .82 .86 .89 ];

CC1=  [0 .14 .27 .37 .44 .5 .54 .58 .62 .65 .68 .69 .7 ];
```

```
Wss0=4;Wsr0=2; %Assumptions

atsc=interp1(BB,H,Bc,'spline'); Dcav=D+2*Hs+Hc;
ATSC=pi*Dcav/P/3*atsc/1e3; Bt30=B13*1.36;
atst=interp1(BB,H,Bt30,'spline');
ATST=atst*Hs/1000;ATS=ATSC+ATST; rat1=Wss0/Lg;
k01=interp1(Ratio,CC,rat1,'spline');

kgs=SPitch/(SPitch-Wss0 *k01);

rat2=Wsr0/Lg;k02=interp1(Ratio,CC,rat2,'spline');spr0=pi*Dr/Sr;
kgr=spr0/(spr0-Wsr0*k02);kg=kgs*kgr;Lgd=Lg*kg;rat3=bvd/Lg;
kv=interp1(Ratio,CC1,rat3,'spline'); if rat3 >=12 kv=0.7;end;

Ld=L-kv*nvd*bvd;Aag=pi*D/P*Ld;Bg=FI*1e6/Aag;B30d=1.36*Bg;
ATg=0.796            *B30d*Lgd*1e3;            Btr30=Brt*1.36;
atrt=interp1(BB,H,Btr30,'spline'); ATRT=atrt*Hsr/1e3;

Dcrav=Dr-2*Hsr-dcr; atrc=interp1(BB,H,Brc,'spline');

ATRC= pi*Dcrav/1e3/P/3 *atrc;  ATR=ATRC+ATRT; ATT=ATS+ATR+ATg;

Im=P/2*ATT/(1.17*Kw*Tph);I0=sqrt(Iw^2+Im^2);I0byI=I0/Iph;

pf0=Iw/I0;if pf0 >=0.16 continue;end;

%(5)<------------Short-Circuit-Current-------------------->');

h2=1.6;h2r=1.6;h3r=HwR;h4r=HLR;ks=1; %Assumptions

h1=Zsh*(Hstrip+insS);h3=Hw;h4=HL;bs=Ws;b0=Wss0;

Lmdss=h1/3/bs   +h2/bs   +2*h3/(bs+b0)   +h4/b0;   h1r=wcR;br=Wsr;
br0=Wsr0;      Lmdsr=h1r/3/br +h2r/br +2*h3r/(br+br0) +h4r/br0;

Lmddsr=Kw^2*S/Sr*Lmdsr;ssp=Lmdss+Lmddsr; gd=S/P/3; p=P/2;

Xs=15.8*f*L*ssp*Tph^2/(p*gd)*1e-9;

L0Lmd0=ks*PP^2/pi/SPitch/1000;

X0=15.8*f*L0Lmd0*Tph^2/(p*gd)*1e-6;

gs=S/P; gr=Sr/P; Xm=Vph/Im; Xz=5/6 *Xm* (1/gs^2+1/gr^2);

X=Xs+X0+Xz;R=Rph+Rr;Z=sqrt(R^2+X^2);Isc=Vph/Z;pfsc=R/Z;
RAT=Isc/Iph;

%(6)<--------------Performance-------------------->');

Pt=PnL+Pcus+Pcur;EFF=KW/(KW+Pt/1000)*100;

if EFF <88.5continue; end;

Rinp=KW*1000+Pfw+Pcur; SFL=Pcur/Rinp*100;
```

```
Tst=(Isc/Ir)^2*SFL/100; Pmax=3*Vph*(Isc-I0)/2/(1+pfsc)*1e-3;

Acool1=(pi*D*(L*2.5)+2*pi*(D+50)*0.04)/1e6;
Acool2=Acool1*(1+0.1*v);Acool3=pi*D0*L/1e6;

AcoolT=Acool2+Acool3; Pst=Pcus+Pit+Pic;

Tr=0.03*Pst/AcoolT;if Tr>55.6 continue;end;

Ars=Wsr*Hsr*Sr;Dri=Dcrav-dcr;

Wri=(pi*(Dr^2-Dri^2)/4-Ars)*L*7.8e-6;

Wtot=1.01*(Wcus+Wt+Wc+Wri+Wcur); KgPkw=Wtot/KW;

if KgPkw >6.3 continue;end;

%--------------------End of Program------------------------->

sn=sn+1;if EFF >=EFFmax  EFFmax=EFF;end;

if abs(EFF-EFFmax)<=1e-3 M2=sn; end; sn,

if KgPkw <=minKgPkw  minKgPkw=KgPkw;end;

if abs(KgPkw-minKgPkw)<= 0.001 M1=sn; end;

if Tr <=minTr  minTr=Tr;end;

if abs(Tr-minTr)<=0.0001 M3=sn; end;

if I0byI <=minI0byI  minI0byI=I0byI;end;

if abs(I0byI-minI0byI)<=0.0001 M4=sn; end;

fprintf(f2,'\n%2d%3d%3d%4.0f%3.0f%4.0f%3d%4.1fX%3.1f%4.1f',sn,P
,Zs,D,v,L,S,Tstrip,Hstrip,CDSW);

fprintf(f2,'%5.1fX%2.0f%6.3f%5.1f%4.0f%6.2f%6.2f%5.2f%5.2f%5.2f
%4.1f%3.0f%4.0f%4.1f%6.2f',Ws,Hs,Btmax,Tr,D0,EFF,KgPkw,I0byI,pf
0,pfsc,SFL,Zr,hcR,cdr,Brtmax);end;end;end;end;end;end;

fprintf(f2,'\n-- -- -- ------ --- -- ------- ---- ------ -----
-- ---- ----- ----- ---- --- ---- ---- -- --- --- -----');

fprintf(f2,'\nSelection of Design Variant based on Optimization
Criteria:');

fprintf(f2,'\nIf  Maximum  Efficiency  is  Required,  Select
Variant(Sn)=%3d  (%5.2f perc)',M2,EFFmax);

fprintf(f2,'\nIf  Minimum  Kg/KW  is  Required,  Select
Variant(Sn)=%3d  (%5.2f)',M1,minKgPkw);

fprintf(f2,'\nIf  Minimum  Temp-Rise  is  Reqd,  Select
Variant(Sn)=%3d  (%4.2f)',M3,minTr);

fprintf(f2,'\nIf Minimum  ratio of I0/I   is Reqd, Select
Variant(Sn)=%3d  (%4.2f)',M4,minI0byI);fclose(f2);
```

7.2.5 (a) Computer Output Results for Optimal Design

Optimal Design of 30 KW, 440 V, 50 HZ 3-Ph Slip-ring Ind Motor

Sn	P	Zs	D	m/s	L	S	wcaXhca	CDSW	WsXHs	Btmax	Tr	D0	Eff	KgPkw	I0/I	pf0	pfsc	Slip	Zr	hcR	cdr	Brtmx
1	4	20	250	20	190	36	1.8×5.5	3.0	12.6×43	1.573	51.0	420	89.94	5.97	0.32	0.16	0.22	2.4	2	12	4.5	1.79
2	4	20	250	20	190	36	1.9×5.5	3.0	13.0×43	1.628	50.4	420	90.01	6.00	0.32	0.16	0.22	2.4	2	12	4.5	1.79
3	4	20	250	20	190	36	2.0×5.0	3.0	13.4×41	1.711	50.5	420	90.00	5.86	0.34	0.15	0.23	2.4	2	12	4.5	1.79
4	4	20	250	20	190	36	2.0×5.0	3.0	13.4×41	1.711	50.5	420	89.82	5.81	0.32	0.15	0.24	2.6	2	11	5.0	1.75
5	4	20	250	20	190	36	2.0×5.0	3.0	13.4×41	1.711	50.5	420	89.60	5.76	0.31	0.16	0.25	2.9	2	10	5.5	1.71
6	4	20	250	20	190	36	1.6×5.5	3.5	11.8×43	1.472	52.5	420	89.75	5.89	0.31	0.16	0.23	2.4	2	12	4.5	1.79
7	4	20	250	20	190	36	1.7×5.0	3.5	12.2×41	1.539	52.6	420	89.74	5.77	0.32	0.16	0.24	2.4	2	12	4.5	1.79
8	4	20	250	20	190	36	1.8×5.0	3.5	12.6×41	1.593	51.8	420	89.84	5.80	0.32	0.15	0.23	2.4	2	12	4.5	1.79
9	4	20	250	20	190	36	1.5×5.0	4.0	11.4×41	1.443	54.6	420	89.50	5.70	0.31	0.16	0.25	2.4	2	12	4.5	1.79
10	4	20	250	20	190	36	1.6×4.5	4.0	11.8×38	1.507	54.8	410	89.50	5.55	0.31	0.15	0.26	2.4	2	12	4.5	1.79
11	4	20	250	20	190	36	1.6×4.5	4.0	11.8×38	1.507	54.8	410	89.32	5.50	0.30	0.16	0.26	2.6	2	11	5.0	1.75
12	4	20	250	20	190	36	1.7×4.5	4.0	12.2×38	1.558	53.8	410	89.63	5.58	0.32	0.15	0.25	2.4	2	12	4.5	1.79
13	4	20	250	20	190	36	1.7×4.5	4.0	12.2×38	1.558	53.8	410	89.45	5.53	0.30	0.16	0.26	2.6	2	11	5.0	1.75
14	6	18	330	17	160	54	2.0×5.0	3.0	10.9×46	1.644	54.6	500	88.61	6.38	0.35	0.14	0.27	3.3	2	10	5.0	1.78
15	6	18	330	17	160	54	1.9×5.0	3.5	10.6×46	1.597	55.5	500	88.51	6.33	0.34	0.14	0.27	3.3	2	10	5.0	1.78

Selection of Design Variant based on Optimization Criteria:

If Maximum Efficiency is Required, Select Variant (Sn) = 2 (90.01 perc)

If Minimum Kg/KW is Required, Select Variant (Sn) = 11 (5.50)

If Minimum Temp-Rise is Required, Select Variant (Sn) = 2 (50.37)

If Minimum ratio of I0/I is Required, Select Variant (Sn) = 11 (0.30).

CHAPTER 8

Single-Phase Induction Motors

8.1 Introduction

Single-phase induction motors are not self-starting type. Based on type of starting they are categorized as (1) Split phase, (2) Capacitor Start, (3) Shaded Pole, and (4) Repulsion Type. Theory portion of design is not given in this book, but necessary formulae, curves and tables given in standard books are made use of.

Computer-aided design for Split-phase type is given in this book. Total design is split into six parts in a proper sequence. Design Calculations are given for a given Rating of motor, followed by Computer Program written in "C" language using MATLAB software for each part. Finally all the Programs are added together to get the total Program by running which we get the total design. Computer output of total design is given.

This design may not be the optimum one. Now optimization objective and Design Constraints are inserted into this total Program. When this program is run we will get various alternative feasible designs from which the selected variant based on the Optimization Criteria can be picked up. Computer output showing the important design parameters for various feasible alternatives is given at the end of this chapter along with a logic diagram.

8.2 Sequential Steps for Design of Each Part and Programming Simultaneously

(a) Calculate Main dimensions of stator core OD, Width of tooth, stator core depth, Slot width and depth, Air gap length

(b) Calculate for the Main Winding: No. of series turns, Turns in each coil, conductor dia, by checking current density. Calculate mean length, Resistance, Copper losses

Calculate Rot dia, no. of Rotor Slots, Cond dia, Core depth, Resistance and Copper Losses, Weight of rotor copper

(c) Calculate Carter Coefft and Ampere-turns for Air gap, Stator tooth, Stator core, Rotor tooth, Rotor core and Total No-load AT, Magnetizing current

(d) Calculate for the Auxiliary Winding: No. of series turns, Turns in each coil, conductor dia, by checking current density. Calculate mean length, Resistance, Copper losses

(e) Calculation of Wt of tooth, core Iron Losses, Total Weight, Kg/hp

(f) Solution by Equivalent Circuit, Slip Vs Torque curve.

Note: By adding programs established for each part sequentially we get the Program for complete design.

Problem: Design a 1.0 HP, 230V, 2850 RPM, 2 Pole, 50Hz, Split-phase, Ind.motor.

Table 8.1 Frequency Constant (Kf)

Freq(HZ)	25	30	40	45	50	60
Kf	0.865	0.885	0.923	0.94	0.96	1.0

Table 8.2 Output Coefficient (C0) at 1 hp.

Poles	2	4	6	8
CO	0.32	0.22	0.19	0.17

Table 8.3 Ratio of Di/D0.

Poles	2	4	6	8
Di/D0	0.5	0.59	0.64	0.67

8.3 Calculation of Main Dimensions (Part-1)

Frequency Constant (Kf) = 0.96 from Table.1

Output coefficient for 2 pole motor (C0) = 0.32 from Table.2

Type constant for Split-phase type (Kt) = 1.42 (assumed)

Now, $\text{D0sqL} = 16.5 \times \text{C0} \times \text{hp/N} \times \text{kf} \times \text{kt} \times 10^6$

$$= 16.5 \times 0.32 \times 1.0/2850 \times 0.96 \times 1.42 \times 10^6 = 2525.5 \text{ cm}^2;$$

Assuming L/D0 ratio (LbyD0) = 0.3,

$$\text{Outer dia of Stator core (D0)} = \sqrt[3]{\frac{\text{DosqL}}{\text{LbyD0}}} = \sqrt[3]{\frac{2525.5}{0.3}} = 20.34 \approx 20 \text{ cm (rounded off)}$$

Stator Core Length (L) = 0.3 × D0 = 0.3 × 20 = 6 cm

Stator core Inner Dia (Di) = 0.5 × 20 = 10 cm (Using Table. 3)

Assuming Stator slots (S1) = 24 and Slot Cross section as Trapezoidal (Teeth will be rectangular),

Slot opening at tip (b10) = 0.068 + 0.0175 × Di

$$= 0.068 + 0.0175 \times 10 = 0.243 \text{ cm (Emperical Formula)}$$

Assuming depth of the tip at slot opening (h10) = 0.07 cm and depth of the mouth (h11) = 1.3 × 0.07 = 0.091 cm,

Using Empirical Formula,

$$\text{Width of tooth (bt1)} = \frac{(1.1 + 0.032 \times \text{Di}) \times \text{Di}}{\text{S1}}$$

$$= \frac{(1.1 + 0.032 \times 10) \times 10}{24} = 0.5917 \approx 0.59 \text{ cm (rounded off)}$$

Assuming Flux density in St.tooth (Bt) = 1.35 T and St.Core (Bc) = 1.15 T,

$$\text{St.Core depth(dc1)} = \left(\frac{\text{Bt}}{\text{Bc}}\right) \times \left(\frac{\text{S1} \times \text{bt1}}{\pi \times \text{P}}\right) = \left(\frac{1.35}{1.15}\right) \times \left(\frac{24 \times 0.59}{\pi \times 2}\right) = 2.6456 \text{ cm}$$

Width of Slot at top section (b11)

$$= \frac{\pi \times [\text{Di} + 2 \times (\text{h10} + \text{h11})]}{\text{S1}} - \text{bt1} = \frac{\pi \times [10 + 2 \times (0.07 + 0.091)]}{24} - 0.59 = 0.7611 \text{ cm}$$

Depth of the Slot (h14) = 0.05 × (D0 – Di) – (h10 + h11 + dc1)

= 0.05 × (20 – 10) – (0.07 + 0.091 + 2.6456) = 2.1934 cm

Half-Slot angle (alfa) = 180/S1 = 7.5°

Slot-width at the bottom (b13) = b11 + 2 × h14 × tan (alfa)

= 0.7611 + 2 × 2.1934 × tan (7.5 × π/180) = 1.3387 cm

Using Empirical Formula,

Length of Air-gap (Lg)

$$= 0.0013 + \frac{0.0042 \times \text{Di}}{\sqrt{\text{P}}} = 0.0013 + \frac{0.0042 \times 10}{\sqrt{2}} = 0.0427 \text{ cm}$$

8.3 *(a) Computer Program in "C" in MATLAB for Part-1*

```
%Design of 1hp,230V,1-ph IM
%<--------Calculation of Main Dimensions(Part-1)------->
hp=1;V=230;P=2;N=2850;hz=50; %Main Data
kt=1.42;LbyD0=0.3;ki=0.93;S1=24;%Assumptions
h10=0.07;h11=1.3*h10;Bt=1.35;Bc=1.15;cdm1=4.5;S2=30;
%Assumptions
HZ=[25 30 40 45 50 60];          KF=[0.865 .885 .923 .94 .96 1];
kf=interp1 (HZ,KF,hz, 'spline');
POLES=[2   4   6   8];           OPTC=[0.32   0.22   .19   .17];
C0=interp1(POLES,OPTC,P,'spline');
DIBYD0=[.5 .59 .64.67];
DibyD0=interp1(POLES,DIBYD0,P,'spline');
```

```
D0sqL=16.5*C0*hp/N*kf*kt*1e6;               D01=(D0sqL/LbyD0)^(1/3);
D0=floor(D01);    L=LbyD0*D0;      Di1=DibyD0*D0;     Di=ceil(Di1);
SP=S1/P;                           if P==2 bt1a=(1.1+0.032*Di)*Di/S1;
bt1=floor(bt1a*100)/100; end;
ifP==4 bt1a=(1.27+0.035*Di)*Di/S1;bt1=floor(bt1a*100)/100; end;
SS=[24 36 48]; B10=[0.068+0.0175*Di 0.038+0.0175*Di 0.0175*Di];
b10=interp1(SS,B10,S1,'spline');
dc1=Bt/Bc*(S1*bt1)/(pi*P);b11=pi*(Di+2*h10+2*h11)/S1-bt1;
h14=DibyD0*(D0-Di)-(h10+h11+dc1);
alfa=180/S1;b13=b11+2*h14*tan(alfa*pi /180);
Lg=0.013+0.0042*Di/sqrt(P);
```

8.4 Design of Main Winding (Part-2)

No. of Coils/pole (Ncpp) = $\frac{S1}{P \times 2} = \frac{24}{2 \times 2} = 6$, Now using Concentric Winding,

Pitch factor for Concentric Winding

Coil No.	Slots spanned (CT)	Factor(y1)	Pitch Factor(Kp) = sin(y1*π/2)
1	1 to13 (12)	12/13	Sin (12/13 × π/2) = 0.9927
2	2 to12 (10)	10/13	Sin (10/13 × π/2) = 0.935
3	3 to11 (8)	8/13	Sin (8/13 × π/2) = 0.8230
4	4 to10 (6)	6/13	Sin (6/13 × π/2) = 0.6631
5	5 to9 (4)	4/13	Sin (4/13 × π/2) = 0.4647
6	6 to8 (2)	2/13	Sin (2/13 × π/2) = 0.2393
			∑Kp = 4.1179

Percentage No. of Turns in Each Coil

Coil No (n)	1	2	3	4	5	6	∑
%Turns(Tn) = Kp(n)/4.1179 × 100	24.11	22.71	19.99	16.10	11.28	5.81	100

Winding factor (Kwm) =∑Kp(n) × T(n)

$$= (0.9927 \times 24.11) + (0.935 \times 22.71) + (0.823 \times 19.99) + (0.6631 \times 16.1) + (0.4647 \times 11.28) + (0.2393 \times 5.81)/100 = 0.7892$$

Considering Iron factor (ki) = 0.93;

Net iron length in core (Li) = ki × L = 0.93 × 6 = 5.58 cm

Slots/Pole (SP) = S1/P = 24/2 = 12

Area of tooth (At) = bt1 × Li × SP = 0.59 × 5.58 × 12 = 39.5064 cm^2;

Flux/Pole (FI) = $0.637 \times At \times Bt \times 10^{-4} = 0.637 \times 39.5064 \times 1.35 \times 10^{-4} = 0.0034$ Wb

Considering Induced EMF (E) = $0.95 \times V$,

No of series turns in Main winding (Tm)

$$= \frac{0.95 \times 230}{4.44 \times hz \times FI \times Kwm} = \frac{0.95 \times 230}{4.44 \times 50 \times 0.0034 \times 0.7892} = 367 \approx 368$$ (rounded off to even integer)

Total Conductors in Main winding (Nm) = $2 \times Tm = 2 \times 368 = 736$

No. of series turns/pole in Main winding (Tppm) = Tm/P = 368/2 = 184

No. of Turns in Each Coil

Coil No (n)	1	2	3	4	5	6	∑
Turns(Tc)n = %Turns × Tppm/100	44	42	37	30	21	10	184

(Example: for 1st coil, Tc1 = $24.11 \times 184/100 = 44.36 \approx 44$(rounded off to nearest integer)

Table 8.4 Efficiency and PF.

Output(HP)	0.05	0.1	0.25	0.5	0.75	1.0
Efficiency(pu)	0.35	0.44	0.58	0.65	0.68	0.7
PF	0.45	0.5	0.56	0.62	0.64	0.66

From Table. 8.4, efficiency (effa) = 0.7pu and

Power factor (pfa) = 0.66 corresponding to 1.0 hp

$$\text{Current in main wdg (Im)} = \frac{hp \times 746}{effa \times V \times pfa} = \frac{1.0 \times 746}{0.7 \times 230 \times 0.66} = 7.02 \text{ A}$$

Assuming a current density (cdm) = 4.5 A/mm^2,

$$\text{Area of Conductor (Amc)} = Im/cdm = \frac{Im}{cdm} = \frac{7.02}{4.5} = 1.5601 \text{ mm}^2,$$

$$\text{Diameter of conductor (dmc)} = \sqrt{\frac{4 \times Amc}{\pi}} = \sqrt{\frac{4 \times 1.5601}{\pi}} = 1.4094 \text{ mm},$$

Table 8.5 Standard Wire Gauge (SWG) for round Conductor as per British Standard Specification.

SWG no	5	6	7	8	9	10	11	12	13	14	15	16	17	18	19
Dia (mm)	5.38	4.88	4.47	4.06	3.66	3.25	2.95	2.64	2.31	2.03	1.83	1.63	1.42	1.22	1.02

From Table. 8.5 suitable dia of bare conductor (dmc) = 1.42 mm

Considering Enamel insulation thickness as 0.075 mm,

Dia of insulated conductor (dmcins) = 1.42 + 0.075 = 1.4975 mm;

Area of CS of copper in the conductor (Amc) = $\dfrac{\pi \times dmc^2}{4} = \dfrac{\pi \times 1.42^2}{4} = 1.5837\ mm^2$

Current density in main winding (cdm) = $\dfrac{Im}{Amc} = \dfrac{7.02}{1.5837} = 4.433 A/mm^2$

Space occupied by 44 insulated conductors in the slot (Socm)

$$= 44 \times \frac{\pi \times dmcins^2}{4 \times 100} = 44 \times \frac{\pi \times 1.495^2}{4 \times 100} = 0.7724\ cm^2$$

Ares of CS of Gross Slot (Asg) = [h11 × (b10 + b11) + h14 × (b11 + b13)]/2

= [0.091 × (0.243 + 0.7611) + 2.1934 × (0.7611 + 1.3387)]/2 = 2.3486 cm^2;

Space factor of the slot (Sfsm) = $\dfrac{Socm}{Asg} = \dfrac{0.7724}{2.3486} = 0.3289$ (< 0.5 and hence OK)

Height of Stator Slot (Hs) = (h11 + h14) = (0.091 + 2.1934) = 2.2844 cm

Length of mean Turn (Lmt) in Each Coil

From Table.8.5 suitable dia of bare conductor (dmc) = 1.42 mm

Considering Enamel insulation thickness as 0.075mm,

Dia of insulated conductor (dmcins) = 1.42 + 0.075 = 1.4975 mm;

Area of CS of copper in the conductor (Amc)

$$= \frac{\pi \times dmc^2}{4} = \frac{\pi \times 1.42^2}{4} = 1.5837\ mm^2$$

Current density in main winding (cdm) = $\dfrac{Im}{Amc} = \dfrac{7.02}{1.5837} = 4.433\ A/mm^2$

Space occupied by 44 insulated conductors in the slot (Socm)

$$= 44 \times \frac{\pi \times dmcins^2}{4 \times 100} = 44 \times \frac{\pi \times 1.495^2}{4 \times 100} = 0.7724\ cm^2$$

Ares of CS of Gross Slot (Asg) = [h11 × (b10 + b11) + h14 × (b11 + b13)] / 2

= [0.091 × (0.243 + 0.7611) + 2.1934 × (0.7611 + 1.3387)] / 2 = 2.3486 cm^2;

Space factor of the slot (Sfsm) = $\dfrac{Socm}{Asg} = \dfrac{0.7724}{2.3486} = 0.3289$ (< 0.5 and hence OK)

Height of Stator Slot (Hs) = (h11 + h14) = (0.091 + 2.1934) = 2.2844 cm

Length of mean Turn (Lmt) in each coil

Coil No (n)	1	2	3	4	5	6
Lmt (cm)	63.59	54.0	46.4	37.8	29.2	20.6

Example for Coil. 1

Lmt (1) = 8.4 × (Di + Hs)/S1 × Slots Spanned + (2 × L)

$= 8.4 \times (10 + 2.2844)/24 \times 12 + (2 \times 6) = 63.59$ cm

Mean Length of turn of main wdg (Lmtm) = $\sum$Lmt(n) × T(n)

$= [(63.59 \times 44) + (54 \times 42) + (46.4 \times 37) + (37.8 \times 30) + (29.2 \times 21) + (20.6 \times 10)]/184 = 47.705$ cm

Resistance of Main wdg at 75° (Rm)

$$= \frac{0.021 \times \text{Lmtm} \times \text{Tm}}{\text{Amc} \times 100} = \frac{0.021 \times 47.705 \times 368}{1.5837 \times 100} = 2.328\,\Omega$$

Copper loss in main wdg (Pcus) = $\text{Im}^2 \times \text{Rm} = 7.02^2 \times 2.328 = 114.74$ W

8.4 *(a) Computer Program in "C" in MATLAB for Part-2*

```
%<--------Design of Main Winding(Part-2)--------->

S1=24;P=2;SP=12;ki=0.93;L=6;bt1=0.59;Bt=1.35;V=230;hz=50;hp=1;
cdm1=4.5;h11=0.091;b10=0.243;b11=0.7611;h14=2.1934; b13=1.3387;
Di=10;%Input Data

Ncpp=S1/P/2;sum1=0;sum2=0;sum3=0;sum4=0;sum5=0;

for   i=1:Ncpp;    y1=(SP-(i-1)*2)/(SP+1);    Kp(i)=sin(y1*pi/2);
sum1=sum1+Kp(i);end;

for i=1:Ncpp;   PTm(i)=Kp(i)/sum1*100;    sum2=sum2+PTm(i);   end;

for i=1:Ncpp; sum3=sum3+Kp(i)*PTm(i); end; Kwm=sum3/sum2;

Li=ki*L;            At=bt1*Li*SP;              FI=0.637*At*Bt*1e-4;
Tm1=0.95*V/(4.44*hz*FI*Kwm);

Tm=ceil(Tm1/2)*2; Nm=Tm*2; Tppm=Tm/P;

for   i=1:Ncpp;   Tc1(i)=Tppm*PTm(i)/100;    Tc(i)=round(Tc1(i));
sum4=sum4+Tc(i);  end; corr=Tppm-sum4;  Tc(Ncpp)=corr+Tc(Ncpp);
sum4=sum4+corr;

HP=[.05  .1  .25  .5  .75  1]; EFF=[.35  .44  .58  .65  .68  .7];
PF=[.45 .5 .56 .62 .64 .66];

effa=interp1(HP,EFF,hp,'spline');
pfa=interp1(HP,PF,hp,'spline');

Im=hp*746/(effa*V*pfa);  Amc1=Im/cdm1;    dmc1=sqrt(4*Amc1/pi);
```

```
DSWG=[5.38 4.88 4.47 4.06 3.66 3.25 2.95 2.64 2.31 2.03 1.83
1.63 1.42 1.22 1.02];

for i=1:15; if dmc1 > = DSWG(i) continue; end; dmc=DSWG(i);end;
dmcins=dmc+0.075;

Amc=pi*dmc^2/4;  cdm=Im/Amc;if cdm<2.8||cdm>4.5 continue; end;

Socm=Tc(1)*pi*dmcins^2/4/100;

Asg=(h11*(b10+b11)+h14*(b11+b13))/2; Sfsm=Socm/Asg;

if Sfsm >0.5 continue;end;  Hs=h11+h14;

for i=1:Ncpp;   Lmt(i)=8.4*(Di+Hs)/S1*(SP-(i-1)*2)+2*L;

sum5=sum5+Lmt(i)*Tc(i);end;

Lmtm=sum5/sum4;Rm=0.021*Lmtm*Tm/(Amc*100)

Pcus=Im^2*Rm;   Wcum=Lmtm*Amc*Tm*8.9e-5;
```

8.5 Design of Rotor (Part-3)

Rotor outer dia (Dr) = Di – 2 × Lg = 10 – 2 × 0.0427 = 9.9146 cm

Assuming Rotor slots (S2) = 30,

Width of tooth at minimum section (bt2)

$$= \frac{0.95 \times S1 \times bt1}{S2} = \frac{0.95 \times 24 \times 0.59}{30} = 0.4484 \text{ cm}$$

Assuming Slot opening (b20) = 0.075cm and depth of tip (h20) = 0.08 cm,

Radius of round slot (r21)

$$= \frac{\pi \times (Dr - 2 \times h20) - S2 \times bt2}{2 \times (S2 + \pi)} = \frac{\pi \times (9.9146 - 2 \times 0.08) - 30 \times 0.4484}{2 \times (30 + \pi)} = 0.2594 \text{ cm}$$

Dia of rotor conductor (drc) = 2 × r21 – 0.038 = 2 × 0.2594 – 0.038 = 0.4808 cm

From Table. 8.5, suitable conductor is SWG:6 with dia (drc) = 4.88 mm

Area of conductor (Ab) = $\frac{\pi \times drc^2}{4} = \frac{\pi \times 4.88^2}{4} = 18.7038 \text{ mm}^2$;

Rotor Core depth (dc2) = 0.95 × dc1 = 0.95 × 2.6456 = 2.5133 cm

Length of Bar /conductor (Lb) = L + 1 = 6 + 1 = 7 cm

Area of CS of end ring (Aer) = $\frac{Nb \times Ab}{\pi \times P} = \frac{30 \times 18.704}{\pi \times 2} = 89.304 \text{ mm}^2$

Resistance of Rotor wdg at 75° referred to stator main wdg (R2md)

$$= P \times Nm^2 \times Kwm^2 \times 0.021 \times \left[\frac{Lb}{Ab \times Nb} + \frac{0.64 \times Dm}{P^2 \times Aer}\right]$$

$$= 2 \times 736^2 \times 0.7892^2 \times 0.021 \times \left[\frac{7}{18.704 \times 30} + \frac{0.64 \times 0.07}{4 \times 89.304}\right] = 3.5359\,\Omega$$

Equivalent Rotor current (I2d) = Im × pfa = 7.02 × 0.66 = 4.6335 A

Rotor Copper Loss (Pcur) = $I2d^2$ × R2md = 4.6335^2 × 3.5359 = 75.92 W

Weight of Rotor Copper (Wcur) = Lb × Ab × Nb × 8.9 × 10^{-5}

= 7 × 18.704 × 30 × 8.9 × 10^{-5} = 0.3496 Kg

Weight or Rotor Steel (Wtr) = $[(\pi \times Dr^2/4) - (S2 \times \pi \times r21^2)] \times L \times 7.8 \times 10^{-3}$

$= [(\pi \times 9.9146^2/4) - (30 \times \pi \times 0.2594^2)] \times 6 \times 7.8 \times 10^{-3} = 3.3164$ Kg

8.5 *(a) Computer Program in "C" in MATLAB for Part-3*

```
%<--------Design of Rotor(Part-3)--------->
b20=0.075;h20=0.08;Dm=0.07; %assumption
Di=10;Lg=0.0427;S1=24;bt1=0.59;S2=30;dc1=2.6456;L=6;P=2;Nm=736;
%Input Data
DSWG=[5.38 4.88 4.47 4.06 3.66 3.25 2.95 2.64 2.31 2.03 1.83
1.63 1.42 1.22 1.02];
Kwm=0.7892;Im=7.02;pfa=0.66; %Input Data
Dr=Di-2*Lg;bt2=0.95*S1*bt1/S2;
r21=(pi*(Dr-2*h20)-S2*bt2)/(2*(S2+pi)); drc1= 2*r21-0.038;
for i=1:15; if drc1*10>= DSWG(i)continue;end; drc=DSWG(i); end;
Ab=pi*drc^2/4;dc2=0.95*dc1;Lb=L+1;Nb=S2;Aer=Ab/pi*Nb/P;
R2md=P*Nm^2*Kwm^2*0.021*(Lb/(100*Ab*Nb)+2/pi*Dm/P^2/Aer);
I2d=Im*pfa;Pcur=I2d^2*R2md; Wcur=Lb*Ab*Nb*8.9e-5;
Wtr=((pi*Dr^2/4)-(S2*pi*r21^2))*L*7.8*1e-3;
```

8.6 Amp-Turns Calculation (Part-4)

$$\frac{\text{Stator Slot Opening}}{\text{Air Gap}} = \frac{b10}{Lg} = \frac{0.243}{0.0427} = 5.6911$$

Carter Coefft. read from Fig. 8.1 corresponding to the ratio = 5.6911 is

(K01) = 0.6445

$$\text{Slot pitch (sp1)} = \frac{\pi \times \text{Di}}{\text{Sl}} = \frac{\pi \times 10}{24} = 1.309 \text{ cm}$$

$$\text{Stator Slot coefft (Kgs)} = \frac{\text{spl}}{\text{spl} - \text{b10} \times \text{k01}} = \frac{1.309}{1.309 - 0.243 \times 0.6445} = 1.1359$$

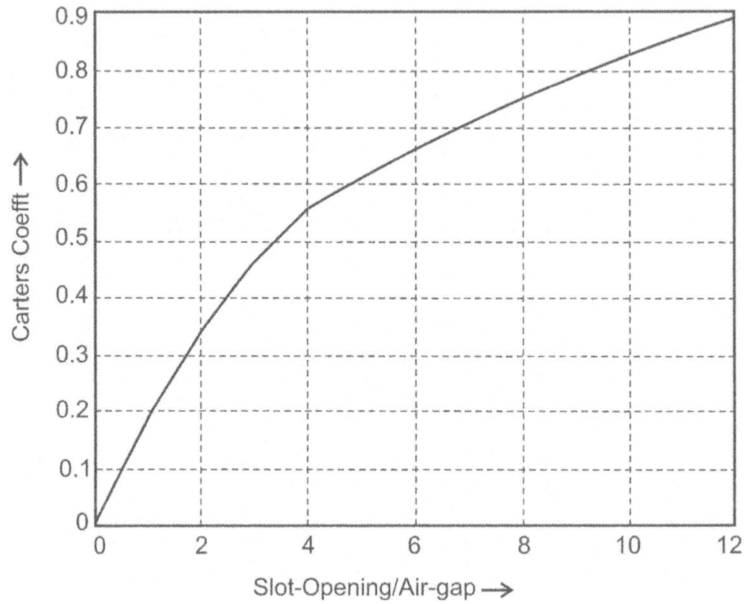

Fig. 8.1 Carter coefft. for semi-closed slot.

$$\frac{\text{Rotor Slot Opening}}{\text{Air Gap}} = \frac{\text{b20}}{\text{Lg}} = \frac{0.075}{0.0427} = 1.7565$$

Carter Coefft. read from Fig. 8.1 corresponding to the ratio

$$= 1.7565 \text{ is (K02)} = 0.3065$$

$$\text{Slot pitch(sp2)} = \frac{\pi \times \text{Dr}}{\text{S2}} = \frac{\pi \times 9.9146}{30} = 1.0383 \text{ cm}$$

$$\text{Stator Slot coefft (Kgr)} = \frac{\text{sp2}}{\text{sp2} - \text{b20} \times \text{k02}} = \frac{1.0383}{1.0383 - 0.075 \times 0.3065} = 1.0226$$

$$\text{Air gap area(Aga)} = \frac{\pi \times \text{Di}}{\text{P} \times \text{L}} = \frac{\pi \times 10}{2 \times 6} = 94.248 \text{ cm}^2$$

Air gap coefft(Kag) = Kgs × Kgr = 1.1359 × 1.0226 = 1.1616

Air Gap length, effective (Lgd) = Kag × Lg = 1.1616 × 0.0427 = 0.0496 cm

$$\text{Flux density in Air gap (Bag)} = \frac{\text{Fi} \times 10^4}{0.637 \times \text{Aga}} = \frac{0.0034 \times 10^4}{0.637 \times 94.248} = 0.5659 \text{ T}$$

$$\text{Amp Turns for the gap (ATag)} = 0.796 \times \text{Bag} \times \text{Kag} \times \text{Lgd} \times 10^4$$

$$= 0.796 \times 0.5659 \times 1.1616 \times 0.0496 \times 10^4 = 259.5$$

$$\text{Area of St.teeth per Pole (At1)} = \frac{\text{Li} \times \text{bt1} \times \text{Sl}}{\text{P}} = \frac{5.58 \times 0.59 \times 24}{2} = 39.5064 \text{ cm}^2;$$

$$\text{Flux density in St tooth (Bt1)} = \frac{\text{Fi} \times 10^4}{0.637 \times \text{At1}} = \frac{0.0034 \times 10^4}{0.637 \times 39.5064} = 1.35 \text{ T}$$

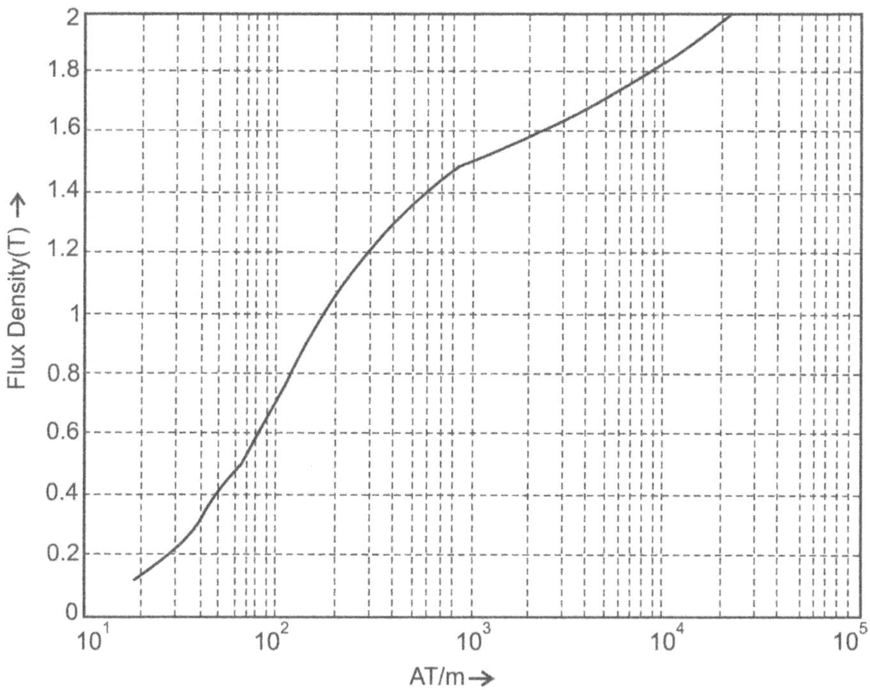

Fig. 8.2 Magnetization curve for Lohys steel.

For (Bt1 = 1.35T), AT/m (att1) = 467.7 AT/m as read from Fig. 8.2

Length of Flux Path in St tooth (Lfpt1) = h14 + h11 + h10

$$= 2.1034 + 0.091 + 0.07 = 2.3544 \text{ cm}$$

Amp-Turns for St-Teeth (ATT1) = att1 × Lfpt1/100 = 467.7 × 2.3544/100 = 11.0

Area for St.Core (Ac1) = Li × dc1 = 5.58 × 2.6456 = 14.7623 cm^2;

$$\text{Flux density in St Core (Bc1)} = \frac{\text{Fi} \times 10^4}{0.637 \times \text{Ac1}} = \frac{0.0034 \times 10^4}{0.637 \times 14.7623} = 1.1507 \text{ T}$$

For (Bc1 = 1.1507 T), AT/m (atc1) = 254.9 AT/m as read from Fig. 8.2

Length of Flux Path in St Core (Lfpc1)

$$= \frac{\pi \times (D0 - dc1)}{2 \times P} = \frac{\pi \times (20 - 2.6456)}{2 \times 2} = 13.63 \text{ cm}$$

Amp-Turns for St-Core (ATC1) = atc1 × Lfpc1/100 = 254.9 × 13.63/100 = 34.74

$$\text{Area of Ror. teeth per Pole (At2)} = \frac{Li \times bt2 \times S2}{P} = \frac{5.58 \times 0.4484 \times 30}{2} = 37.5311 \text{ cm}^2;$$

$$\text{Flux density in Rotor tooth (Bt2)} = \frac{Fi \times 10^4}{0.637 \times At2} = \frac{0.0034 \times 10^4}{0.637 \times 37.5311} = 1.4211 \text{ T}$$

For (Bt2 = 1.4211T), AT/m (att2) = 640.4 AT/m as read from Fig. 8.2

Length of Flux Path in Rot tooth (Lfpt2) = 2 × r21 + h20

= (2 × 0.2594) + 0.08 = 0.6 cm

Amp-Turns for Rot-Teeth (ATT2) = att2 × Lfpt2/100 = 640.4 × 0.6/100 = 3.83

CS Area for Rot Core (Ac2) = Li × dc2 = 5.58 × 2.5133 = 14.024 cm^2;

$$\text{Flux density in Rot Core (Bc2)} = \frac{Fi \times 10^4}{0.637 \times Ac2} = \frac{0.0034 \times 10^4}{0.637 \times 14.0242} = 1.2112 \text{ T}$$

For (Bc2 = 1.2112 T), AT/m (atc2) = 304.15 AT/m as read from Fig. 8.2

Length of Flux Path in Rot Core (Lfpc2)

$$= \frac{\pi \times (2 \times r21 + h20 + dc2)}{2 \times P} = \frac{\pi \times (2 \times 0.2594 + 0.08 + 2.5133)}{2 \times 2} = 2.444 \text{ cm}$$

Amp-Turns for Rot-Core (ATC2) = atc2 × Lfpc2 /100 = 304.15 × 2.444/100 = 7.43

Total AT required (ATT) = ATag + ATT1 + ATC1 + ATT2 + ATC2

= 259.5 + 11 + 34.74 + 3.83 + 7.43 = 316.5

$$\text{Saturation factor (satf)} = \frac{ATT}{ATag} = \frac{316.5}{259.5} = 1.22$$

8.6 *(a) Computer Program in "C" in MATLAB for Part-4*

```
%<--------Amp-Turns Calculation (Part-4)--------->

b10=0.243;Lg=0.0427;Di=10;S1=24;b20=0.075;Dr=9.9146;S2=30;P=2;

L=6;FI=0.0034; Li=5.58;bt1=0.59;h14=2.1034;h11=0.091; h10=0.07;
dc1=2.6456; D0=20;bt2=0.4484;%input Data

r21=0.2594;h20=0.08;dc2=2.5133;%input Data

SobyAg1=b10/Lg;    SOBYAG=[0 1 2 3 4 5 6 7 8 9 10 11 12];
```

```
CC=[0 .19 .34 .46 .55 .61 .66 .71 .75 .79 .82 .86 .89];%semi-
closed-slot plot(SOBYAG,CC);grid;
xlabel('Slot-Opening/Air-gap-->');ylabel('Carters Coefft-->');
title('Carter coefft for Semi closed Slot');
K01=interp1(SOBYAG,CC,     SobyAg1,'spline');     sp1=pi*Di/S1;
Kgs=sp1/(sp1-b10*K01);    SobyAg2=b20/Lg;
K02=interp1(SOBYAG,CC,SobyAg2,'spline');          sp2=pi*Dr/S2;
Kgr=sp2/(sp2-b20*K02);Kag=Kgs*Kgr;Lgd=Kag*Lg; Aga=pi*Di/P*L;
Bag=FI*1e4/0.637/Aga;  ATag=0.796*Bag*Kag*Lgd*1e4;
BB=[.1  .2  .3  .4  .5  .6  .7  .8  .9  1   1.1  1.2  1.3  1.4
1.5  1.6  1.7  1.8  1.9  2.0 ];%Tesla
HH=[18  28  38  48  67  80  95 120 140 180  220  295  390  580
1000 2200 5000  9000 16000 24000];%for Lohys steel
semilogx(HH,BB);grid;xlabel('AT/m-->');
ylabel('Fluxdensity(T)-->');
title('Magnetization curve for Lohys steel');
At1=Li*bt1*S1/P;  Bt1=FI*1e4/0.637/At1;
if Bt1 <1.3 || Bt1 >1.7 continue ;  end;
att1=interp1(BB,HH,Bt1,'spline');Lfpt1=h14+h11+h10;
ATT1=att1* Lfpt1/100;     Ac1=Li*dc1;          Bc1=FI*1e4/2/Ac1;
atc1=interp1(BB,HH,Bc1,'spline');
Lfpc1=pi*(D0-dc1)/2/P;   ATC1=atc1*Lfpc1/100;
At2=Li*bt2*S2/P;Bt2=FI*1e4/0.637/At2;
att2=interp1(BB,HH,Bt2,'spline');Lfpt2=2*r21+h20;
ATT2=att2*Lfpt2/100;Ac2=Li*dc2;Bc2=FI*1e4/2/Ac2;
atc2=interp1(BB,HH,Bc2,'spline');
Lfpc2=pi*(2*r21+h20+dc2)/2/P;  ATC2=atc2*Lfpc2/100;
ATT=ATag+ATT1+ATC1+ATT2+ATC2;  satf=ATT/ATag;
```

8.7 Leakage Reactance Calculation (Part-5)

$$\mathbf{Cx} = \frac{\sum_{n=1}^{n=6} Tc^2(n)}{[Tppm]^2} \times \frac{1}{Kwm^2} \times \frac{S1}{4 \times P}$$

$$= \frac{(44^2 + 42^2 + 37^2 + 30^2 + 21^2 + 10^2)}{184^2} \times \frac{1}{0.7892} \times \frac{24}{4 \times 2} = 0.9261$$

$$\text{Ratio (b11byb13)} = \frac{b11}{b13} = \frac{0.7611}{1.3387} = 0.5686 \text{ and}$$

Constant(A) from Fig. 8.3, (constA) = 0.558

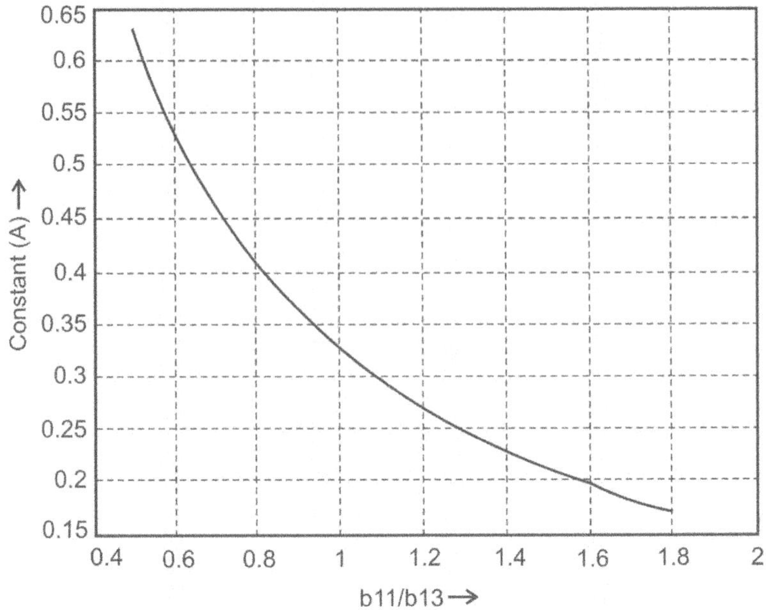

Fig. 8.3 Ratio of b11/b13 vs. constant (A).

Stator Slot constant (Ks1)

$$= A \times \frac{h14}{b13} + \frac{h10}{b10} + \frac{2 \times h11}{b10 + b11} = 0.558 \times \frac{2.1934}{1.3387} + \frac{0.07}{0.243} + \frac{2 \times 0.0910}{0.243 + 0.7611} = 1.3836$$

$$\text{Rotor Slot constant (Ks2)} = \frac{h20}{b20} + 0.623 = \frac{0.08}{0.075} + 0.623 = 1.6897$$

Slot leakage Constant (Ks)

$$= Ks1 \times Cx + \frac{S1}{S2} \times Ks2 = 1.3836 \times 0.9261 + \frac{24}{30} \times 1.6897 = 2.6331$$

$$\text{Factor (Kx)} = 2 \times \pi \times hz \times (Nm \times Kwm)^2 \times 10^{-8}$$

$$= 2 \times \pi \times 50 \times (736 \times 0.7892)^2 \times 10^{-8} = 1.0601$$

Slot Leakage Reactance (Xs)

$$= \frac{Kx \times 2.512 \times Ks \times L}{S1} = \frac{1.0601 \times 2.512 \times 2.6331 \times 6}{24} = 1.7529\ \Omega$$

Stator tooth face (bt10) = sp1 – b10 = 1.309 – 0.243 = 1.066 cm

Rotor tooth face (bt20) = sp2 – b20 = 1.0383 – 0.075 = 0.9633 cm

$$\text{ZigZag constant (Kzz)} = \frac{(bt10 + bt20)^2}{4 \times (sp1 + sp2)} = \frac{(1.066 + 0.9633)^2}{4 \times (1.309 + 1.0383)} = 0.4386$$

ZigZag Leakage Reactance (Xzz)

$$= \frac{0.838 \times L}{Sl \times Lg} \times Kx \times Kzz = \frac{0.838 \times 6}{24 \times 0.0427} \times 1.0601 \times 0.4386 = 2.2811$$

Average Coil Throw of the coils of stator winding (ACT)

$$= \frac{\sum_{n=1}^{n=6} Tc(n) \times CT(n)}{Tppm}$$

$$= \frac{(44 \times 12) + (42 \times 10) + (37 \times 8) + (30 \times 6) + (21 \times 4) + (10 \times 2)}{184}$$

$$= 8.3 \approx 8 \text{ (Integer)}$$

Dia at centre of St slot (De) = Di + h14 + 2 × (h10 + h11)

= 10 + 2.1934 + 2 × (0.07 + 0.091) = 12.5154 cm

End Leakage Reactance (Xe)

$$= Kx \times \frac{1.236 \times De \times ACT}{Sl \times 2} = \frac{1.236 \times 12.5154 \times 8}{24 \times 2} = 2.733\,\Omega$$

$$\text{Factor (Km)} = \frac{Aga}{Lgd \times satf \times P} = \frac{94.2478}{0.0496 \times 1.2197 \times 2} = 778.96$$

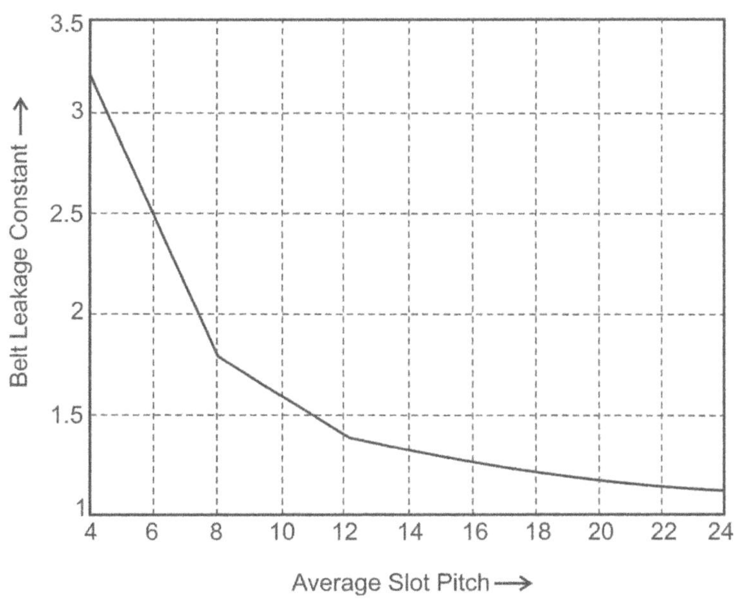

Fig. 8.4 Av. Slot pitch vs. belt leakage constant.

$$\text{Average Slot-Pitch (Nsp)} = \frac{Sl + S2}{2 \times P} = \frac{24 + 30}{2 \times 2} = 13.5$$

Corresponding to (Nsp) = 13.5, From Fig. 8.4 Belt Leakage constant (Kb) = 1.3346

Belt Leakage Reactance (Xb) = 0.000929 × Km × Kb × Kx

= 0.000929 × 778.96 × 1.3346 × 1.0601 = 1.0237 Ω

Assuming Rotor slot skew angle (skew) = 15° and factor (Kp) = 0.97,

$$\text{Factor (Q)} = 0.25 \times \left(\frac{\text{skew}}{100}\right)^2 = 0.25 \times \left(\frac{15}{100}\right)^2 = 0.0056$$

Skew Leakage reactance (Xsk) = 0.2546 × Km × Kp × Q × Kx

= 0.2546 × 778.96 × 0.97 × 0.0056 × 1.0601 = 1.1471 Ω

Total Leakage Reactance of Main Wdg (XLm) = Xs + Xzz + Xe + Xb + Xsk

= 1.7529 + 2.2811 + 2.733 + 1.0237 + 1.1471 = 8.9378 Ω

$$\text{Skew factor (Csk)} = \frac{\sin(\text{skew}/2 \times \pi/180)}{\pi \times (\text{skew}/360)} = \frac{\sin(15/2 \times \pi/180)}{\pi \times (15/360)} = 0.9971$$

Magnetizing Reactance (Xm) = Kx × 0.2546 × Km × Csk

= 1.0601 × 0.2546 × 778.96 × 0.9971 = 209.63 Ω

$$\text{Open Circuit Reactance (X0)} = \text{Xm} + \frac{\text{XLm}}{2} = 209.63 + \frac{8.9378}{2} = 214.1\,\Omega$$

$$\text{Magnetizing Current (Imu)} = \frac{V}{X0} = \frac{230}{214.1} = 1.0743\,\Omega$$

8.7 *(a) Computer Program in "C" in MATLAB for Part-5*

```
%<--------Leakage Reactance(Part-5)--------->

skew=15;Kp=0.97;  %Assumption

Tc=[44  42  37  30  21  10];  Tppm=184;b11=0.7611;  b13=1.3387;
h14=2.1934;  h10=0.07;  b10=0.243;h11=0.091;h20=0.08;  b20=0.075;
S2=30;Nm=736;sp1=1.309;sp2=1.0383;        Lg=0.0427;Aga=94.2478;
Lgd=0.0496;  satf=1.2197;Ncpp=6;S1=24;Kwm=0.7892;P=2;%Input Data

hz=50;L=6;SP=12;Di=10;V=230;%Input Data

sum6=0;for i=1:Ncpp;sum6=sum6+Tc(i)^2;end;

Cx=sum6/Tppm^2*S1/(Kwm^2*4*P);

B11byB13=[.5 .6 .7 .8  .9   1 1.1 1.2 1.3 1.4 1.5 1.6 1.7 1.8];

CONSTANT=[0.63 .53 .46 .41 .36 .33 .3 .27 .25 .23 .21 .2 .18
.17];

plot(B11byB13,CONSTANT);grid;xlabel('b11/b13-->');
ylabel('Constant(A)-->');

title('Ratio of b11/b13 vs Constant(A)');

b11byb13=b11/b13;
```

```
constA=interp1(B11byB13,CONSTANT,b11byb13, 'spline');
Ks1=constA*h14/b13+h10/b10+2*h11/(b10+b11);Ks2=h20/b20+0.623;
Ks=Ks1*Cx+S1/S2*Ks2; Z=Nm;  Kx=2*pi*hz*(Z*Kwm)^2*1e-8;
Xs=Kx*2.512*Ks*L/S1; bt10=sp1-b10;bt20=sp2-b20;
Kzz=(bt10+bt20)^2/(4*(sp1+sp2)); Xzz=Kx*Kzz*0.838 *L/(S1*Lg);
sum7=0;   for i=1:Ncpp;   sum7=sum7+Tc(i)*(SP-(i-1)*2);   end;
ACT1=sum7/Tppm;          ACT=floor(ACT1);De=Di+h14+2*(h10+h11);
Xe=Kx*1.236*De*ACT/S1/P;Km= Aga/(Lgd*satf*P);
AvSP=[4 8 12 16 20 24];BLC=[3.2 1.8 1.4 1.25 1.15 1.1];
plot(AvSP,BLC);grid;xlabel('Average    Slot    Pitch-->');
ylabel('Belt  Leakage  constant)-->');title('Av.Slot  pitch  Vs
Belt Leakage Constant');
Nsp=(S1+S2)/2/P; Kb=interp1(AvSP,BLC,Nsp,'spline');
Xb=0.000929*Km*Kb*Kx; Q=0.25*(skew/100)^2;
Xsk=0.2546*Km*Kp*Q*Kx;  XLm=Xs+Xzz+Xe+Xb+Xsk;
Csk=sin(skew/2*pi/180)/(pi*skew/360);Xm=Kx*0.2546*Km*Csk;
X0=Xm+XLm/2; Imu=V/X0;
```

8.8 Design of Auxiliary Winding (Part-6)

Assuming Transmission ratio (K) = 1.6,

winding factor for Auxiliary Wdg (Kwa) = 0.83, and

Using Concentric Winding, No. of Turns in Aux.wdg (Ta)

$$= K1 \times Tm \times \frac{Kwm}{Kwa} = 1.6 \times 368 \times \frac{0.7892}{0.83} = 559.9 \approx 560 \text{ (Rounded off)}$$

No. of conductors (Na) = Ta × 2 = 560 × 2 = 1120

Turns per pole in Aux.wdg (Tppa) $= \frac{Ta}{P} = \frac{560}{2} = 280$

Pitch factor for Concentric Winding (connected in quadrature to main winding)

Coil No.	Slots spanned	Factor(y1)	Pitch Factor(Kp) = sin(y1 × π/2)
1	13	13/13	Sin(13/13 × π/2) = 1.0
2	11	11/13	Sin(11/13 × π/2) = 0.9709
3	9	9/13	Sin(9/13 × π/2) = 0.8855
4	7	7/13	Sin(7/13 × π/2) = 0.7485
5	5	5/13	Sin(5/13 × π/2) = 0.5681
6	3	3/13	Sin(3/13 × π/2) = 0.3546
			∑Kp = 4.5276

No. of Turns in each coil

Coil No (n)	1	2	3	4	5	6	Σ
Turns(Tn) = Tppa × Kp(n)/4.5276	62	60	55	46	35	22	280

(Example: for 1st coil, Tca1 = 280 × 1.0/4.5276

= 61.84 ≈ 62 (rounded off to nearest integer)

$$\text{Winding factor (Kwa)} = \sum_{n=1}^{n=6} \text{Kp(n)} \times \text{Ta(n)}$$

= [(0.1 × 62) + (0.9709 × 60) + (0.8855 × 55) + (0.7485 × 46) + (0.5681 × 35) + (0.3546 × 22)]/280

= 0.8253 (lies between 0.78 to 0.84 and hence OK)

$$\text{Actual Transmission ratio (K)} = \frac{\text{Ta}}{\text{Tm}} \times \frac{\text{Kwa}}{\text{Kwm}} = \frac{280}{368} \times \frac{0.8253}{0.7892} = 1.5912$$

Length of mean Turn (Lmt) in each coil

Coil No. (n)	1	2	3	4	5	6
Lmt (cm)	67.89	59.3	50.7	42.1	33.5	24.9

Example for Coil. 1

Lmt(1) = 8.4 × (Di + Hs)/S1 × Slots Spanned + (2 × L)

= 8.4 × (10 + 2.2844)/24 × 13 + (2 × 6) = 67.89 cm

$$\text{Mean Length of turn of Auxiliary wdg (Lmta)} = \sum_{n=1}^{n=6} \text{Lmt(n)} \times \text{Ta(n)}$$

= [(67.89 × 62) + (59.3 × 60) + (50.7 × 55) + (42.1 × 46) + (33.5 × 35) + (24.9 × 22)]/280 = 50.76 cm

Assuming area of CS of conductor of Aux. Wdg is 12% of that of main winding, (mf) = 0.12

Area of CS (Aac) = 0.12 × Amc = 0.12 × 1.5837 = 0.19 mm^2;

$$\text{Diameter of conductor (dac1)} = \sqrt{\frac{4 \times \text{Aac}}{\pi}} = \sqrt{\frac{4 \times 0.19}{\pi}} = 0.4919 \text{ mm},$$

Table 8.6 Standard Wire Gauge (SWG) for round Conductor as per British Standard Specification.

SWG no	19	20	21	22	23	24	25	26	27	28	29	30
Dia(mm)	1.02	.914	.813	.711	0.61	.559	.508	.457	.417	.376	.345	.315

From Table. 8.6 suitable dia of bare conductor in aux.wdg (dac) = 0.508 mm

Area of CS of copper in the conductor (Aac) = $\frac{\pi \times dac^2}{4} = \frac{\pi \times 0.508^2}{4} = 0.2027\ mm^2$

Resistance of Aux. wdg at 75° (R1a)

$$= \frac{0.021 \times Lmta \times Ta}{Aac \times 100} = \frac{0.021 \times 50.76 \times 560}{0.2027 \times 100} = 29.4503\ \Omega$$

Rotor Resistance in terms of Aux.Wdg (Rdr) = $K^2 \times R2md$

$$= 1.5912^2 \times 3.5359 = 8.9524\ \Omega$$

Total resistance in terms of Aux.wdg (Rat)

$$= R1a + Rdr = 29.4503 + 8.9524 = 38.4027\ \Omega$$

Total resistance in terms of Main.wdg(Rmt) = Rm + R2md

$$= 2.3279 + 3.5359 = 5.8638\ \Omega$$

Total Leakage Reactance in terms of Aux. wdg (XLa) = $K^2 \times XLm$

$$= 1.5912^2 \times 8.9378 = 22.6292\ \Omega$$

Locked Impedance of main winding (ZLm)

$$= \sqrt{Rmt^2 + XLm^2} = \sqrt{5.8638^2 + 8.9378^2} = 10.6897\ \Omega$$

Locked Impedance of Aux winding (ZLa)

$$= \sqrt{Rat^2 + XLa^2} = \sqrt{38.4027^2 + 22.6292^2} = 44.574\ \Omega$$

Starting Torque (Tqst) = $V^2 \times \frac{P \times K \times R2md}{2 \times \pi \times hz} \times \left(\frac{Rat \times XLm - Rmt \times XLa}{(ZLm \times ZLa)^2}\right)$

$$= 230^2 \times \frac{2 \times 1.5912 \times 3.5359}{2 \times \pi \times 50} \times \left(\frac{38.4027 \times 8.9378 - 5.8638 \times 22.6292}{(10.6897 \times 44.574)^2}\right)$$

$$= 1.7571\ Nw - m$$

Locked Rotor Current in Starting/Aux.Wdg (Isa) = $\frac{V}{ZLa} = \frac{230}{44.574} = 5.16A$

Locked Rotor Current in Main.Wdg (Ism) = $\frac{V}{ZLm} = \frac{230}{10.6897} = 21.516A$

Current Density in Aux.Wdg (cda) = $\frac{Isa}{Aac} = \frac{5.16}{0.2027} = 25.46\ A/mm^2$ (Less than 30 and hence OK)

Weight of Aux Wdg copper (Wcua) = Lmta × Aac × Ta × 8.9×10^{-5}

$$= 50.7574 \times 0.2027 \times 560 \times 8.9 \times 10^{-5} = 0.5127\ Kg$$

8.8 (a) Computer Program in "C" in MATLAB for Part-6

```
%<--------Design of Auxiliary Wdg(Part-6)--------->

K1=1.6;Kwa1=0.83;mf=0.12; %Assumptions

Tm=368;Kwm=0.7892;P=2;Ncpp=6;SP=12;Di=10;Hs=2.2844;S1=24;L=6;

Amc=1.5837;R2md=3.5359;Rm=2.3279;XLm=8.9378;V=230;hz=50;
%Input Data

Ta1=K1*Tm*Kwm/Kwa1;  Ta=ceil(Ta1/2)*2;  Na=Ta*2;  Tppa=Ta/P;

sum11=0; sum12=0;sum13=0;sum14=0; sum15=0;

for  i=1:Ncpp;  y1=(SP+1-(i-1)*2)/(SP+1);  Kp(i)=sin(y1*pi/2);
sum11=sum11+Kp(i); end; sum12=0;

for i=1:Ncpp; Tca1(i)=Tppa*Kp(i)/sum11;  Tca(i)=round(Tca1(i));

sum12=sum12+Tca(i);          end;          corr=Tppa-sum12;
Tca(Ncpp)=corr+Tca(Ncpp); sum12=sum12+corr;

for  i=1:Ncpp;  sum13=sum13+Kp(i)*Tca(i);  end;  Kwa=sum13/Tppa;
K=Ta*Kwa/ Tm/Kwm; sum15=0;

for i=1:Ncpp;  Lmta1(i)=8.4*(Di+Hs)/S1*(SP+1-(i-1)*2)+2*L;

sum15=sum15+Lmta1(i)*Tca(i); end;   Lmta=sum15/Tppa;

Aac1=mf*Amc;  dac1=sqrt(4*Aac1/pi);

DSWG1=[1.02 .914 .813 .711 .610 .559 .508 .457 .417 .376 .345
.315];

for i=1:12;   if dac1 >= DSWG1(i) continue;end;

dac=DSWG1(i);end; Aac=pi*dac^2/4;  R1a=0.021*Lmta*Ta/(Aac*100);

Rdr=K^2*R2md; Rat=R1a+Rdr; Rmt=Rm+R2md; XLa=K^2*XLm; ZLm=sqrt
(Rmt^2+XLm^2); ZLa=sqrt(Rat^2+XLa^2);
Tqst=P*K*R2md*V^2/(2*pi*hz)*(Rat*XLm-Rmt*XLa) /ZLm^2/ZLa^2;

Isa=V/ZLa;      Ism=V/ZLm;      Ist=Isa+Ism;      cda=Isa/Aac;
if cda >30 continue;end;    Wcua=Lmta*Aac*Ta*8.9e-5;
```

8.9 Weights, Losses and Performance using Eq.Ckt (Part-7)

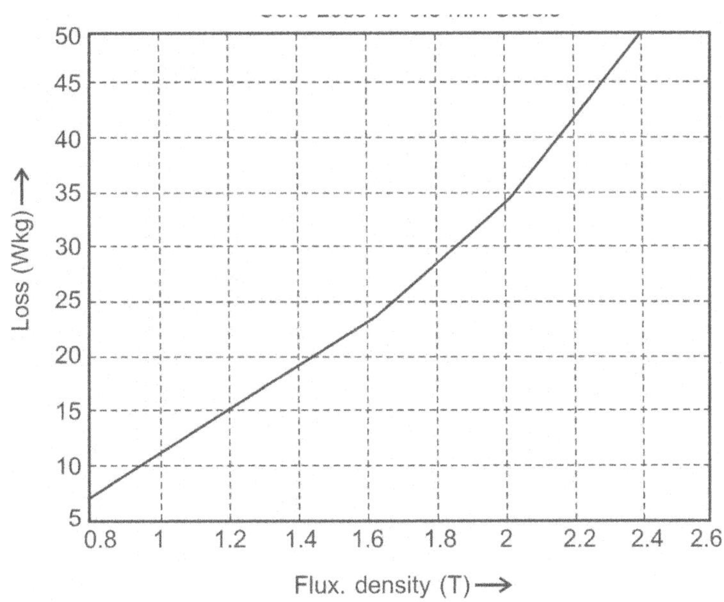

Fig. 8.5 Core loss for 0.5 mm steels.

Stator Tooth

For a flux density (Bt1 = 1.35T), Losses read from Fig. 8.5 (WpKgt) = 17.8254 W/Kg

Wt. of teeth (WtT) = At1 × P × Lfpt1 × 7.8 × 10^{-3}

$$= 39.5064 \times 2 \times 2.3544 \times 7.8 \times 10^{-3} = 1.451 \text{ Kg}$$

Iron loss in Tooth (Pist) = WpKgt × WtT= 17.8254 × 1.451 = 25.8653 W.

Stator Core

For a flux density (Bc1 = 1.1507T),

Losses read from Fig. 8.5 (WpKgc) = 14.0795 W/Kg

Mean Diameter of Core (Dmc) = D0 – dc1 = 20 – 2.6456 = 17.3544 cm

Wt of core (WtC) = π × Dmc × Ac1 × 7.8 × 10^{-3}

$$= \pi \times 17.3544 \times 14.7623 \times 7.8 \times 10^{-3} = 6.2788 \text{ Kg}$$

Iron loss in Core (Pist) = WpKgc × WtC = 14.0705 × 6.2788 = 88.3885 W

Assuming Surface Losses as 90%,

Total Iron Losses (Pi) = 1.9 × (25.8653 + 88.3885) = 217.08 Kg

Assuming 2% of output, Friction and Windage Losses (Pfw)

$$= 0.02 \times \text{hp} \times 746 = 0.02 \times 1 \times 746 = 14.92 \text{ W}$$

Assuming 1 % Insulation Wt, Tot Weight of Active materials (TotWt)

$$= 1.01\times (Wcum + Wcua + WtT + WtC + Wcur + Wtr)$$

$$= 2.474 + 0.513 + 1.451 + 6.28 + 0.35 + 3.3164 = 14.526 \text{ Kg;}$$

Specific Wt. of active material (KgPhp) $= \dfrac{TotWt}{hp} = \dfrac{14.526}{1} = 14.526 Kg/HP$.

Solution by Equivalent Ckt

Synchronous speed (Ns) $= \dfrac{120\times hz}{2} = \dfrac{120\times 50}{2} = 3000 \text{ RPM}$

At rated Speed of 2850 RPM, Slip(s) $= \dfrac{Ns - N}{Ns} = \dfrac{3000 - 2850}{3000} = 0.05 \text{ pu}$

$$Z1 = Rm + j\,(XLm/2) = 2.3279 + j(8.9378/2) = 2.3279 + j(4.4689)$$

$$Z2a = \frac{R2md}{2\times s} + j\frac{XLm}{4} = \frac{3.5359}{2\times 0.05} + j\frac{8.9378}{4} = 35.3593 + j2.2345$$

$$Z2b = j\frac{Xm}{2} = j\frac{209.632}{2} = j104.82$$

$$Z3a = \frac{R2md}{2\times(2-s)} + j\frac{XLm}{4} = \frac{3.5359}{2\times(2-0.05)} + j\frac{8.9378}{4} = 0.9066 + j2.2345$$

$$Z3b = Z2b = j\frac{Xm}{2} = j\frac{209.632}{2} = j104.82$$

$$Z2 = \frac{Z2a\times Z2b}{Z2a + Z2b} = \frac{(35.3593 + j2.2345)\times j104.82}{35.3593 + j2.2345 + j104.82} = 30.564 + j12.2833$$

$$Z3 = \frac{Z3a\times Z3b}{Z3a + Z3b} = \frac{(0.9066 + j2.2345)\times j104.82}{0.9066 + j2.2345 + j104.82} = 0.8691 + j2.1952$$

$$Zeq = Z1 + Z2 + Z3$$

$$= (2.3279 + j4.4689) + (30.564 + j12.2833) + (0.8691 + j2.1952)$$

$$= 33.761 + j18.9474$$

$$I1 = \frac{V}{Zeq} = \frac{230}{33.761 + j18.7494} = 5.1808 - j2.9076 = 5.9409\angle -29.3^{\circ}$$

∴ Total Current (I1m) = 5.9409 A and Power factor (pfx) = cos (– 29.3°) = 0.8721

Current in Forward rotor (If)

$$= I1 \times \frac{Z2b}{Z2a + Z2b} = (5.1808 - j2.9076)$$

$$\times \frac{j104.82}{35.3593 + j2.2345 + j104.82} = 5.4215 - j1.0561$$

Magnitude of If (Ifm) = $\sqrt{5.4215^2 + 1.0561^2} = 5.5234$ A

Current in Back ward rotor (Ib) =

$$I1 \times \frac{Z3b}{Z3a + Z3b} = (5.1808 - j2.9076)$$

$$\times \frac{j104.82}{0.9066 + j2.2345 + j104.82} = 5.0964 - j2.8037$$

Magnitude of Ib (Ibm) = $\sqrt{5.0964^2 + 2.8037^2} = 5.8167$A

Forward Torque (Tf) = $Ifm^2 \times \left(\frac{R2md}{2}\right) \times \frac{1}{s} = 5.5234^2 \times \left(\frac{3.5359}{2}\right) \times \frac{1}{0.05} = 1078.7$

sync-watts Back-ward Torque (Tb)

$$= Ibm^2 \times \left(\frac{R2md}{2}\right) \times \frac{1}{2-s} = 5.8167^2 \times \left(\frac{3.5359}{2}\right) \times \frac{1}{2-0.05} = 30.6757$$

sync-watts

Resultant Torque (Tr) = Tf – Tb = 1078.7 – 30.6 = 1048.1 sync-watts

Net motor Torque (TqFL) = Tr – (Pi + Pfw)

= 1048.1 – (217.08 + 14.92) = 816.1 Watts

Motor Net output (Popt) = TqFL × (1 – s) = 816.1 × (1 – 0.05) = 775.3 W

Efficiency (eff) = $\frac{Popt}{Pinp} = \frac{775.26}{1191.6} \times 100 = 65.06\%$

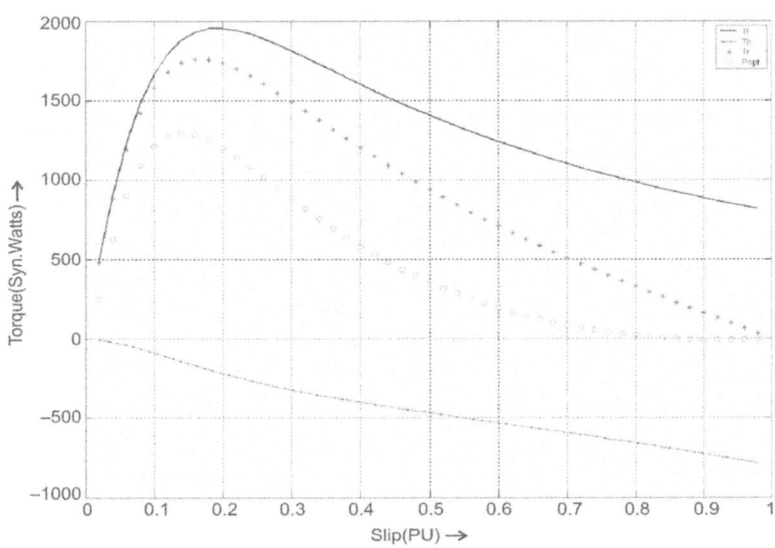

Fig. 8.6 Slip vs. torque curve.

8.9 *(a) Computer Program in "C" in MATLAB for Part-7*

```
%<-----Weights,LossesandPerformance using Eq.Ckt(Part-7)-------
>

Bt1=1.35;At1=39.5064;P=2;Lfpt1=2.3544;Bc1=1.1507;D0=20;
dc1=2.6456;  Ac1=14.7623;hp=1;Wcum=2.474;Wcua=0.513; Wcur=0.35;
Wtr=3.3164;hz=50;       N=2850;Rm=2.3279;XLm=8.9378;R2md=3.5359;
Xm=209.632;V=230;%Input Data

FD=[0.8 1.2 1.6 2 2.4]; WPKG=[7 15 23 34 50]; % for 0.5 mm
plot(FD,WPKG);grid;xlabel('Flux.Density(T)-->');
ylabel('Loss(W/Kg)-->');title('Core Loss for 0.5 mm Steels');

WpKgt=interp1(FD,WPKG,Bt1,'spline');

WtT=At1*P*Lfpt1*7.8*1e-3; Pist=WpKgt* WtT;

WpKgc=interp1(FD,WPKG,Bc1,'spline'); Dmc=D0-dc1;
WtC=pi*Dmc*Ac1*7.8 *1e-3;   Pisc=WpKgc*WtC;

Pi=1.9*(Pist+Pisc);  Pfw=0.02*hp*746;

TotWt=1.01*(Wcum+Wcua+WtT+WtC+Wcur+Wtr);  KgPhp=TotWt/hp;

%------Solution by Equivalent Circuit------->

Ns=120*hz/P;s=(Ns-N)/Ns;

Z1=Rm+XLm/2*j;  Z2a=R2md/(2*s)+XLm/4*j;  Z2b=Xm/2*j;

Z3a=R2md/(2*(2-s))+XLm/4*j;Z3b=Xm/2*j;Z2=Z2a*Z2b/(Z2a+Z2b);
Z3=Z3a*Z3b/(Z3a+Z3b);            Zeq=Z1+Z2+Z3;            I1=V/Zeq;
ang=angle(I1)*180/pi;   pfx=   cos(ang*pi/180);    I1m=abs(I1);
If=I1*Z2b/(Z2a+Z2b);Ifm=abs(If);Ib=I1*Z3b/(Z3a+Z3b);
Ibm=abs(Ib);   Tf=Ifm^2*R2md/(2*s);     Tb=Ibm^2*R2md/(2*(2-s));
Tr=Tf-Tb; TqFL=Tr-(Pi+Pfw);

Popt=TqFL*(1-s); Pinp=V*I1m*pfx; eff=Popt/Pinp*100;

%---Extra Program for Plotting Slip vs Torque ------------->

for          kk=1:49;           s=kk*0.02;           Z1=Rm+XLm/2*j;
Z2a=R2md/(2*s)+XLm/4*j;Z2b=Xm/2*j;Z3a=R2md/(2*(2-s))+  XLm/4*j;
Z3b=Xm/2*j;      Z2=Z2a*Z2b/(Z2a+Z2b);      Z3=Z3a*Z3b/(Z3a+Z3b);
Zeq=Z1+Z2+Z3;I1=V/Zeq;ang=angle(I1)*180/pi;
pfx=cos(ang*pi/180);      I1m=abs(I1);       If=I1*Z2b/(Z2a+Z2b);
Ifm=abs(If);          Ib=I1*Z3b/(Z3a+Z3b);            Ibm=abs(Ib);
Tf=Ifm^2*R2md/(2*s); Tb=Ibm^2*R2md/(2*(2-s));          Tr=Tf-Tb;
TqFL=Tr-(Pi+Pfw);       Popt=TqFL*(1-s);SS(kk)=s;       TF(kk)=Tf;
TB(kk)=-Tb;TR(kk)=Tr;PO(kk)=Popt;end;

plot(SS,TF,'-',SS,TB,'-.',SS,TR,'*',SS,PO,'o');grid;
legend('Tf','Tb','Tr','Popt');xlabel('Slip(PU)-->');

ylabel('Torque(syn.Watts)-->');title('Slip vs Torque Curve');

%---------End of Program--------------->
```

8.10 Computer Program in "C" in MATLAB for Complete Design

```
%Design of 1-ph IM

hp=1;V=230;P=2;N=2850;hz=50; %Main Data

kt=1.42;LbyD0=0.3;ki=0.93;S1=24;S2=30;Bt=1.35;%Assumptions

h10=0.07;h11=1.3*h10;Bt=1.35;Bc=1.15;cdma1=4.5; %Assumptions

f2=fopen('Total_1hp_Output.m','w');

fprintf(f2,'Design of 1-ph Ind.Motor(Split-Phase)of %3.1f
HP,%3.0f Volts,%2.0f HZ:-\n',hp,V,hz);

fprintf(f2,'==================================================
============\n');

fprintf(f2,'<----------Calculation of Main Dimensions(Part-1)--
----------->\n');

HZ=[25 30 40 45 50 60]; KF=[0.865 .885 .923 .94 .96 1];
kf=interp1 (HZ,KF,hz,'spline');

PO LES=[2 4 6 8]; OPTC=[0.32 0.22 .19 .17];
C0=interp1(POLES,OPTC,P,'spline');

DIBYD0=[.5 .59 .64 67];DibyD0=interp1(POLES,DIBYD0,P,'spline');

D0sqL=16.5*C0*hp/N*kf*kt*1e6; D01=(D0sqL/LbyD0)^(1/3);
D0=floor(D01); L=LbyD0*D0; Di1=DibyD0*D0; Di=ceil(Di1);SP=S1/P;

if P==2 bt1a=(1.1+0.032*Di)*Di/S1;bt1=floor(bt1a*100)/100;end;

SS=[24 36 48];B10=[0.068+0.0175*Di 0.038+0.0175*Di 0.0175*Di];
b10=interp1(SS,B10,S1,'spline'); dc1=Bt/Bc*(S1*bt1)/(pi*P);
b11=pi*(Di+2*h10+2*h11)/S1-bt1;

h14=DibyD0*(D0-Di)-(h10+h11+dc1);alfa=180/S1;
b13=b11+2*h14*tan(alfa*pi/ 180); Lg=0.013+0.0042*Di/sqrt(P);

fprintf(f2,'No of Poles                      =%6d\n',P);

fprintf(f2,'Output Coefft                    =%6.3f\n',C0);

fprintf(f2,'Freq.Constant                    =%6.3f\n',kf);

fprintf(f2,'Type.Constant(Split-phase)  =%6.3f\n',kt);

fprintf(f2,'D0sqL (Cu.cm)               =%6.1f\n',D0sqL);

fprintf(f2,'Outer dia of Stator (cm)    =%6.1f\n',D0);

fprintf(f2,'Inner dia of Stator (cm)    =%6.1f\n',Di);

fprintf(f2,'Gross Length of Stator(cm)  =%6.1f\n',L);

fprintf(f2,'No of Stator Slots          =%6d\n',S1);

fprintf(f2,'St.Slot opening (cm)        =%6.3f\n',b10);

fprintf(f2,'St.Slot depth of tip (cm)   =%6.3f\n',h10);
```

```
fprintf(f2,'St.Slot depth of mouth(cm)   =%6.3f\n',h11);
fprintf(f2,'St.Slot tooth width (cm)     =%6.3f\n',bt1);
fprintf(f2,'St.Core depth (cm)           =%6.3f\n',dc1);
fprintf(f2,'Slot width at topandbottom(cm)
  =%6.3fand%6.3f\n',b11,b13);
fprintf(f2,'St.Slot depth (cm)           =%6.3f\n',h14);
fprintf(f2,'Air-gap Length (cm)          =%6.3f\n',Lg);
fprintf(f2,'<-----------Design of Main Winding(Part-2)---------
----------->\n');
Ncpp=S1/P/2;sum1=0;sum2=0;sum3=0;sum4=0;sum5=0;
For   i=1:Ncpp;y1=(SP-(i-1)*2)/(SP+1);      Kp(i)=sin(y1*pi/2);
sum1=sum1+Kp(i); end;
for i=1:Ncpp; PTm(i)=Kp(i)/sum1*100; sum2=sum2+PTm(i);end;
for i=1:Ncpp;sum3=sum3+Kp(i)*PTm(i);end; Kwm=sum3/sum2;
Li=ki*L;At=bt1*Li*SP;FI=0.637*At*Bt*1e-4;
Tm1=0.95*V/(4.44*hz*FI*Kwm);Tm=ceil(Tm1/2)*2;Nm=Tm*2;
Tppm=Tm/P;
for i=1:Ncpp;Tc1(i)=Tppm*PTm(i)/100;Tc(i)=round(Tc1(i));
sum4=sum4+Tc(i);end;corr=Tppm-sum4;      Tc(Ncpp)=corr+Tc(Ncpp);
sum4=sum4+corr;
HP=[.05 .1 .25 .5 .75 1];EFF=[.35 .44 .58 .65 .68 .7];
PF=[.45 .5 .56 .62 .64 .66];effa=interp1(HP,EFF,hp,'spline');
pfa=interp1(HP,PF,hp,'spline');          Im=hp*746/(effa*V*pfa);
Amc1=Im/cdma1; dmc1=sqrt(4*Amc1/pi);
DSWG=[5.38 4.88 4.47 4.06 3.66 3.25 2.95 2.64 2.31 2.03 1.83
1.63 1.42 1.22 1.02];
for i=1:15; if dmc1 >= DSWG(i) continue; end; dmc=DSWG(i); end;
dmcins=dmc+0.075; Amc=pi*dmc^2/4; cdm=Im/Amc;
if cdm <2.8 || cdm >4.5 continue;end;
Socm=Tc(1)*pi*dmcins^2/4/100;Asg=(h11*(b10+b11)+h14*(b11+b13))
/2;
Sfsm=Socm/Asg; if Sfsm >0.5 continue;end;
Hs=h11+h14;for i=1:Ncpp;Lmt(i)=8.4*(Di+Hs)/S1*(SP-(i-1)*2)+2*L;
sum5=sum5+Lmt(i)*Tc(i);end;Lmtm=sum5/sum4;
Rm=0.021*Lmtm*Tm/(Amc*100);Pcus=Im^2*Rm;
Wcum=Lmtm*Amc*Tm*8.9e-5;
```

```
fprintf(f2,'Pitch-Factors                    of                    %2d
coils:%5.3f,%5.3f,%5.3f,%5.3f,%5.3fand%5.3f\n', Ncpp,Kp);

fprintf(f2,'Pecentage-Turns                    of                    %2d
coils:%4.1f+%4.1f+%4.1f+%4.1f+%4.1f+%4.1f=
%4.0f(perc)\n',Ncpp,PTm,sum2);

fprintf(f2,'Winding Factor =                    %6.3f\n',Kwm);

fprintf(f2,'Flux/Pole (Wb) =                    %6.4f\n',FI);

fprintf(f2,'No          of          Turns          in          %2d
coils:%3.0f+%3.0f+%3.0f+%3.0f+%3.0f+%3.0f=
%3.0f\n',Ncpp,Tc,Tppm);

fprintf(f2,'Current in Main Wdg   (A) =         %6.3f\n',Im);

fprintf(f2,'Conductor: dia=%6.3fcmandCS area(total)=
%5.3fsq.mm\n',dmc,Amc);

fprintf(f2,'Current dens in Main wdg(A/sq.mm)=
%6.3f(desirable:<4.5)\n',cdm);

fprintf(f2,'Gross Slot Area(sq.cm) =%6.3f\n',Asg);

fprintf(f2,'Space       Occupied       in       Coil:1       by
%2.0fstrands(sq.cm)=%6.4f\n',Tc(1),Socm);

fprintf(f2,'Space factor of the slot =           %6.3f(
desirable: < 0.5)\n',Sfsm);

fprintf(f2,'Mean        Turn        Lengths        of        %2d
coils(cm):%4.1f,%4.1f,%4.1f,%4.1f,%4.1fand% 4.1f\n',Ncpp,Lmt);

fprintf(f2,'Mean Turn Lengths of Total Wdg(cm):%5.2f\n',Lmtm);

fprintf(f2,'Resistance of Main Wdg(ohm)=       %6.4f\n',Rm);

fprintf(f2,'Copper Losses of Main Wdg(W)=      %6.3f\n',Pcus);

fprintf(f2,'Wt of Main-Wdg-Copper   (Kg)=      %6.3f\n',Wcum);

fprintf(f2,'<----------Design of Rotor(Part-3)-----------------
----------->\n');

b20=0.075;h20=0.08;Dm=0.07;                  %assumption

Dr=Di-2*Lg;bt2=0.95*S1*bt1/S2;

r21=(pi*(Dr-2*h20)-S2*bt2)/(2*(S2+pi)); drc1=2* r21-0.038;

for i=1:15;if drc1*10>=DSWG(i) continue; end; drc=DSWG(i); end;

Ab=pi*drc^2/4;dc2=0.95*dc1;Lb=L+1;Nb=S2;Aer=Ab/pi*Nb/P;

R2md=P*Nm^2*Kwm^2*0.021*(Lb/(100*Ab*Nb)+2/pi*Dm/P^2/Aer);

I2d=Im*pfa;Pcur=I2d^2*R2md;                 Wcur=0Lb*Ab*Nb*8.9e-5;
Wtr=((pi*Dr^2/4)-(S2*pi*r21^2))*L*7.8*1e-3;

fprintf(f2,'Outer dia of Rotor (cm) =           %6.4f\n',Dr);
```

```
fprintf(f2,'Gross Length of Rotor (cm) =          %6.1f\n',L);
fprintf(f2,'No of Rotor Slots =                   %6d\n',S2);
fprintf(f2,'Tooth width at min section(cm)=       %6.3f\n',bt2);
fprintf(f2,'Rt.Slot opening (cm) =                %6.3f\n',b20);
fprintf(f2,'Rt.Slot depth of tip  (cm)  =         %6.3f\n',h20);
fprintf(f2,'Radius of round slot  (cm) =          %6.3f\n',r21);
fprintf(f2,'Conductor dia of Rotor(cm) =          %6.4f\n',drc);
fprintf(f2,'Rt.Core depth (cm) =                  %6.3f\n',dc2);
fprintf(f2,'Rot.Resist   refered   to   Main   Wdg(ohm)   =
%6.4f\n',R2md);
fprintf(f2,'Equivalent Rot.Current(A) =           %6.3f\n',I2d);
fprintf(f2,'Rotor Copper Losses (W) =             %6.3f\n',Pcur);
fprintf(f2,'Wt of Rot-Copper (Kg) =               %6.3f\n',Wcur);
fprintf(f2,'<---------------Amp-Turns Calculation (Part-4)-----
----------->\n');
SobyAg1=b10/Lg;SOBYAG=[0 1 2 3 4 5 6 7 8 9 10 11 12];
CC=[0 .19 .34 .46 .55 .61 .66 .71 .75 .79 .82 .86 .89];
%semi-closed-slot
K01=interp1(SOBYAG,CC,SobyAg1,'spline');
sp1=pi*Di/S1;Kgs=sp1/(sp1-b10* K01);
SobyAg2=b20/Lg;K02=interp1(SOBYAG,CC,SobyAg2,'spline');
sp2=pi*Dr/S2;Kgr=sp2/(sp2-b20*K02);  Kag=Kgs*Kgr;  Lgd=Kag*Lg;
Aga=pi*Di/P*L;Bag=FI*1e4/0.637/Aga; ATag=0.796*Bag*Kag*Lgd*1e4;
BB= [ .1  .2  .3  .4  .5  .6  .7  .8  .9  1  1.1  1.2  1.3
1.4  1.5  1.6  1.7  1.8  1.9  2.0  ]; %Tesla
HH= [ 18  28  38  48  67  80  95 120 140 180  220  295  390
580 1000 2200 5000  9000 16000 24000];%forLohys
%semilogx(HH,BB,'-',H,BB,'-.',Hcs,BB,'*');grid;
xlabel('AT/m-->');ylabel('Fluxdensity(T)-->');
%title('Magnetization curves');
At1=Li*bt1*S1/P;Bt1=FI*1e4/0.637/At1;
if Bt1 <1.3 || Bt1 >1.7 continue;end;
att1=interp1(BB,HH,Bt1,'spline');Lfpt1=h14+h11+h10;
ATT1=att1*Lfpt1/100;
Ac1=Li*dc1;Bc1=FI*1e4/2/Ac1;atc1=interp1(BB,HH,Bc1,'spline');
Lfpc1=pi*(D0-dc1)/2/P;ATC1=atc1*Lfpc1/100;
```

```
At2=Li*bt2*S2/P;Bt2=FI*1e4/0.637/At2;

att2=interp1(BB,HH,Bt2,'spline');Lfpt2=2*r21+h20;

ATT2=att2*Lfpt2/100;

Ac2=Li*dc2;Bc2=FI*1e4/2/Ac2;atc2=interp1(BB,HH,Bc2,'spline');

Lfpc2=pi*(2*r21+h20+dc2)/2/P;ATC2=atc2*Lfpc2/100;

ATT=ATag+ATT1+ATC1+ATT2+ATC2;satf=ATT/ATag;

fprintf(f2,'Stator Gap Coefft  (Kgs) =        %6.4f\n',Kgs);

fprintf(f2,'Rotor Gap Coefft  (Kgr) =         %6.4f\n',Kgr);

fprintf(f2,'Air Gap Coefft (Kag) =            %6.4f\n',Kag);

fprintf(f2,'Air-Gap Effectivelength(cm) =     %6.4f\n',Lgd);

fprintf(f2,'Air-Gap:               Flux-density(T)=%6.4f   and
AmpTurns=%6.1f\n',Bag,ATag);

fprintf(f2,'Stator      tooth:Flux-density(T)=%6.4f      and
AmpTurns=%6.1f\n',Bt1,ATT1);

fprintf(f2,'Stator     Core     :Flux-density(T)=%6.4f     and
AmpTurns=%6.1f\n',Bc1,ATC1);

fprintf(f2,'Rotor     tooth:     Flux-density(T)=%6.4f     and
AmpTurns=%6.1f\n',Bt1,ATT2);

fprintf(f2,'Rotor    Core    :    Flux-density(T)=%6.4f    and
AmpTurns=%6.1f\n',Bc1,ATC2);

fprintf(f2,'Total   AmpTurns          =%6.1f   and   Saturation
factor=%6.3f\n',ATT,satf);

fprintf(f2,'<-----------------Leakage Reactances(Part-5)-------
----------->\n');

skew=15;Kp=0.97;  %Assumption

sum6=0;fori=1:Ncpp;sum6=sum6+Tc(i)^2;end;

Cx=sum6/Tppm^2*S1/(Kwm^2*4 *P);

B11byB13=[.5 .6 .7 .8  .9  1 1.1 1.2 1.3 1.4 1.5 1.6 1.7 1.8];

CONSTANT=[0.63 .53 .46 .41 .36 .33 .3 .27 .25 .23 .21 .2 .18
.17];                                     b11byb13=b11/b13;
constA=interp1(B11byB13,CONSTANT,b11byb13,'spline');

Ks1=constA*h14/b13+h10/b10+2*h11/(b10+b11);Ks2=h20/b20+0.623;
Ks=Ks1*Cx+S1/S2*Ks2;      Z=Nm;      Kx=2*pi*hz*(Z*Kwm)^2*1e-8;
Xs=Kx*2.512*Ks*L/S1; Lmd1=pi*Di/S1; Lmd2=pi*Dr/S2;

bt10=sp1-b10;bt20=sp2-b20;   Kzz=(bt10+bt20)^2/(4*(Lmd1+Lmd2));
Xzz=Kx*Kzz*0.838*L/ (S1*Lg);
```

```
sum7=0;fori=1:Ncpp;sum7=sum7+Tc(i)*(SP-(i-1)*2);end;
ACT1=sum7/    Tppm;ACT    =floor(ACT1);    De=Di+h14+2*(h10+h11);
Xe=Kx*1.236*De*ACT/S1/P; Km=Aga/(Lgd*satf*P);

AvSP=[4 8 12 16 20 24];BLC=[3.2 1.8 1.4 1.25 1.15 1.1];
Nsp=(S1+S2)/2/P; Kb=interp1(AvSP,BLC,Nsp,'spline');
Xb=0.000929*Km*Kb*Kx; Q=0.25*(skew/100)^2;
Xsk=0.2546*Km*Kp*Q*Kx; XLm=Xs+Xzz+Xe+Xb+Xsk;

Csk=sin(skew/2*pi/180)/(pi*skew/360);       Xm=Kx*0.2546*Km*Csk;
X0=Xm+XLm/2;Imu=V/X0;

fprintf(f2,'Slot-Leakage Reactance(Xs) =   %6.4f ohms\n',Xs);

fprintf(f2,'Zig-Zag      Reactance(Xzz)=   %6.4f ohms\n',Xzz);

fprintf(f2,'End-Leakage  Reactance(Xe) =   %6.4f ohms\n',Xe);

fprintf(f2,'Belt-Leakage Reactance(Xb) =   %6.4f ohms\n',Xb);

fprintf(f2,'Skew-Leakage Reactance(Xsk) =  %6.4f ohms\n',Xsk);

fprintf(f2,'Total-LeakageReactance(XLm) =  %6.4f ohms\n',XLm);

fprintf(f2,'Magnetizing  Reactance(Xm) =   %6.3f ohms\n',Xm);

fprintf(f2,'Open-Circuit Reactance(Xe) =   %6.3f ohms\n',X0);

fprintf(f2,'Magnetizing  Current (Imu) =   %6.4f amps\n',Imu);

fprintf(f2,'<--------------Design of Auxiliary Winding(Part-6)-
----------->\n');

K1=1.6;Kwa=0.83;mf=0.12; %Assumptions

Ta1=K1*Tm*Kwm/Kwa;Ta=ceil(Ta1/2)*2;Na=Ta*2;Tppa=Ta/P;

sum11=0; sum12=0;sum13=0;sum14=0;sum15=0;

fori=1:Ncpp;y1=(SP+1-(i-1)*2)/(SP+1);Kp(i)=sin(y1*pi/2);
sum11=sum11+Kp(i); end;

sum12=0;for i=1:Ncpp;Tca1(i)=Tppa*Kp(i)/sum11;

Tca(i)=round(Tca1 (i));

sum12=sum12+Tca(i);end;corr=Tppa-sum12;
Tca(Ncpp)=corr+Tca(Ncpp);sum12=sum12+corr;

fori=1:Ncpp;sum13=sum13+Kp(i)*Tca(i);end;Kwa=sum13/Tppa;
K=Ta*Kwa/Tm /Kwm;sum15=0;

for i=1:Ncpp; Lmta1(i)=8.4*(Di+Hs)/S1*(SP+1-(i-1)*2)+2*L;

sum15=sum15+Lmta1(i)*Tca(i);end;Lmta=sum15/Tppa;

Aac1=mf*Amc;dac1=sqrt(4*Aac1/pi);

DSWG1=[1.02 .914 .813 .711 .610 .559 .508 .457 .417 .376 .345
.315];

for i=1:12; if dac1 >= DSWG1(i) continue;end;
```

```
dac=DSWG1(i);end;Aac=pi*dac^2/4; R1a=0.021*Lmta*Ta/(Aac*100);

Rdr=K^2*R2md;Rat=R1a+Rdr;Rmt=Rm+R2md;XLa=K^2*XLm;
ZLm=sqrt(Rmt^2+XLm^2);                    ZLa=sqrt(Rat^2+XLa^2);
Tqst=P*K*R2md*V^2/(2*pi*hz)*(Rat*XLm-Rmt*XLa)/ ZLm^2/ZLa^2;

Isa=V/ZLa;  Ism=V/ZLm;  Ist=Isa+Ism;  cda=Isa/Aac;if  cda  >30
continue; end; Wcua=Lmta*Aac*Ta*8.9e-5;

fprintf(f2,'Pitch-Factors                of               %2d
coils:%5.3f,%5.3f,%5.3f,%5.3f,%5.3fand%5.3f\n', Ncpp,Kp);

fprintf(f2,'No        of        Turns        in           %2d
coils:%3.0f+%3.0f+%3.0f+%3.0f+%3.0f+%3.0f=      %3.0f\n',Ncpp,
Tca,Tppa);

fprintf(f2,'Winding Factor %6.3f(desirable:0.78to0.84)\n',Kwa);

fprintf(f2,'Transformation   Ratio(K)   =   %6.3f(desirable:1.5
to2)\n',K);

fprintf(f2,'Mean        Turn        Lengths        of        %2d
coils(cm):%4.1f,%4.1f,%4.1f,%4.1f,%4.1f                     and
%4.1f\n',Ncpp,Lmta1);

fprintf(f2,'Mean Turn Lengths of Total Wdg(cm) =%5.2f\n',Lmta);

fprintf(f2,'Conductor: of dia=%6.3fcm and CS area(total)=%6.3f
sq.mm\n',dac,Aac);

fprintf(f2,'Resistance of Aux  Wdg(ohm) =      %6.4f\n',R1a);

fprintf(f2,'Rot.Res in terms of aux wdg(ohm)=  %6.4f\n',Rdr);

fprintf(f2,'Tot.Res in terms of aux wdg(ohm) = %6.4f\n',Rat);

fprintf(f2,'Tot.Res in terms of Main wdg(ohm)= %6.4f\n',Rmt);

fprintf(f2,'Tot.Leak.React in terms of aux wdg(ohm) =
%6.4f\n',XLa);

fprintf(f2,'Locked Imped of main Wdg(ohm) =    %6.4f\n',ZLm);

fprintf(f2,'Locked Imped of Aux  Wdg(ohm) =    %6.4f\n',ZLa);

fprintf(f2,'Starting Torque(N-m) =             %6.3f\n',Tqst);

fprintf(f2,'Locked Rot current in aux wdg(A) = %6.2f\n',Isa);

fprintf(f2,'Locked Rot current in Main wdg(A)= %6.2f\n',Ism);

fprintf(f2,'Total Starting Current (A) =       %6.3f\n',Ist);

fprintf(f2,'Conductor Dia (mm) =               %6.3f\n',dac);

fprintf(f2,'Current dens in Aux wdg(A/sq.mm) =
%6.3f(desirable:<30)\n',cda);

fprintf(f2,'Wt of Aux-Wdg-Copper   (Kg) =      %6.3f\n',Wcua);
```

```
fprintf(f2,'<-------Weights,LossesandPerformance         using
Eq.Ckt(Part-7)------->\n');
FD=[0.8 1.2 1.6 2 2.4]; WPKG=[7 15 23 34 50]; % for 0.5 mm
WpKgt=interp1(FD,WPKG,Bt1,'spline');WtT=At1*P*Lfpt1*7.8*1e-3;
Pist=WpKgt* WtT; WpKgc=interp1(FD,WPKG,Bc1,'spline');
Dmc=D0-dc1;WtC=pi*Dmc*Ac1* 7.8*1e-3;
Pisc=WpKgc*WtC;Pi=1.9*(Pist+Pisc);Pfw=0.02*hp*746;
TotWt=1.01*(Wcum+Wcua+WtT+WtC+Wcur+Wtr);KgPhp=TotWt/hp;
fprintf(f2,'Stator     tooth:Weight(Kg)=%6.4f     and     Iron
Losses=%6.2f\n',WtT,Pist);
fprintf(f2,'Stator    Core:    Weight(Kg)=%6.4f    and     Iron
Losses=%6.2f\n',WtC,Pisc);
fprintf(f2,'Total Iron  Losses(W) =                 %6.2f\n',Pi);
fprintf(f2,'Friction and Windage Losses(W) =  %6.2f\n',Pfw);
fprintf(f2,'Total       Weight(Kg)=%6.3f      and      Specific
Wt(Kg/HP)=%6.2f\n',TotWt,KgPhp);
%------Solution by Equivalent Circuit------->
Ns=120*hz/P;s=(Ns-N)/Ns;
Z1=Rm+XLm/2*j;Z2a=R2md/(2*s)+XLm/4*j;Z2b=Xm/2*j;Z3a=R2md/(2*(2-
s))+XLm/4*j;Z3b=Xm/2*j;
Z2=Z2a*Z2b/(Z2a+Z2b);Z3=Z3a*Z3b/(Z3a+Z3b);Zeq=Z1+Z2+Z3;
I1=V/Zeq;
ang=angle(I1)*180/pi;pfx=cos(ang*pi/180);I1m=abs(I1);If=I1*Z2b/
(Z2a+Z2b);Ifm=abs(If);
Ib=I1*Z3b/(Z3a+Z3b);Ibm=abs(Ib);Tf=Ifm^2*R2md/(2*s);Tb=Ibm^2*R2
md/(2*(2-s));
Tr=Tf-Tb;TqFL=Tr-(Pi+Pfw);Popt=TqFL*(1-s);Pinp=V*I1m*pfx;
eff=Popt/Pinp*100;
fprintf(f2,'Sync.Speed(RPM) =%4.0f   and Rated Slip(pu) =
%6.3f\n',Ns,s);
fprintf(f2,'Line Current (Amps)                = %6.2f\n',I1m);
fprintf(f2,'Forward Torque  (syn.Watts)        = %6.2f\n',Tf);
fprintf(f2,'Backword Torque (syn.Watts)        = %6.2f\n',Tb);
fprintf(f2,'Output at Rated Speed (Watts)      = %6.2f\n',Popt);
fprintf(f2,'PF at Rated Speed                  = %6.3f\n',pfx);
fprintf(f2,'Efficiency at Rated Speed(perc)   = %6.2f\n',eff);
fprintf(f2,'<-------------------------End of output------------
-------->\n\n');
fclose(f2);
```

8.10 *(a) Computer Output Results for Complete Design*

Design of 1-ph Ind.Motor(Split-Phase)of 1.0 HP,230 Volts,50 HZ:

==

<----------Calculation of Main Dimensions(Part-1)------------->

No. of Poles	=	2
Output Coefft.	=	0.320
Freq.Constant	=	0.960
Type.Constant(Split-phase)	=	1.420
D0sqL (Cu.cm)	=	2525.5
Outer dia of Stator (cm)	=	20.0
Inner dia of Stator (cm)	=	10.0
Gross Length of Stator(cm)	=	6.0
No. of Stator Slots	=	24
St.Slot opening (cm)	=	0.243
St.Slot depth of tip (cm)	=	0.070
St.Slot depth of mouth(cm)	=	0.091
St.Slot tooth width (cm)	=	0.590
St.Core depth (cm)	=	2.646
Slot width at top and bottom(cm)	=	0.761and 1.339
St.Slot depth (cm)	=	2.193
Air-gap Length (cm)	=	0.043

<-----------Design of Main Winding(Part-2)------------------->

Pitch-Factors of 6 coils: 0.993, 0.935, 0.823, 0.663, 0.465 and 0.239

Pecentage-Turns of 6 coils:24.1+22.7+20.0+16.1+11.3+ 5.8= 100(perc)

Winding Factor	=	0.789
Flux/Pole (Wb)	=	0.0034
No. of Turns in 6 coils: 44+ 42+ 37+ 30+ 21+ 10	=	184
Current in Main Wdg (A)	=	7.021
Conductor: dia = 1.420 cm and CS area (total)	=	1.584 sq.mm
Current dens in Main wdg(A/sq.mm)	=	4.433(desirable:<4.5)
Gross Slot Area(sq.cm)	=	2.349
Space Occupied in Coil:1 by 44strands(sq.cm)	=	0.7724
Space factor of the slot	=	0.329(desirable: < 0.5)

Mean Turn Lengths of 6 coils(cm):63.6,55.0,46.4,37.8,29.2and20.6

Mean Turn Lengths of Total Wdg(cm):47.70

Resistance of Main Wdg(ohm) = 2.3279

Copper Losses of Main Wdg(W) = 114.737

Wt. of Main-Wdg-Copper (Kg) = 2.474

<----------Design of Rotor(Part-3)--------------------------->

Outer dia of Rotor (cm) = 9.9146

Gross Length of Rotor (cm) = 6.0

No. of Rotor Slots = 30

Tooth width at min section(cm) = 0.448

Rt.Slot opening (cm) = 0.075

Rt.Slot depth of tip (cm) = 0.080

Radius of round slot (cm) = 0.259

Conductor dia of Rotor(cm) = 4.8800

Rt.Core depth (cm) = 2.513

Rot.Resist refered to Main Wdg(ohm) = 3.5359

Equivalent Rot. Current(A) = 4.634

Rotor Copper Losses (W) = 75.915

Wt of Rot-Copper (Kg) = 0.350

<--------------Amp-Turns Calculation (Part-4)--------------->

Stator Gap Coefft (Kgs) = 1.1359

Rotor Gap Coefft (Kgr) = 1.0226

Air Gap Coefft (Kag) = 1.1616

Air-Gap Effectivelength(cm) = 0.0496

Air-Gap: Flux-density(T) = 0.5659 and AmpTurns = 259.5

Stator tooth:Flux-density(T) = 1.3500 and AmpTurns = 11.0

Stator Core :Flux-density(T) = 1.1507 and AmpTurns = 34.7

Rotor tooth: Flux-density(T) = 1.3500 and AmpTurns = 3.8

Rotor Core : Flux-density(T) = 1.1507 and AmpTurns = 7.4

Total AmpTurns = 316.5 and Saturation factor = 1.220

<----------------Leakage Reactances(Part-5)----------------->

Slot-Leakage Reactance(Xs) = 1.7529 ohms

Zig-Zag Reactance(Xzz) = 2.2811 ohms

End-Leakage Reactance(Xe) = 2.7330 ohms

Belt-Leakage Reactance(Xb) = 1.0237 ohms

Skew-Leakage Reactance(Xsk) = 1.1471 ohms

Total-LeakageReactance(XLm) = 8.9378 ohms

Magnetizing Reactance(Xm) = 209.632 ohms

Open-Circuit Reactance(Xe) = 214.101 ohms

Magnetizing Current (Imu) = 1.0743 amps

<--------------Design of Auxiliary Winding(Part-6)------------>

Pitch-Factors of 6 coils:1.000,0.971,0.885,0.749,0.568and0.355

No of Turns in 6 coils: 62+ 60+ 55+ 46+ 35+ 22=280

Winding Factor = 0.825(desirable:0.78to0.84)

Transformation Ratio(K) = 1.591(desirable:1.5to2)

Mean Turn Lengths of 6 coils(cm):67.9,59.3,50.7,42.1,33.5and24.9

Mean Turn Lengths of Total Wdg(cm) = 50.76

Conductor: of dia= 0.508cm and CS area(total) = 0.203 sq.mm

Resistance of Aux Wdg(ohm) = 29.4503

Rot.Res in terms of aux wdg(ohm) = 8.9524

Tot.Res in terms of aux wdg(ohm) = 38.4027

Tot.Res in terms of Main wdg(ohm) = 5.8638

Tot.Leak.React in terms of aux wdg(ohm) = 22.6292

Locked Imped of main Wdg(ohm) = 10.6897

Locked Imped of Aux Wdg(ohm) = 44.5740

Starting Torque(N-m) = 1.757

Locked Rot current in aux wdg(A) = 5.16

Locked Rot current in Main wdg(A) = 21.52

Total Starting Current (A) = 26.676

Conductor Dia (mm) = 0.508

Current dens in Aux wdg(A/sq.mm) = 25.458(desirable:<30)

Wt. of Aux-Wdg-Copper (Kg) = 0.513

<-------Weights,LossesandPerformance using Eq.Ckt(Part-7)------->

Stator tooth:Weight(Kg)=1.4510 and Iron Losses	=	25.87
Stator Core: Weight(Kg)=6.2778 and Iron Losses	=	88.39
Total Iron Losses(W)	=	217.08
Friction and Windage Losses(W)	=	14.92
Total Weight(Kg)=14.526 and Specific Wt(Kg/HP)	=	14.53
Sync.Speed(RPM) =3000 and Rated Slip(pu)	=	0.050
Line Current (Amps)	=	5.94
Forward Torque (syn.Watts)	=	1078.75
Backword Torque (syn.Watts)	=	30.68
Output at Rated Speed (Watts)	=	775.27
PF at Rated Speed	=	0.872
Efficiency at Rated Speed(perc)	=	65.06

<------------------------End of output-------------------->

8.11 Modifications to be done in the above Program to get Optimal Design

1. Insert "for" Loops for the following parameters to iterate the total program between min and max permissible limits for selecting the feasible design variants:
   (a) No. of Poles
   (b) Stator Slots
   (c) Main Wdg Current Densities
   (d) Tooth Flux density
   (e) Core Flux density
   (f) Transmission ratio.
2. Insert also minimum or maximum range of required objective functional values as constraint values, for example (a) Efficiency, (b) Kg/HP, (c) Starting Torque (d) Max output.
3. Run the program to get various possible design variants.

Note: From the feasible design variants printed in the output, select that particular design fulfilling the objective of optimal parameter for the design.

8.12 Computer Program in "C" in MATLAB for Optimal Design

```
%Design of 1-ph IM :
hp=1;V=230;N=2850;hz=50; %Main Data
f2=fopen('Optimal_1hp_Output.m','w');
fprintf(f2,' Design of 1-ph Ind.Motor of %3.1f HP,%4.0f
Volts,%4.0f HZ:-\n',hp,V,hz);
fprintf(f2,' Sn  P  D0   L   Di  S1  Tm  dmc  cdm S2  Bt1  Bc1
K  Ta  dac  cda  Ist  Tqst  pfx  Popt  eff  KgPhp\n');
fprintf(f2,'
==================================================================
========================================\n');
%<--------Calculation of Main Dimensions(Part-1)------->
kt=1.42;LbyD0=0.3;ki=0.93; %Assumptions
h10=0.07;h11=1.3*h10;cdm1=4.5; %Assumptions
Sn=0; M1=0; M2=0; M3=0; EFFmax=50; maxTqst=1.5; minKgPhp=16;
maxPopt=700; %Assumptions
for P=2:2:4; for S1=24:12:48; for cdm1=3.5:0.1:4.5;
HZ=[25  30  40  45  50  60];KF=[0.865  .885  .923  .94  .96  1];
kf=interp1(HZ,KF,hz,'spline');
POLES=[2   4   6   8];      OPTC=[0.32   0.22   .19   .17];
C0=interp1(POLES,OPTC,P,'spline');
DIBYD0=[.5 .59.64 .67];DibyD0=interp1(POLES,DIBYD0,P,'spline');
D0sqL=16.5*C0*hp/N*kf*kt*1e6;D01=(D0sqL/LbyD0)^(1/3);
D0=floor(D01);L=LbyD0*D0; Di1=DibyD0*D0;Di=ceil(Di1);SP=S1/P;
if P==2 bt1a=(1.1+0.032*Di)*Di/S1;bt1=floor(bt1a*100)/100;end;
if P==4 bt1a=(1.27+0.035*Di)*Di/S1;bt1=floor(bt1a*100)/100;end;
SS=[24 36 48];B10=[0.068+0.0175*Di 0.038+0.0175*Di 0.0175*Di];
b10=interp1 (SS,B10,S1,'spline');
for Bt=1.3:0.05:1.7;for Bc=0.8:0.05:1.15; btbybc=Bt/Bc;
dc1=btbybc*(S1*bt1)/(pi*P);b11=pi*(Di+2*h10+2*h11)/S1-bt1;
h14=DibyD0*(D0-Di)-(h10+h11+dc1);              alfa=180/S1;
b13=b11+2*h14*tan(alfa*pi/ 180);
Lg=0.013+0.0042*Di/sqrt(P);
%<--------Design of Main Winding(Part-2)--------->
Ncpp=S1/P/2;     if     ceil(Ncpp)>     Ncpp     continue,end;
sum1=0;sum2=0;sum3=0;sum4=0; sum5=0;
for i=1:Ncpp;y1=(SP-(i-1)*2)/(SP+1);Kp(i)=sin(y1*pi/2);
sum1= sum1+Kp(i); end;
for i=1:Ncpp;PTm(i)=Kp(i)/sum1*100;sum2=sum2+PTm(i);end;
for i=1:Ncpp;sum3=sum3+Kp(i)*PTm(i);end;Kwm=sum3/sum2;
Li=ki*L;At=bt1*Li*SP; FI=0.637*At*Bt*1e-4;
```

```
Tm1=0.95*V/(4.44*hz*FI*Kwm); Tm=ceil(Tm1/2)*2;
Nm=Tm*2; Tppm=Tm/P;
for i=1:Ncpp;Tc1(i)=Tppm*PTm(i)/100;Tc(i)=round(Tc1(i));
sum4=sum4+Tc(i);end;corr=Tppm-sum4;      Tc(Ncpp)=corr+Tc(Ncpp);
sum4=sum4+ corr; HP=[.05 .1 .25 .5 .75 1];
EFF=[.35 .44 .58 .65 .68 .7];  PF=[.45 .5 .56 .62 .64 .66];
effa=interp1(HP,EFF,hp,'spline');
pfa=interp1(HP,PF,hp,'spline');
Im=hp*746/(effa*V*pfa);Amc1=Im/cdm1;dmc1=sqrt(4*Amc1/pi);
DSWG=[5.38 4.88 4.47 4.06 3.66 3.25 2.95 2.64 2.31 2.03 1.83
1.63 1.42 1.22 1.02 .914 .813 .711 .61];
for i=1:19; if dmc1 >= DSWG(i) continue; end; dmc=DSWG(i);
end;dmcins=dmc +0.075;Amc=pi*dmc^2/4;cdm=Im/Amc; if cdm <2.8 ||
cdm    >4.5    continue;end;    Socm=Tc(1)*pi*dmcins^2/4/100;
Asg=(h11*(b10+b11)+h14*(b11+b13))/2;Sfsm=Socm/Asg;
if Sfsm >0.5 continue;end; Hs=h11+h14;
for i=1:Ncpp; Lmt(i)=8.4*(Di+Hs)/S1*(SP-(i-1)*2)+2*L;
sum5=sum5+Lmt(i)*Tc(i);end;Lmtm=sum5/sum4;
Rm=0.021*Lmtm*Tm/(Amc*100);Pcus=Im^2*Rm;
Wcum=Lmtm*Amc*Tm*8.9e-5;
%<--------Design of Rotor(Part-3)--------->
b20=0.075;h20=0.08;Dm=0.07;     %assumption    for    ij=1:2;
S2=S1+(ij+2)*P; Dr=Di-2*Lg;bt2=0.95*S1*bt1/S2;
r21=(pi*(Dr-2*h20)-S2*bt2)/(2*(S2+pi));drc1=2* r21-0.038;
for i=1:19; if drc1*10 >= DSWG(i)continue;end;drc=DSWG(i); end;
Ab=pi*drc^2/4;dc2=0.95*dc1;Lb=L+1;Nb=S2;Aer=Ab/pi*Nb/P;
R2md=P*Nm^2*Kwm^2*0.021*(Lb/(100*Ab*Nb)+2/pi*Dm/P^2/Aer);
I2d=Im*pfa;Pcur=I2d^2*R2md;Wcur=Lb*Ab*Nb*8.9e-5;
Wtr=((pi*Dr^2/4)-(S2*pi*r21^2))*L*7.8*1e-3;
%<--------Amp-Turns Calculation (Part-4)--------->
SobyAg1=b10/Lg;SOBYAG=[0 1 2 3 4 5 6 7 8 9 10 11 12];
CC=[0 .19 .34 .46 .55 .61 .66 .71 .75 .79 .82 .86 .89];
%semi-closed-slot, K01=interp1(SOBYAG,CC,SobyAg1,'spline');
sp1=pi*Di/S1;Kgs=sp1/(sp1-b10*K01);
SobyAg2=b20/Lg;K02=interp1(SOBYAG,CC,SobyAg2,'spline');
sp2=pi*Dr/S2;Kgr=sp2/(sp2-b20*K02);  Kag=Kgs*Kgr;  Lgd=Kag*Lg;
Aga=pi*Di/P*L;Bag=FI*1e4/0.637/Aga; ATag=0.796*Bag*Kag*Lgd*1e4;
```

```
BB=[.1 .2 .3 .4 .5 .6 .7  .8  .9  1   1.1  1.2  1.3  1.4  1.5
1.6  1.7   1.8  1.9   2.0  ]; %Tesla

HH=[18 28 38 48 67 80 95 120 140 180  220  295  390  580 1000
2200 5000  9000 16000 24000];% for Lohys

At1=Li*bt1*S1/P;Bt1=FI*1e4/0.637/At1;

if Bt1 <1.3 || Bt1 >1.7 continue;end;

att1=interp1(BB,HH,Bt1,'spline');Lfpt1=h14+h11+h10;

ATT1=att1*Lfpt1/100; Ac1=Li*dc1; Bc1=FI*1e4/2/Ac1;

if Bc1 <0.8 || Bc1 >1.2 continue;end;

BtbyBc=Bt1/Bc1;if BtbyBc >1.2 continue;end;

atc1=interp1(BB,HH,Bc1,'spline');

Lfpc1=pi*(D0-dc1)/2/P;ATC1=atc1*Lfpc1/100;

At2=Li*bt2*S2/P;Bt2=FI*1e4/0.637/At2;

att2=interp1(BB,HH,Bt2,'spline');Lfpt2=2*r21+h20;

ATT2=att2*Lfpt2/100;       Ac2=Li*dc2;        Bc2=FI*1e4/2/Ac2;
atc2=interp1(BB,HH,Bc2,'spline');

Lfpc2=pi*(2*r21+h20+dc2)/2/P; ATC2=atc2*Lfpc2/100;

ATT=ATag+ATT1+ATC1+ATT2+ATC2; satf=ATT/ATag;

%<--------Leakage Reactance(Part-5)--------->

skew=15;Kp=0.97;         %Assumption          sum6=0;          for
i=1:Ncpp;sum6=sum6+Tc(i)^2;end; Cx=sum6/Tppm^2*S1/(Kwm^2*4*P);

B11byB13=[.5 .6 .7 .8  .9   1 1.1 1.2 1.3 1.4 1.5 1.6 1.7 1.8];

CONSTANT=[0.63 .53 .46 .41 .36 .33  .3 .27 .25 .23 .21  .2 .18
.17];

b11byb13=b11/b13;
constA=interp1(B11byB13,CONSTANT,b11byb13,'spline');

Ks1=constA*h14/b13+h10/b10+2*h11/(b10+b11);Ks2=h20/b20+0.623;
Ks=Ks1*Cx+S1/S2*Ks2;               Z=Nm;Kx=2*pi*hz*(Z*Kwm)^2*1e-8;
Xs=Kx*2.512*Ks*L/S1; Lmd1=pi*Di/S1; Lmd2=pi*Dr/S2;

bt10=sp1-b10;bt20=sp2-b20;    Kzz=(bt10+bt20)^2/(4*(Lmd1+Lmd2));
Xzz=Kx*Kzz* 0.838*L/(S1*Lg);

sum7=0;fori=1:Ncpp;       sum7=sum7+Tc(i)*(SP-(i-1)*2);       end;
ACT1=sum7/Tppm; ACT=floor(ACT1);

De=Di+h14+2*(h10+h11);                    Xe=Kx*1.236*De*ACT/S1/P;
Km=Aga/(Lgd*satf*P);

AvSP=[4 8 12 16 20 24];BLC=[3.2 1.8 1.4 1.25 1.15 1.1];

Nsp=(S1+S2)/2/P;Kb=interp1(AvSP,BLC,Nsp,'spline');
Xb=0.000929*Km*Kb*Kx;

Q=0.25*(skew/100)^2;Xsk=0.2546*Km*Kp*Q*Kx;XLm=Xs+Xzz+Xe+Xb+Xsk;
```

```
Csk=sin(skew/2*pi/180)/(pi*skew/360);Xm=Kx*0.2546*Km*Csk;
X0=Xm+XLm/2;Imu=V/X0;

%<--------Design of Auxiliary Wdg(Part-6)--------->

K1=1.5;Kwa1=0.83; %Assumptions

for K1=1.5:0.1:2;

%for ij=1:3;mf=(0.10+ij*0.01);Aac1=mf*Amc;dac1=sqrt(4*Aac1/pi);

%DSWG1=[1.63 1.42 1.22 1.02 .914 .813 .711 .610 .559 .508 .457
.417 .376 .345 .315];

%for i=1:15; if dac1 >= DSWG1(i) continue;end; dac=DSWG1(i);
end;Aac = pi*dac^2/4;Ta1=K1*Tm*Kwm/Kwa1; Ta=ceil(Ta1/2)*2;
Na=Ta*2;Tppa=Ta/P; sum11=0;sum12=0;sum13=0;sum14=0;sum15=0;

fori=1:Ncpp;y1=(SP+1-(i-1)*2)/(SP+1);
Kp(i)=sin(y1*pi/2);sum11=sum11+Kp(i); end;

sum12=0;for i=1:Ncpp;Tca1(i)=Tppa*Kp(i)/sum11;

Tca(i)= round(Tca1(i));sum12=sum12+Tca(i); end;

corr=Tppa-sum12;Tca(Ncpp)=corr+ Tca(Ncpp);sum12= sum12+corr;

for 1:Ncpp; sum13=sum13+Kp(i)*Tca(i); end; Kwa=sum13/Tppa;
K=Ta*Kwa/Tm/Kwm; sum15=0; for i=1:Ncpp;
Lmta1(i)=8.4*(Di+Hs)/S1*(SP+1-(i-1)*2)+2*L;

sum15=sum15+Lmta1(i)*Tca(i);end;Lmta=sum15/Tppa;

for ij=1:3;mf=(0.10+ij*0.01);Aac1=mf*Amc;dac1=sqrt(4*Aac1/pi);

DSWG1=[1.02 .914 .813 .711 .610 .559 .508 .457 .417 .376 .345
.315];

for i=1:12; if dac1 >= DSWG1(i) continue;end;

dac=DSWG1(i);end;Aac=pi*dac^2/4; R1a=0.021*Lmta*Ta/(Aac*100);

Rdr=K^2*R2md;Rat=R1a+Rdr;Rmt=Rm+R2md;XLa=K^2*XLm;
ZLm=sqrt(Rmt^2+XLm^2);ZLa=sqrt(Rat^2+XLa^2);
Tqst=P*K*R2md*V^2/(2*pi*hz)*(Rat*XLm-Rmt*XLa)/ ZLm^2/ZLa^2;

if Tqst < 1.64 continue;end;Isa=V/ZLa;Ism=V/ZLm;Ist=Isa+Ism;

cda=Isa/Aac; if cda >30 continue;end; Wcua=Lmta*Aac*Ta*8.9e-5;

%<----Weights,LossesandPerformance using Eq.Ckt(Part-7)--------
->

FD=[0.8 1.2 1.6 2 2.4]; WPKG=[7 15 23 34 50]; % for 0.5 mm

WpKgt=interp1(FD,WPKG,Bt1,'spline');WtT=At1*P*Lfpt1*7.8*1e-3;
Pist=WpKgt* WtT; WpKgc=interp1(FD,WPKG,Bc1,'spline');

Dmc=D0-dc1;WtC=pi*Dmc*Ac1*7.8*1e-3; Pisc=WpKgc*WtC;
Pi=1.9*(Pist+Pisc); Pfw=0.02*hp*746;

TotWt=1.01*(Wcum+Wcua+WtT+WtC+Wcur+Wtr);

KgPhp=TotWt/hp;if KgPhp >14.7 continue;end;
```

```
%------Solution by Equivalent Circuit------->
Ns=120*hz/P;s=(Ns-N)/Ns;%for kk=20:100; s=kk*1e-3;
Z1=Rm+XLm/2*j; 2a=R2md/(2*s)+XLm/4*j;Z2b=Xm/2*j;
Z3a=R2md/(2*(2-s))+XLm/4*j; Z3b=Xm/2*j;Z2=Z2a*Z2b/(Z2a+Z2b);
Z3=Z3a*Z3b/(Z3a+Z3b);Zeq=Z1+Z2+Z3;I1=V/Zeq;
ang=angle(I1)*180/pi;pfx=cos(ang*pi/180);I1m=abs(I1);
If=I1*Z2b/(Z2a+Z2b);Ifm=abs(If);Ib=I1*Z3b/(Z3a+Z3b);
Ibm=abs(Ib);Tf=Ifm^2*R2md/(2*s);Tb=Ibm^2*R2md/(2*(2-s));
Tr=Tf-Tb;TqFL=Tr-(Pi+Pfw);Popt=TqFL*(1-s);
if Popt < 700 continue;end;
Pinp=V*I1m*pfx;eff=Popt/Pinp*100;if eff < 64.8 continue;end;
Sn=Sn+1; Sn,mf, if eff >=EFFmax EFFmax=eff;end;if abs(eff-
EFFmax)<=2e-3 M1=Sn; end;
if KgPhp <=minKgPhp minKgPhp=KgPhp;end;if abs(KgPhp-
minKgPhp)<=0.001 M2=Sn; end;
if Tqst >=maxTqst maxTqst=Tqst;end;if abs(Tqst-
maxTqst)<=0.0001 M3=Sn; end;
if Popt >=maxPopt maxPopt=Popt;end;if abs(Popt-
maxPopt)<=0.0001 M4=Sn; end;
fprintf(f2,'%3d%3d%5.1f%5.1f%5.1f%3d%4d%6.3f%4.1f%3d%5.2f%5.2f%
4.1f%4d%6.3f%5.1f%5.1f%5.2f%6.3f%5.0f%6.2f%6.2f\n',...
Sn, P, D0, L, Di, S1,Tm,dmc, cdm, S2, Bt1, Bc1, K, Ta,
dac,cda, Ist, Tqst, pfx,Popt, eff,KgPhp);
end;end;end;end;end;end;end;end;
fprintf(f2,'
================================================================
========================================\n');
fprintf(f2,'Selection of Design Variant based on Optimization
Criteria:');
fprintf(f2,'\nIf Maximum Efficiency is Required, Select
Variant(Sn)=%3d (%5.2f perc)',M1,EFFmax);
fprintf(f2,'\nIf Minimum Kg/HP is Required, Select
Variant(Sn)=%3d (%5.2f)',M2,minKgPhp);
fprintf(f2,'\nIf Maximum Starting Torque is Reqd, Select
Variant(Sn)=%3d (%4.2f)',M3,maxTqst);
fprintf(f2,'\nIf Max Output at Rated speed Reqd, Select
Variant(Sn)=%3d (%4.0f)',M4,maxPopt);
fclose(f2);
```

8.12 (a) Computer Output Results for Optimal Design

Design of 1-ph Ind.Motor of 1.0 HP, 230 Volts, 50 HZ

Sn	P	D0	L	Di	S1	Tm	dmc	cdm	S2	Bt1	Bc1	K	Ta	dac	cda	Ist	Tqst	pfx	Popt	eff	KgPhp
1	2	20.0	6.0	10.0	24	382	1.420	4.4	30	1.30	1.10	1.5	546	0.508	26.4	25.4	1.77	0.874	721	65.09	14.62
2	2	20.0	6.0	10.0	24	382	1.420	4.4	30	1.30	1.10	1.5	546	0.508	26.4	25.4	1.77	0.874	721	65.09	14.62
3	2	20.0	6.0	10.0	24	382	1.420	4.4	30	1.30	1.10	1.6	582	0.508	24.1	24.9	1.65	0.874	721	65.09	14.66
4	2	20.0	6.0	10.0	24	382	1.420	4.4	30	1.30	1.10	1.6	582	0.508	24.1	24.9	1.65	0.874	721	65.09	14.66
5	2	20.0	6.0	10.0	24	382	1.420	4.4	32	1.30	1.10	1.5	546	0.508	26.8	25.9	1.76	0.878	774	65.90	14.66
6	2	20.0	6.0	10.0	24	382	1.420	4.4	32	1.30	1.10	1.5	546	0.508	26.8	25.9	1.76	0.878	774	65.90	14.66
7	2	20.0	6.0	10.0	24	382	1.420	4.4	32	1.30	1.10	1.6	582	0.508	24.5	25.5	1.64	0.878	774	65.90	14.70
8	2	20.0	6.0	10.0	24	382	1.420	4.4	32	1.30	1.10	1.6	582	0.508	24.5	25.5	1.64	0.878	774	65.90	14.70
9	2	20.0	6.0	10.0	24	382	1.420	4.4	32	1.30	1.15	1.5	546	0.508	26.7	25.9	1.76	0.875	764	65.19	14.52
10	2	20.0	6.0	10.0	24	382	1.420	4.4	32	1.30	1.15	1.5	546	0.508	26.7	25.9	1.76	0.875	764	65.19	14.52
11	2	20.0	6.0	10.0	24	382	1.420	4.4	32	1.30	1.15	1.5	546	0.559	24.5	26.5	1.71	0.875	764	65.19	14.63
12	2	20.0	6.0	10.0	24	368	1.420	4.4	30	1.35	1.15	1.5	526	0.508	28.0	27.2	1.89	0.872	775	65.06	14.49
13	2	20.0	6.0	10.0	24	368	1.420	4.4	30	1.35	1.15	1.5	526	0.508	28.0	27.2	1.89	0.872	775	65.06	14.49
14	2	20.0	6.0	10.0	24	368	1.420	4.4	30	1.35	1.15	1.5	526	0.559	25.7	27.8	1.84	0.872	775	65.06	14.60
15	2	20.0	6.0	10.0	24	368	1.420	4.4	30	1.35	1.15	1.6	560	0.508	25.5	26.7	1.76	0.872	775	65.06	14.53
16	2	20.0	6.0	10.0	24	368	1.420	4.4	30	1.35	1.15	1.6	560	0.508	25.5	26.7	1.76	0.872	775	65.06	14.53
17	2	20.0	6.0	10.0	24	368	1.420	4.4	30	1.35	1.15	1.6	560	0.559	23.3	27.2	1.69	0.872	775	65.06	14.63
18	2	20.0	6.0	10.0	24	368	1.420	4.4	32	1.35	1.15	1.5	526	0.508	28.4	27.8	1.88	0.876	832	65.85	14.53
19	2	20.0	6.0	10.0	24	368	1.420	4.4	32	1.35	1.15	1.5	526	0.508	28.4	27.8	1.88	0.876	832	65.85	14.53
20	2	20.0	6.0	10.0	24	368	1.420	4.4	32	1.35	1.15	1.5	526	0.559	26.1	28.4	1.83	0.876	832	65.85	14.63
21	2	20.0	6.0	10.0	24	368	1.420	4.4	32	1.35	1.15	1.6	560	0.508	25.9	27.3	1.75	0.876	832	65.85	14.56
22	2	20.0	6.0	10.0	24	368	1.420	4.4	32	1.35	1.15	1.6	560	0.508	25.9	27.3	1.75	0.876	832	65.85	14.56
23	2	20.0	6.0	10.0	24	368	1.420	4.4	32	1.35	1.15	1.6	560	0.559	23.7	27.8	1.69	0.876	832	65.85	14.67

Selection of Design Variant based on Optimization Criteria:

If Maximum Efficiency is Required, Select Variant (Sn) = 8 (65.90 perc)

If Minimum Kg/HP is Required, Select Variant (Sn) = 13 (14.49)

If Minimum Starting Torque is Required, Select Variant (Sn) = 13 (1.89)

If Minimum Output at Rated speed is Required, Select Variant (Sn) = 23 (832).

Bibliography

Performance & Design of Alternating Current Machines by M.G. Say

Electrical Engineering Design Manual by M.G. Say

Alternating Current Machines by A.F. Puchstein, T.C. Lloyd & A.G. Conrod

Electrical Machinery by A.E. Fitzgerald, Charles Kingsley, Jr & Stephen D. Umans

Electrical Technology by B.L. Theraja & A.K. Theraja

Design of Electrical Machines by V.N. Mittle & A. Mittal

Getting Started with MATLAB by Rudrapratap

Author's Design Experience of 29 years in BHEL

Relevant Indian Standards

Index

A

Air gap 22, 148, 182, 257, 310
Armature reaction 65, 183, 213
Armature tooth 22, 23
Armature windings 33, 34, 37, 44
AT per pole 60
Aux winding 295
Av loss factor 145, 186
Axial length 124, 234

B

Back-ward torque 299
Bar current 230
Bearing loss 192, 215
Brush contact loss 72
Brush friction 71
Brush loss 160
Brushes 39, 44, 71

C

Carter coefficient 150, 153
Coil throw 291
Cold rolled steels 96, 106
Commutator 4, 5, 44, 71, 72,
Commutator segment 4, 36, 59
Compensating winding 73
Cond/slot 143, 269
Conducting materials 15, 16
Conductor 17, 34, 38, 60, 227, 270
Conductor width 60, 85
Constraints 4, 93, 277
Cooling 24, 25, 30
Cooling area 104, 110, 133, 238
Cooling surface 24, 63, 70, 127
Cooling tubes 5, 94, 134
Copper loss 16, 74, 101, 124, 186, 231, 310
Core 2, 26, 60, 67, 93, 99, 144, 153, 185, 195, 312
Core type transformer 99, 101
Current density 16, 71, 100, 134, 182, 270
Current volume 4, 57, 60

D

DC copper loss 145, 186, 188
Depth of winding 70
Design of poles 65, 68,
Design of rotor 147, 190, 284
Dissipation coefficient 162, 163

Double layer winding 38, 41, 47, 48
Dummy coils 44

E

Eddy current loss 4, 62, 186
Efficiency 1, 112, 159, 224, 281
Electric circuit 20
Electrical materials 13, 15
EMF 21
End leakage reactance 291, 306, 311
End ring 230, 248, 260
Equivalent circuit 278
Equivalent rotor current 258, 285
Exciter voltage 159

F

Field coil 57, 65, 141, 163
Field regulator 69
Flux in core 62, 146, 186, 228
Flux density 2, 22, 62, 73, 110, 153, 186, 234, 288
Flux/Pole 22, 59, 142, 212
Forward rotor 298
Free length 62
Frequency constant 4, 278

G

Gap coefficient 22, 68, 151
Gross Iron length 184, 225

H

Heat dissipation 24, 25
Heating 24, 25
Height of core 57, 62, 223
HV Winding 93, 100, 124
Hysteresis loss 2, 18, 19
Hysteresis loop 18

I

Insulating materials 13, 16, 17
Insulation 17, 123, 227, 298
Inter pole 72, 73, 84
Inter pole winding 73
Inter turn insulation 98
Iron factor 58, 119, 142, 183, 280
Iron length 172, 225, 268, 280
Iron loss 63, 95, 120, 142, 161, 229
Iron path 66

K

Kg/KVA 1, 5, 142, 215
Kg/KW 57, 85, 224, 250, 257

L

L/D0 ratio 278
Laminations 18, 62, 189
Lap winding 35, 37-39
Leakage flux 146, 188
Leakage reactance 4, 146, 187, 289, 311
Length of bar 230, 248, 284
Length of pole 65, 148
Lip 73, 227, 258
Locked impedance 295
Locked rotor current 295
Loss factor 4, 145, 186
Losses 5, 16, 94, 145, 159, 215, 297
Lowest cost 5
LV winding 98, 99, 123

M

Magnetic circuit 20, 150, 192, 195
Magnetic frame 94, 118
Magnetic path 22, 152, 153
Magnetization curves 67, 69, 304
Magnetizing current 97, 223, 232, 257, 311

Main core 119, 133
Main dimensions 4, 58, 142, 182, 224, 278
Main pole 66
Main winding 280, 309
Maximum efficiency 90, 91, 117, 180, 276, 317
Mean length of turn 69, 111, 124, 134, 186, 228, 283
Mechanical loss 182, 192
Minimum section 284

N

Net iron length 23, 184, 247, 268
Net motor torque 299
No load loss 249, 259, 269
No-load current 96, 121, 133, 259

O

Objective function 2, 5, 85, 135
Oil duct 99, 101
Open circuit characteristic 142, 154, 182, 197
Optimal design 5, 32, 84, 85, 113, 135, 174, 215, 216, 250, 312
Output coefft 58, 247, 268
Overhang 28, 146, 267, 270

P

P.U. impedance 103, 126
P.U. resistance 103, 126
Performance 73, 75, 102, 126, 238, 297
Periphoral velocity 57, 225
Permeance 145, 237, 270
Pitch 36, 38, 41, 184, 257, 291
Pitch factor 45, 280, 307, 309
Pole arc 4, 58
Pole pitch 4, 34, 46, 85, 174, 268

R

Radial width 98, 111, 125, 134
Reactances 5, 311
Regulation 4, 94, 118, 159, 173, 201, 214
Resistance 16, 20, 70, 103, 112, 145, 160, 283, 311
Rotor 2, 48, 147, 182, 190, 230, 257, 284
Rotor slots 4, 248, 310
Round slot 310

S

Saturation factor 4, 288
Series field winding 69, 70
Shell type transformer 123
Short circuit current 236
Shunt field current 60
Shunt field winding 4, 69
Silicon steel 2, 18, 118, 189
Single layer winding 45
Skew factor 292
Skew leakage reactance 292, 306, 311
Slack 101, 134
Slip 2, 224, 257, 270
Slip at full load 238
Slip rings 160, 257
Slip-ring rotor 257
Slot 4, 33, 34, 38, 41, 191, 249, 286, 310
Slot balance 57, 182, 191, 257
Slot loading 141, 143
Slot opening 233, 284, 310

Slot pitch 23, 60, 144, 185, 226, 286
Slot width 144, 185, 213, 248, 309
Slots spanned 283, 293, 294
Specific electrical loading 15
Specific magnetic loading 4, 14, 15, 142, 182, 183
Specific permeance 145, 187, 236
Spindle 72, 84
Split-phase type 277, 278
Squirrel cage 2, 223, 230
Standard ratings 25
Standards 1
Starting torque 4, 224, 257, 295, 307, 311
Strand 98, 100
Stray losses 102, 124, 125, 145, 186

T

Tank design 104, 112, 127, 134
Temperature rise 1, 13, 29, 104, 127
Total weight 75, 128, 163, 201, 311
Transmission ratio 293, 294
Tube dia 104, 127
Turns/ph 172, 226, 247, 268

V

Ventilating ducts 26, 27, 58, 142, 183
Ventilation loss 192, 215
Ventilation schemes 26
Voltage regulation 159, 173, 182, 201, 214
Volts/turn 94, 118, 133

W

Wattful current 97, 121, 249, 269
Wave winding 35, 41, 42, 44
Wedge 61, 212, 213, 227, 258
Width of bar 230
Width of pole 65, 148
Winding factor 142, 224, 257, 293, 311
Window area 94, 110, 119, 133
Window space factor 4, 94, 110, 119, 129, 133

Y

Yoke 57, 67, 84, 95, 112, 133

Z

Zig zag reactance 237, 310

www.ingramcontent.com/pod-product-compliance
Ingram Content Group UK Ltd.
Pitfield, Milton Keynes, MK11 3LW, UK
UKHW060019200726
13845UKWH00046B/93

9 789352 300662